Open-Source Electronics Platforms

Open-Source Electronics Platforms

Development and Applications

Special Issue Editor

Trung Dung Ngo

MDPI • Basel • Beijing • Wuhan • Barcelona • Belgrade

MDPI

Special Issue Editor
Trung Dung Ngo
The More-Than-One Robotics Laboratory
Faculty of Sustainable Design Engineering
University of Prince Edward Island
Canada

Editorial Office
MDPI
St. Alban-Anlage 66
4052 Basel, Switzerland

This is a reprint of articles from the Special Issue published online in the open access journal *Electronics* (ISSN 2079-9292) from 2018 to 2019 (available at: https://www.mdpi.com/journal/electronics/special_issues/opensource_elec_platforms).

For citation purposes, cite each article independently as indicated on the article page online and as indicated below:

LastName, A.A.; LastName, B.B.; LastName, C.C. Article Title. *Journal Name* **Year**, *Article Number*, Page Range.

ISBN 978-3-03897-972-2 (Pbk)
ISBN 978-3-03897-973-9 (PDF)

Cover image courtesy of Trung Dung Ngo.

Contents

About the Special Issue Editor

Trung Dung Ngo is an Associate Professor at the Faculty of Sustainable Design Engineering, University of Prince Edward Island (UPEI), Canada. He has been the Founding Director of the More-Than-One Robotics Laboratory since 2008 and the Lead Researcher for the UPEI Centre for Excellence in Robotics and Industrial Automation, where he has coordinated a number of research projects and contributed technical consultation to government sectors and industrial companies in robotics, automation, and seafood processing. He has been interested in the development and usage of open-source electronics for teaching and research activities. He was the Guest Editor for a number of Special Issues and an edited Book. He is a Senior Member of IEEE.

Preface to "Open-Source Electronics Platforms"

Open-source electronics are becoming very popular, and are integrated with our daily educational and developmental activities. At present, the use open-source electronics for teaching science, technology, engineering, and mathematics (STEM) has become a global trend. Off-the-shelf embedded electronics such as Arduino- and Raspberry-compatible modules have been widely used for various applications, from do-it-yourself (DIY) to industrial projects. In addition to the growth of open-source software platforms, open-source electronics play an important role in narrowing the gap between prototyping and product development. Indeed, the technological and social impacts of open-source electronics in teaching, research, and innovation have been widely recognized.

This book is a collection of 12 selected chapters from authors around the world. This collection represents the diversity as well as impact of open-source electronics through numerous system developments and applications. A summary of the chapters in this book can be overviewed as follows.

Chapter 1: Trilles et al. developed Internet of Things (IoT)-based sensor nodes named SEnviro for smart agriculture. The wireless sensor node enables the integration of temperature, air humidity, barometer, soil moisture, and weather sensors, as well as a 3G wireless communication module and a sonar panel. A state machine-based software platform consisting of logic control, basic configuration, communication, and energy consumption modules was implemented to govern the behavioral operations of the nodes. The wireless sensor nodes were deployed and tested with the application of monitoring and detecting diseases in vineyards.

Chapter 2: Zhang et al. presented a platform for the real-time data transmission and analysis of livestock. The platform was an integration of wireless sensor nodes mounted on livestock and repeaters for data relay and processing. The developed system was deployed and examined with the processes of feeding and breeding management in grazing on a real field.

Chapter 3: Hang et al. proposed a sensor-cloud-based platform capable of virtually representing physical sensors in the Cloud of Things (CoT) environment. The design and implementation procedures of the sensor-cloud platform governing different types of wireless sensor nodes with the faulty sensor detection capability was addressed and verified through comparison analysis with the existing systems.

Chapter 4: Kashevnick et al. proposed an ontological approach of blockchain-based coalition formation for cyber-physical systems (CPSs). A model of abstract and operational context management for interaction and ontology of multi-agent cyber-physical systems was developed and evaluated through the collaboration of a heterogenous system of mobile robots.

Chapter 5: Rooney et al. presented a method of hardware trojan creation and detection using FPGAs and off-the-shelf components. They demonstrated that using off-the-shelf components could reduce the cost of integrated circuit design and the fabrication of trojan detection in different settings.

Chapter 6: Ferrari et al. proposed an experimental methodology for examining the impact of quality of service (QoS) on the communication delay between production lines and cloud platforms using open platform communication unified architecture (OPC UA) gateways. The experiment results measuring the overall time delay between a machine with an OPC UA interface and a cloud platform demonstrate the impact of the QoS parameters on the communication delay, which is an important factor to guarantee the real-time processing of industrial IoTs.

Chapter 7: Merenda et al. presented the design and implementation of smart converters for metering and testing the current and voltage of renewable energy systems. Using smart convertors, a developer can focus on the development of software algorithms for controlling and managing ecosystems. The key features of the developed converters, such as system-level management, real-time diagnostics, and on-the-flight parameter change, were tested and verified with a solar simulator as well as photovoltaic generators.

Chapter 8: Yang et al. developed single-channel bio-signal-based human–computer interfaces (HCIs) to estimate the horizontal position of the eyeballs of disabled people. Input signals from electrooculograms (EOGs) and electromyograms (EMGs) were processed in real time through a modified sliding window algorithm using piecewise linear approximation (PLA), and the eyeball position was identified through a curve-fitting model using the support vector regression (SVR) method. The advantages of the proposed method were evaluated in comparison with the conventional EOG-based HCI.

Chapter 9: Ariza presented an open-source hardware and software platform for a learning embedded system named DSCBlocks. Using algorithm visualizations with a graphical building block of embedded C codes for dsPIC, a learner can focus on designing an algorithm to program digital signal controllers and observe the configuration at the hardware level.

Chapter 10: Ngo et al. introduced an open platform for multiple-disciplinary teaching methodology in terms of the problem-based learning (PBL) and the Engineering Projects in Community Service (EPICS) course for engineering students. The open platform consists of low-cost automated guided vehicles built of off-the-shelf components including ARM Cortex M4 32-bit, WiFi module, proximity sensors, camera, cylinders, and is equipped with open-source libraries. It was demonstrated that the open platform was productively used by students in mechatronics, computer science, and mechanics in their collaborative projects.

Chapter 11: Vega et al. introduced an open low-cost robotics platform named PiBot for STEM education. The platform was developed by the integration of off-the-shelf components including the Raspberry Pi 3 controller board, Raspberry PiCamera, ultrasonic sensor, infrared ranging sensors, and motor encoders under the Do-It-Yourself (DIY) philosophy. A simulated robot under the Gazebo simulator was also developed to provide an alternative learning platform for students. The robotic platforms were examined and evaluated through a series of exercises of robot programming, control, and vision.

Chapter 12: Costa et al. provided a comprehensive review of open-source electronics platforms as enabling technologies for smart cities. The key features, including advantages and disadvantages of using off-the-shelf electronics and computing boards (e.g., Raspberry Pi, BeagleBoard, and Arduino) and open-source software platforms (e.g., Arduino, Raspbian OS) are discussed through numerous smart-city applications.

In summary, the diversity of open-source electronic platforms as well as their applications presented in this collection demonstrates that there is no restriction or limitation in using open-source electronics for our education, research, and product development. We are delighted to deliver up-to-date knowledge and technology of open-source electronics to our readers. In particular, we believe that this book could be used as a good reference for engineering and science students for their education and research activities.

I would like to thank all the authors who contributed their great work to this collection. I am very grateful to the reviewers for their time and effort to provide useful comments and suggestions to the submitted papers, which yielded the polished chapters of this book. Last but not least, I would like to thank all the staff of the Electronics Editorial Office for our collaboration through this project. Without your contributions and hard work, this book would not be as excellent as it is.

Trung Dung Ngo
Special Issue Editor

electronics

MDPI

Editorial

Open-Source Electronics Platforms: Development and Applications

Trung Dung Ngo

The More-Than-One Robotics Laboratory, Faculty of Sustainable Design Engineering,
University of Prince Edward Island, Charlottetown, PE C1A 4P3, Canada; tngo@upei.ca; Tel.: +1-902-566-6078

Received: 8 April 2019; Accepted: 9 April 2019; Published: 12 April 2019

1. Introduction

Open-source electronics are becoming very popular with our daily educational and developmental purposes. Currently, using open-source electronics for teaching Science, Technology, Engineering and Mathematics (STEM) is becoming the global trend. Off-the-shelf embedded electronics such as Arduino- and Raspberry-compatible modules have been widely used for various applications, from Do-It-Yourself (DIY) to industrial projects. In addition to the growth of open-source software platforms, open-source electronics play an important role in narrowing the gap between prototyping and product development. Indeed, the technological and social impacts of open-source electronics in teaching, research and innovation have been widely recognized.

2. Summary of the Special Issue

This Special Issue is a collection of 11 technical papers and one review article selected through the peer-review process. This collection represents the diversity, as well as the impact of open-source electronics through numerous system developments and applications. The contributions in this Special Issue can be summarized as follows.

Trilles et al. [1] developed an Internet of Things (IoT)-based sensor node, named SEnviro, for smart agriculture. The wireless sensor node resulted in the integration of temperature, air humidity, barometer, soil moisture and weather sensors, as well as a 3G wireless communication module and a sonar panel. A state machine-based software platform consisting of logic control, basic configuration, communication and energy consumption modules was implemented to govern the behavioural operations of the nodes. The wireless sensor nodes were deployed and tested with the application of monitoring and detecting diseases in vineyards.

Similarly, Zhang et al. [2] presented a platform for real-time data transmission and analysis of livestock. The platform is an integration of wireless sensor nodes mounted on livestock and repeaters for data relay and processing. The developed system was deployed and examined with the process of feeding and breeding management in grazing in the real field.

Hang et al. [3] proposed a sensor cloud-based platform that is capable of virtually representing physical sensors in the Cloud of Things (CoT) environment. The design and implementation procedures of the sensor-cloud platform governing different types of wireless sensor nodes with faulty sensor detection capability were addressed and verified through comparison analysis with existing systems.

Kashevnick et al. [4] proposed an ontological approach of blockchain-based coalition formation for Cyber-Physical Systems (CPS). A model of abstract and operational context management for the interaction and ontology of multi-agent cyber-physical systems was developed and evaluated through the collaboration of a heterogeneous system of mobile robots.

Rooney et al. [5] presented a method of hardware trojan creation and detection using FPGAs and off-the-shelf components. They demonstrated that by using off-the-shelf components, they were able to reduce the cost of integrated circuit design and fabrication for trojan detection in different settings.

Ferrari et al. [6] proposed an experimental methodology of examining the impact of Quality of Service (QoS) on the communication delay between the production line and the cloud platforms using the Open Platform Communication Unified Architecture (OPC UA) gateways. The experiment results of measuring the overall time delay between a machine with an OPC UA interface and a cloud platform demonstrated the impact of the QoS parameters on the communication delay, which is an important factor to guarantee real-time processing of industrial IoTs.

Merenda et al. [7] presented the design and implementation of smart converters for metering and testing the current and voltage of renewable energy systems. Using smart converters, a developer can focus on developing software algorithms for controlling and managing ecosystems. The key features of the developed converters such as system-level management, real-time diagnostics and on-the-fly parameter change were tested and verified with a solar simulator, as well as photovoltaic generators.

Yang et al. [8] developed a single-channel bio-signal-based Human–Computer Interface (HCIs) to estimate the horizontal position of the eyeballs of disabled people. Input signals from Electrooculograms (EOG) and Electromyograms (EMG) were processed in real-time through a modified sliding window algorithm using Piecewise Linear Approximation (PLA), and the eyeball position was identified through the curve-fitting model using the Support Vector Regression (SVR) method. The advantages of the proposed method were evaluated in comparison with the conventional EOG-based HCI.

Ariza [9] presented an open-source hardware and software platform for learning embedded system, named DSCBlocks. Using algorithm visualizations with the graphical building block of embedded C codes for dsPIC, a learner can focus on designing an algorithm to program digital signal controllers and observe the configuration at the hardware level.

In a similar way, Ngo et al. [10] introduced an open platform for the multi-disciplinary teaching methodology in terms of Problem Base Learning (PBL) and the Engineering Projects in Community Service course (EPICS) for engineering students. The open platform is a low-cost automated guided vehicle built of off-the-shelf components including the ARM Cortex M4 32-bit, a WiFi module, proximity sensors, a camera, cylinders and equipped with open-source libraries. It was demonstrated and surveyed that the open platform has been productively used for students in mechatronics, computer science and mechanics in their collaborative projects.

In addition, Vega et al. [11] introduced an open low-cost robotics platform, named PiBot, for STEM education. The platform was developed by the integration of off-the-shelf components including the Raspberry Pi 3 controller board, the Raspberry PiCamera, n ultrasonic sensor, infrared ranging sensors and motor encoders under the Do-It-Yourself (DIY) philosophy. A simulated robot under the Gazebo simulator was also developed to provide an alternative learning platform for students. The robotic platforms were examined and evaluated through a series of exercises of robot programming, control and vision.

Lastly, Costa et al. [12] provided a comprehensive review of open-source electronics platforms as enabling technologies for smart cities. The key features including the advantages and disadvantages of using off-the-shelf electronics and computing boards, e.g., Raspberry Pi, BeagleBoard and Arduino, and open-source software platforms, Arduino and Raspbian OS, were discussed through numerous smart-city applications.

The diversity of open-source electronic platforms, as well as their applications presented in this collection [1–12] demonstrates that there is no restriction nor limitation in using open-source electronics for education, research and product development. We are delighted to offer this Special Issue in order to deliver up-to-date knowledge and technology of open-source electronics to our readers. In particular, we believe that this Special Issue can be a good reference for engineering and science students in their education and research activities.

Funding: I hereby acknowledge the support of NSERC (RGPIN-2017-05446), MITACS (IT11073) and DND-IDEaS (IDEaS-1-1A-CP0726) for my research activities related to this Special Issue.

Electronics **2019**, *8*, 428

Acknowledgments: I would like to thank all the authors of this Special Issue for their great contributions. I am grateful to the reviewers for their time and efforts in providing useful comments and suggestions to the submitted papers. Finally, I would like to thank the Editor-in-Chief, as well as all the staff of the Editorial Office of Electronics for our productive collaboration through this Special Issue.

Conflicts of Interest: The author declares no conflict of interest.

References

1. Trilles, S.; González-Pérez, A.; Huerta, J. A Comprehensive IoT Node Proposal Using Open Hardware. A Smart Farming Use Case to Monitor Vineyards. *Electronics* **2018**, *7*, 419. [CrossRef]
2. Zhang, L.; Kim, J.; LEE, Y. The Platform Development of a Real-Time Momentum Data Collection System for Livestock in Wide Grazing Land. *Electronics* **2018**, *7*, 71. [CrossRef]
3. Hang, L.; Jin, W.; Yoon, H.; Hong, Y.; Kim, D. Design and Implementation of a Sensor-Cloud Platform for Physical Sensor Management on CoT Environments. *Electronics* **2018**, *7*, 140. [CrossRef]
4. Kashevnik, A.; Teslya, N. Blockchain-Oriented Coalition Formation by CPS Resources: Ontological Approach and Case Study. *Electronics* **2018**, *7*, 66. [CrossRef]
5. Rooney, C.; Seeam, A.; Bellekens, X. Creation and Detection of Hardware Trojans Using Non-Invasive Off-The-Shelf Technologies. *Electronics* **2018**, *7*, 124. [CrossRef]
6. Ferrari, P.; Flammini, A.; Rinaldi, S.; Sisinni, E.; Maffei, D.; Malara, M. Impact of Quality of Service on Cloud Based Industrial IoT Applications with OPC UA. *Electronics* **2018**, *7*, 109. [CrossRef]
7. Merenda, M.; Iero, D.; Pangallo, G.; Falduto, P.; Adinolfi, G.; Merola, A.; Graditi, G.; Della Corte, F. Open-Source Hardware Platforms for Smart Converters with Cloud Connectivity. *Electronics* **2019**, *8*, 367. [CrossRef]
8. Yang, J.J.; Gang, G.; Kim, T. Development of EOG-Based Human Computer Interface (HCI) System Using Piecewise Linear Approximation (PLA) and Support Vector Regression (SVR). *Electronics* **2018**, *7*, 38. [CrossRef]
9. Álvarez Ariza, J. DSCBlocks: An Open-Source Platform for Learning Embedded Systems Based on Algorithm Visualizations and Digital Signal Controllers. *Electronics* **2019**, *8*, 228. [CrossRef]
10. Ngo, H.Q.T.; Phan, M.H. Design of an Open Platform for Multi-Disciplinary Approach in Project-Based Learning of an EPICS Class. *Electronics* **2019**, *8*, 200. [CrossRef]
11. Vega, J.; Cañas, J. PiBot: An Open Low-Cost Robotic Platform with Camera for STEM Education. *Electronics* **2018**, *7*, 430. [CrossRef]
12. Costa, D.; Duran-Faundez, C. Open-Source Electronics Platforms as Enabling Technologies for Smart Cities: Recent Developments and Perspectives. *Electronics* **2018**, *7*, 404. [CrossRef]

electronics

MDPI

Article

A Comprehensive IoT Node Proposal Using Open Hardware. A Smart Farming Use Case to Monitor Vineyards

Sergio Trilles *, Alberto González-Pérez and Joaquín Huerta

Institute of New Imaging Technologies, Universitat Jaume I, Av. Vicente Sos Baynat s/n, 12071 Castellón de la Plana, Spain; algonzal@uji.es (A.G.-P.); huerta@uji.es (J.H.)
* Correspondence: strilles@uji.es; Tel.: +34-964-387686

Received: 17 November 2018; Accepted: 4 December 2018; Published: 10 December 2018

Abstract: The last decade has witnessed a significant reduction in prices and an increased performance of electronic components, coupled with the influence of the shift towards the generation of open resources, both in terms of knowledge (open access), programs (open-source software), and components (open hardware). This situation has produced different effects in today's society, among which is the empowerment of citizens, called makers, who are themselves able to generate citizen science or build assembly developments. Situated in the context described above, the current study follows a Do-It-Yourself (DIY) approach. In this way, it attempts to define a conceptual design of an Internet of Things (IoT) node, which is reproducible at both physical and behavioral levels, to build IoT nodes which can cover any scenario. To test this conceptual design, this study proposes a sensorization node to monitor meteorological phenomena. The node is called *SEnviro* (node) and features different improvements such as: the possibility of remote updates using Over-the-Air (OTA) updates; autonomy, using 3G connectivity, a solar panel, and applied energy strategies to prolong its life; and replicability, because it is made up of open hardware and other elements such as 3D-printed pieces. The node is validated in the field of smart agriculture, with the aim of monitoring different meteorological phenomena, which will be used as input to disease detection models to detect possible diseases within vineyards.

Keywords: Internet of Things; open hardware; smart farming

---

1. Introduction

More than 15 years ago, initiatives such as Arduino [1] constituted the first project to enable citizens to make their own prototypes. Subsequently numerous initiatives appeared, such as Raspberry PI [2], Beaglebone [3] and PCduino [4], among others. Similar to Arduino, all these projects were characterized by making their schematics available; this is known as open hardware [5]. In addition to the open-hardware movement, there has also been a significant drop in the price of these types of hardware platforms [6], thanks to advances in semiconductor manufacturing technology. These platforms have become more affordable and have been distributed efficiently, due to the open-source distribution policy.

All of this means that these open-hardware platforms are very well known in our day-to-day activities [5]. A large number of projects have been developed, which bring end users closer to electronics in a fast and straightforward way [7,8]. This approach is summarized in the Do-It-Yourself (DIY) initiative, where the end user becomes the consumer and creator of these technologies and projects, thus eliminating structural, technological, and economic obstacles [9]. In recent years, initiatives such as Instructables [10], Make Magazine [11], OpenMaterials [12],

Adafruit [13] and Sparkfun [14] have appeared, offering tutorials and instructions on how to use these open-hardware components.

The spread of this movement has aided the proliferation of devices which are always connected to the Internet, either directly or through a gateway, called Internet of Things (IoT) devices [15,16]. This proliferation has led to a real revolution, within environments such as industry, which has generated a new industrial model, called Industry 4.0 [17], where everything is connected to everything. Many projects with open-hardware devices have been used in the industrial domain and in others, such as smart cities [18], healthcare [19], agriculture [20] and the domotics [21], among others.

To connect these IoT devices to a server to manage them and handle all their functionalities, wire or wireless communication is required. For this purpose, all open-hardware platforms have been adapted to support technologies such as Bluetooth, Zigbee, Wi-Fi and 3-5G, among others [22]. To establish these communications, protocols are required. Although HTTP through RESTful interfaces is commonly used, other protocols such as Constraint Application Protocol (CoAP) and Message Queuing Telemetry Transport (MQTT) are suggested to replace HTTP [23]. The most widely used connectivity protocol in IoT and Machine-to-Machine (M2M) applications is MQTT [24]. MQTT is a lightweight protocol designed to connect physical devices [25] to IoT middleware due to it offering better energy performance. This last consideration is significant because IoT solutions are usually installed in harsh environments without an electrical connection.

This study focuses on providing a solution to design an IoT node using open-hardware components. More specifically, the main goals are (a) to propose an IoT node architecture design, both at physical and logical levels; (b) to guide a step-by-step example of how to build an IoT node with open-hardware components and provide a replicable research; and (c) to validate the proposal within smart farming by proposing effective M2M communication.

The balance of this paper is organized as follows. Section 2 presents the background which positions the current study. Section 3 details the agnostic-technology IoT node. Section 4 presents a technological solution to develop the agnostic IoT node approach and reveals some energy tests. Section 5 validates the solution in a smart farming scenario. Section 6 enumerates and compares similar related work. The paper ends in Section 7 with conclusions and future work.

2. Background

In this section, we first present some different open-hardware microcontrollers. Then, to locate the IoT node approach, we define the IoT architecture. Finally, we detail the IoT protocols used to establish an Internet connection.

2.1. Open Hardware

As already mentioned, the cost reduction and the increase of open-hardware popularity have triggered different options for open-hardware microcontroller-based platforms [26]. In this study, the selected IoT scenario requires 3G connectivity, following this requirement this subsection presents some different platforms that support this kind of connectivity. The most notable platforms are: *Particle Electron, Adafruit Feather 32u4 FONA, Hologram Dash, Arduino GPRS shield, LinkIt ONE* and *GOBLIN 2*. All these options are completely or partially open hardware. Below, we provide a short description of each of them. Table 1 shows a more specific comparison.

Table 1. Comparison between different open-hardware microcontroller-based platforms with 3G connectivity.

	Particle Electron	Adafruit Feather 32u4 FONA	Hologram Dash	Arduino GPRS Shield	LinkIt ONE	GOBLIN 2
Microprocessor	ARM Cortex M3	ATmega32u4	ARM Cortex M4	-	MT2502A, ARM7EJ-S	ATmega328P
Architecture	32 Bits	32 Bits	32 Bits	-	32 Bits	32 Bits
Clock speed	120 MHz	120 MHz	120 MHz	-	260 MHz	16 MHz
RAM	128 KB	2 KB	128 KB	-	4 MB	2 KB
Flash	32 KB	32 KB	1 MB	-	16 MB	32 KB
Min. power	42 mA	500 mA	700 mA	-	3 mA	300 mA
Cellular modem	3G/2G	2G	3G/2G	2G	2G	3G/2G
PINS	36 Pins (28 GPIOs)	28 pins (20 GPIO)	25 GPIO	12 GPIO	Arduino pin-out	36 Pins (25 GPIOs)
Fuel Guauge	Yes	No	No	No	No	No
Over-the-air (OTA)	Yes	No	Yes	No	No	No
Cost	$69.00	$44.95	$59.00	$4.30	$59.00	$89.99
Measures (mm)	8 × 6.5 × 20.5	61 × 23 × 7	20.32 × 56.68 × 16.53	68.33 × 53.09	83.82 × 53.34	65.5 × 82.2
Open hardware	Completely	Completely	Completely	Completely	Completely	Completely

- **Particle Electron** uses the STM32F205 microcontroller. It presents 36 total pins, such as UART, SPI, I2C, and CAN bus. Electron provides 1 MB of Flash and 128 k of RAM. If we compare Electron with Arduino, the first one is a competent board. The hardware design for the Electron is open source. It includes a SIM card, with a global cellular network for connectivity in 100+ countries, and cloud services. All Electron family products can be set up in minutes using the Particle mobile app or browser-based setup tools.
- **Adafruit Feather 32u4 Fona** is created by Adafruit and is an Arduino-compatible microcontroller plus audio/SMS/data capable cellular board. It is powered by a Li-Po battery for mobile use and micro-USB port when stationary. Feather is a flexible, portable, and light microcontroller. The SIM800 is the heart of this board and supports GSM cellular connectivity.
- **Hologram Dash** allows for interaction with devices by easily routing incoming and outgoing messages via a secure and scalable API. Hologram offers the Hologram Dash, compatible with Arduino IDE. The board is pre-certified for end-use and equipped with the Hologram's networking firmware and OTA code updates. It offers a cloud-friendly cellular connection, a SIM is included to connect and send messages for free (up to 1 MB) for life.
- **Arduino GPRS shield** connects an Arduino to the Internet using the GPRS wireless network. The Shield is compatible with all boards which have the same form factor (and pin-out) as a standard Arduino Board. This shield is configured and controlled via its UART using simple AT commands and is based on the SIM900 module from SIMCOM. It has 12 GPIOs, 2 PWMs, and an ADC. Moreover, as always with Arduino, every element of the platform (hardware & software) makes it easy to get started.
- **LinkIt ONE** includes a development board, Wi-Fi and Bluetooth antenna, GSM (2G) antenna, and a GPS/GLONASS antenna, all powered by a Li-battery. LinkIt ONE uses hardware and an API that is similar to Arduino boards. It uses MediaTek MT2502A SoC to get some features such as communications and media options, with support for GSM, GPRS, Bluetooth 2.1 and 4.0, SD Cards, and MP3/AAC Audio, Wi-Fi and GNSS.
- **GOBLIN2** uses a high-performance ATmega328P microcontroller to develop IoT projects. It is compatible with Arduino. GOBLIN2 is built with a module to control the charge of a Li-Po battery from 3.7V to 4.2V. The GOBLIN2 charges using a solar cell or a Micro-USB.

All the cellular microcontrollers listed work similarly, they use a mobile network to transmit data to and from the cloud. All of them work correctly; the characteristics are shown in Table 1 can be used to select the microcontroller for a specific use case.

For the use case presented (Section 4), we have chosen the Particle Electron microcontroller. When compared with other platforms, Particle Electron is more appropriate for autonomous work, since it features different sleep mode functionalities. Currently, Particle has a vast user community [27] that can help to resolve specific issues. All Particle microcontrollers are easier to use and have lower prices than the others presented above. It offers a complete solution concerning hardware, network, and cloud management.

2.2. IoT Architecture

IoT devices establish any communication network using a set of rules (protocols) for data transmissions. The TCP/IP architecture is the framework that underpins the communication rules within the Internet. More specifically, this architecture describes four layers: the Perception layer, the Network layer, the Middleware layer, and the Application and Business layer [28].

- **Perception layer**: in the same level as the physical layer in the TCP/IP model. Perception layer known as the "Device Layer" contains sensor devices and physical objects belong in it. Its role is to *Capture* information from each sensor device (state, temperature, and location, among others).

- **Network layer**: transfers the information from the sensors to the upper layer. It can use 2G/3G, Wi-Fi, Bluetooth, Zigbee, infrared technology, and so on. It corresponds with the *Communication* stage.
- **Middleware layer**: receives data from the network layer, stores and analyzes data from a sensor and sends it for its visualization.
- **Application layer**: manages all the processed information from the Middleware layer in a particular IoT application,
- **Business layer**: responsible for the management of the whole IoT system and applies a business model to a smart scenario, such as smart health, smart home, smart mobility, and smart agriculture, among others.

This study is focused on the first two layers (Perception and Network). Both layers are implemented at the IoT node itself. In what follows, both layers are defined in depth.

The Perception layer is the origin of the information and the first layer of the IoT model. It contains the components themselves and how they operate at a logical level to meet a required functionality. The main goal of this layer is to observe and act. It can obtain any type of information from the physical world through sensors, such as Quick Response (QR) or tags (RFID), pictures (camera) and location (GPS), among others. It also enables the ability to act on some type of actuator, such as switches, motors, etc.

This layer is defined by the following: (1) that it is composed of devices at the hardware level. It must adapt to the final application where the node will be integrated; (2) that it can be deployed in a huge range of environments (wireless/wire, power supply with batteries, and so on); (3) that heterogeneity is one of the main virtues of this layer, since it offers a vast variety of hardware; (4) that it must offer a hardware component to establish a communication connection (Bluetooth, ZigBee, 3G, and so on); and (5) that it is organized following a strategy such as star, cluster tree or mesh, among others.

The network layer is responsible for transmitting data in a transparent way to a neighboring node or to a gateway with Internet access. The layer performs an important role in the secure transfer and keeps information going from the sensor devices to the central server confidential. Depending on the type of connectivity technology included in the previous layer, it establishes a connection to the Internet using 3G, 4G, UMTS, Wi-Fi, WiMAX, RFTD, infrared or satellite, among others. The capabilities of this layer are: (1) network management technologies (wire, wireless and mobile networks); (2) energy efficiency of the network; (3) Quality of Service (QoS) requirements; (4) processing of data and signals; and (5) security and privacy.

2.3. IoT Protocols

Due to the proliferation of IoT devices, different protocols to manage these IoT nodes have been developed to meet their features [29]. These protocols define a set of rules that both endpoints in each part of the communication must use and know to make a successful connection. These are used by the network layer and are called M2M protocols. The main M2M protocols are MQTT and CoAP, which are defined below:

- **MQTT** follows a client-server paradigm, where each part publishes or subscribes a messaging transport agreement. It is agile, open, simple, and designed to be easy to deploy. The program can run over the TCP/IP network or another. The main features are: (1) use of the publish/subscribe message pattern that provides one-to-many message distribution; (2) an agnostic messaging transport to payload content; and (3) the offer of three quality services for message delivery: "At most once", "At least once" and "Exactly one time";
- **CoAP** is a dedicated web-streaming Protocol for use with limited nodes and a limited network. The protocol provides a request-response interaction model between nodes endpoints, embedded support services, and discovery resources, and includes key Web concepts such as Uniform

Resource Identifiers (URIs) and Internet media types. CoAP is designed to be an HTML-friendly interface for integration with the Web while at the same time addressing specific requirements, such as support for multicast with small overhead.

Both MQTT and CoAP are designed to meet the requirements of M2M applications in restricted environments. The CoAP communication model is similar to the HTTP client/server model and is based on the REST architecture. This requires the client to have prior knowledge of the server preferences to establish a connection. To avoid the intensive resources required by HTTP [30], it is preferable to use CoAP in edge-based devices. As already mentioned, CoAP runs on UDP with support for multicast addressing. Similar to HTTP, CoAP provides secure communications based on the security of the datagram transport layer (DTLS).

Unlike CoAP, MQTT runs through TCP/IP [24]. This protocol has been recognized as particularly suitable for devices with IoT restrictions and unreliable mesh networks. Due to its lightweight realization, MQTT has been used in a variety of sensor network scenarios: smart homes, telemetry, remote monitoring, warning systems and health care, among others. Unlike CoAP, MQTT uses a publish/subscribe architecture. The MQTT interface differs from the REST architecture in the sense that there is an intermediary between the publisher and the subscriber, called Broker. MQTT uses SSL/TLS protocols for secure communications. This last protocol has been selected to establish a M2M connection in the current study, due to lower traffic generation and reduction in energy consumption [31].

3. A Generic IoT Node Architecture

The current IoT node architecture proposal aims to follow a modular design at all levels, both at the hardware components and at the behavioral level. In this way, it is intended to design a transversal architecture over any proprietary or specific solution of particular hardware or application. At the hardware level, blocks with a specific functionality are defined. These blocks will be materialized by any open-hardware component that exists in the market. Regarding these blocks, they are classified into four different groups to define the IoT node concerning physical components.

At the behavioral level, general modules are also defined, so that they abstract the technology that will be used after their development. Seven different modules are defined that include all the IoT applications in any IoT scenario.

In this way, this section presents a conceptual design of an IoT node. The section is divided into two sections. The first one shows all conceptual physical parts and how they are structured and connected. The second subsection exhibits the different logic modules needed to meet all IoT abilities.

3.1. Physical Conceptual Parts

At the level of physical components (or functionalities), the node follows the same detailed composition in [7], although, in this new version, some modifications are added to increase the possibilities. The new composition is shown in Figure 1. Similar to [7], the current approach is also composed of four well-differentiated groups depending on their function. These are *Core*, *Sensors/Actuators*, *Power supply*, and *Communication*. This approach constitutes an extension of [7] in that it features the coalescence of actuators within the *Sensors* category. Consequently, the group has been renamed as *Sensors/Actuators*. Below, each category is described in the form in which it will be distributed in the current study.

The first group, *Core*, does not present physical modifications from the previous one presented in [7]. It is divided into four subgroups (*Microcontroller*, *Connectors*, *Memory*, and *Clock*) and its functionality remains the same as that presented in the cited work.

One of the main changes is focused on the *Power supply*. That is because the new design proposal puts more emphasis on offering an energy-autonomous solution. In this way, the *Power Supply* group has been divided into four elements. They are detailed below:

- **Battery** offers energetic autonomy to keep the IoT node power up. The capacity will vary depending on the consumption of the node.
- **Solar panel** functions to generate energy to recharge the battery and maintain its charge.
- **Gauge counter** provides the current status of the battery, which will be used to establish an energy strategy that will favor energy saving and the life cycle of the node.
- **MPPT** optimizes the match between the solar panel and the battery. MPPT can convert the higher voltage DC output from the solar panels down to the lower voltage needed to charge the batteries.

Sensors and Actuators compose the third group. The former can observe a physical phenomenon, such as temperature, particulate, gas, or humidity, and take and transform a signal produced by the sensor into a quantitative or qualitative value. Actuators allow interaction with an IoT node environment to produce a change, such as turning off a light or deactivating a watering function. Each node can be composed with different configurations of sensors and/or actuators.

The last group (*Communication*) centralizes all telecommunication technologies to connect devices with other nodes, a gateway, or a central server. Different (wired or wireless) network technologies are used to establish a connection (see Section 2.2).

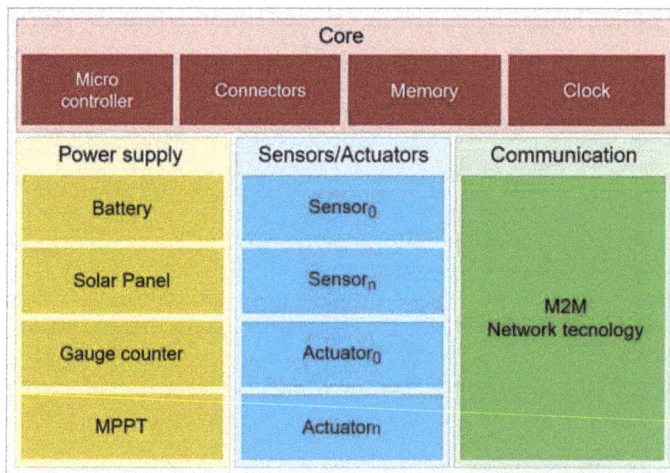

Figure 1. Different hardware concepts that compose the IoT node.

3.2. Behavior Conceptual Parts

After detailing how the IoT node is built at the physical level, this section describes the modules that are included for its functionality at the logical level. Just as it is defined at the physical level, IoT node also follows a modular design at behavioral level. In this way, seven modules are defined (Figure 2). These modules are *Logic control*, *Basic config*, *Communication*, *Sensing*, *Acting*, *Energy savings*, and *Update mode*. Below, each module is detailed.

- **Logical control** is in charge of defining all the IoT node logic and connects with the other six logical modules.
- **Basic configuration** is responsible for saving the configuration of settings in the node, such as how it can establish a connection, what sensors it should observe and if the node has an update, among others.
- **Communication** establishes the connection to other nodes or a server. It knows the features of the hardware component to connect to, as well as the M2M communication protocols.
- **Sensing** collects all data from the available sensors. Some transformations will be required to transform a sensor signal to a useful value.

- **Acting** will be responsible for carrying out the management of the available actuators and the actions that are taken.
- **Energy saving** defines strategies for energy saving and prolongs the battery life to its maximum.
- **Update mode** updates the node when an update is pending.

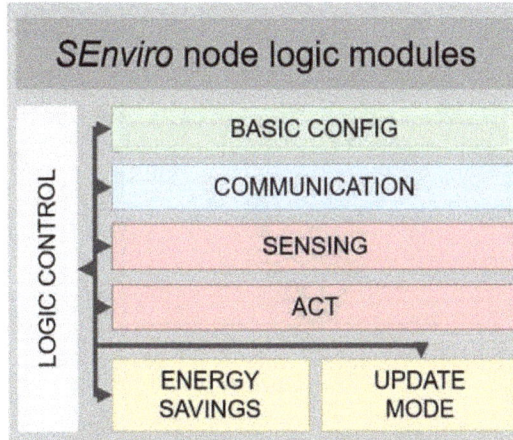

Figure 2. Different behavior conceptual parts.

Figure 3 reveals a state machine that IoT node arises to fulfil with its functionality. A state machine can be a theoretical mechanism, or a digital device in our case, that defines a set of states, and movements between states which are performed according to specific criteria [32].

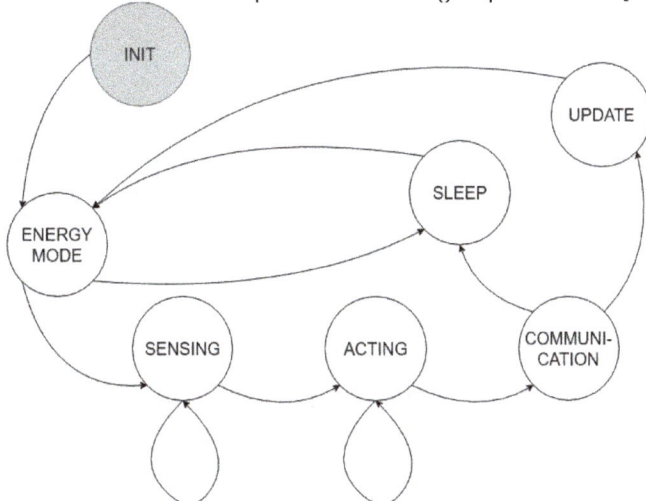

Figure 3. A state machine defined using the logic modules defined.

The two main components of a state machine are states and transitions. The states are points on the machine where it performs some action, in our case the defined logical modules. Transitions are a set of conditions that must be met for a machine to pass from the current state to another, these transitions will depend on the input of sensors, battery, or elapsed time, for instance.

It is important to note that the state machine can only be in one status position at a time. That means that the machine only must check the conditions of the current state (depending on the IoT node state, battery, sensing values, and so on) and verify the new state to which it will subsequently move. The state machines require a persistent memory (provided by the IoT node) and the ability to perform logical operations. The logical operations are those that determine the actions of a state and calculate the conditions that must be met to move to the next state.

More specifically, the defined state machine is adapted to the previous behavior shown in Figure 2, although some differences appear that are detailed below. The *Logic control* module corresponds to the state machine itself. The state machine connects and determinants the different states depending on the transitions.

The first difference is the *Basic Configuration* module, which does not appear directly, but does so through the INIT state. The INIT state is where the configuration variables will be defined, and the first values will be set. These variables will be updated throughout the procedure of the state machine and will be used to discern between states.

The states *energy mode, sensing, acting, communication,* and *update* correspond to the modules that bear the same name and perform the same function. The *sensing* and *acting* states may return to their own state depending on the number of sensors and actuators.

The *Sleep* state supposes another difference from the previous version. It is responsible for putting the machine into rest mode and for applying the energy strategies determined in the *energy mode* state.

The *energy mode* state is the first to be activated after initialization; this is because the first step will be to determine the amount of battery, and thereby adapt to the best possible energy strategy to ensure the IoT node life cycle.

4. A Technological Solution to Develop IoT Nodes Using Open Hardware: *SEnviro* Node

In this section, an example of the IoT node (called *SEnviro* node) which follows the proposed architecture is presented. The section is divided into three subsections. The first subsection shows how the IoT node is built at the hardware level. The second subsection details how the IoT node works at the software level. Finally, the last subsection analyses the energy consumption both theoretically and practically.

4.1. Building a SEnviro Node

The main objective of the current work is to propose an example of IoT node which follows the architecture made using open-hardware components, as detailed in the previous section. In this way, the development below is composed intimately of these types of elements.

Following the physical components defined in Section 3, the IoT node is composed as shown in Figure 4. We now detail how each defined group (*Core, Sensing/Acting, Power Supply* and *Communication*) is developed. The same categorization is used to describe each component in Table 2. The total price per *SEnviro* node is € 256.45. In what follows, each component is detailed and classified in each fixed group.

- **Particle** (Figure 5): This component has been chosen to be part of the core of the IoT node. As introduced in Section 2, the Particle Electron microcontroller follows an open-source design. As shown in Figure 5, this microcontroller is present in different blocks defined in Figure 1, such as *Core, Power Supply* and *Communication*. In the first group, *Core,* the Electron acts as a microcontroller and is responsible for bringing the node to life and implementing all business models so that the node functions correctly. Unlike a conventional PC, Electron can only run one program, which will be detailed in the next subsection. It incorporates a STM32F205RGT6 ARM Cortex M3 chip which works at a frequency of 120 MHz. It can be updated using OTA updates. This functionality adds a considerable expansion regarding keeping each node updated and supporting new functionalities or behaviors in the future without the need to physically go to where the IoT node is deployed. The Electron will also be responsible for storing all the variables to support a normal operation by using RAM (128 KB RAM) and ROM (1 MB Flash). Within the

Core group, it will also function to keep the current time; this is possible because Electron offers a Real-Time Operating System (RTOS) module.

The second group in which the Electron microcontroller is present is the *Power supply*. The Electron provides a chip called MAX17043, which can measure the energy consumed by the microcontroller (and all IoT node components). This microchip has an ultra-compact design, and its cost is meagre.

Finally, in the last group, *Communication*, the announced microcontroller provides 2G and 3G (or any new technology such as 5–6 G) connectivity. The cellular module, U-blox SARA-U270, allows an Internet connection. To do this, it has a cellular antenna, which is essential for the microcontroller to establish a connection to a cellular tower. One difference from the version presented in [7] can be found in the communication group. This new version provides 2G and 3G connectivity instead of Wi-Fi. This feature increases the possibility of installing a node anywhere with mobile data coverage.

- **Weather shield**: This component is an easy-to-use circuit compatible with the Particle microcontrollers (Photon and Electron). The Electron can interact with the world through pins (30 Mixed-Signal GPIO), which will be used to connect this shield. The weather shield has built-in barometric pressure sensors (MPL3115A2), relative humidity and temperature (Si7021). Besides, the shield contains some RJ11 connectors to plug external sensors such as an anemometer and a rain gauge. Table 3 describes all the features of the sensors. The weather shield is present in the *Core* (Connectors) and *Sensors* groups.

- **Solar panel**: This component is waterproof, scratch resistant, and UV resistant. It uses a high-efficiency monocrystalline cell. The output is 6V at 530 mA via 3.5 mm × 1.1 mm DC jack connector. The solar panel will be used to charge the battery and offers an uninterrupted IoT node operation. The solar panel is included in the *Power supply*.

- **Lithium ion battery**: This component has a 2000 mA capacity and has been used to offer an energetically autonomous platform. It supplies an output voltage of 3.7 V and is charged using the power generated by the solar panel. The battery is included in the *Power supply*.

- **Sunny Buddy solar charger**: This component is a MPPT solar charger and is included in the *Power supply*. It is responsible for monitoring current flow and is limited to a specific value to prevent damage to the battery. It is also an open-hardware component.

- **Soil Moisture**: This component also has an open-hardware design, and through a simple breakout it can measure the humidity of the soil. The two pads act as a variable resistor; the more water there is in the soil, the better the conductivity between them. This results in less resistance and higher output. It is considered to be a part of the *Sensor group*.

- **Weather meters** provide the three components of weather measurement: wind speed, wind direction, and rainfall. It offers RJ11 connectors, which facilitates installation with the above-mentioned weather shield. The weather meters are included in the *Sensor group*.

An enclosure has been designed to house all the components and to protect the electronic components. The box has been purposely designed to meet this end (Figure 6), it has been printed using a 3D printer, and is valid for printing using polylactic acid (PLA) filament. Each node is identified by a QR code. This code is used to claim a device and start to collect data from the node.

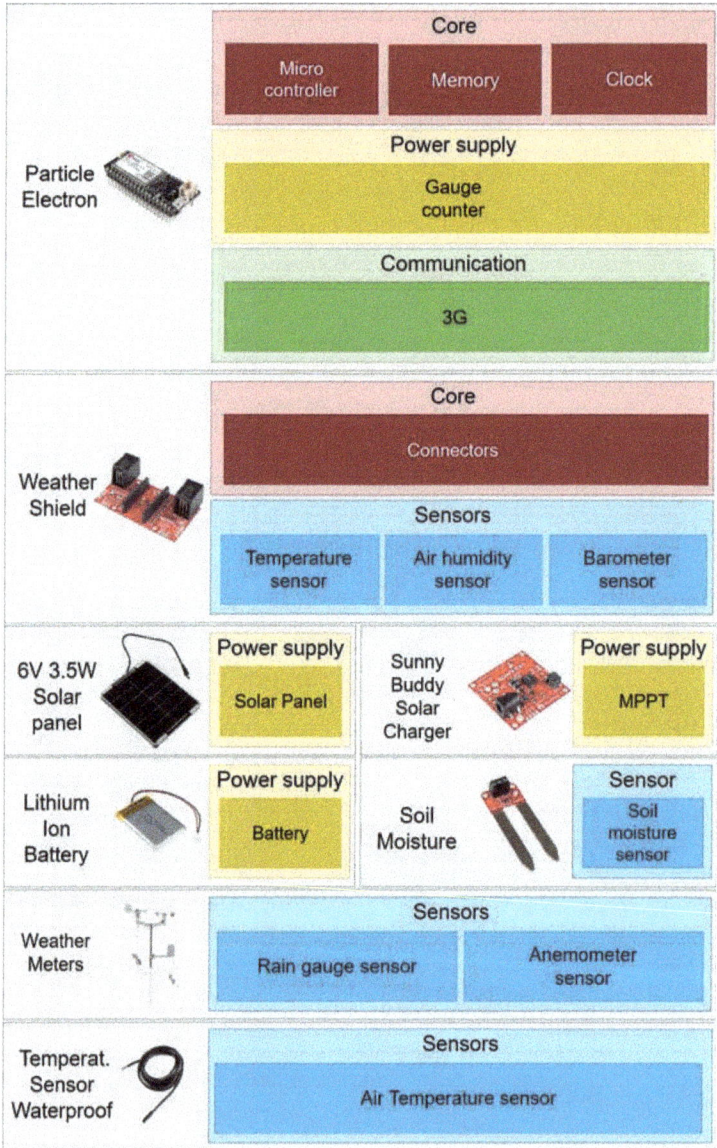

Figure 4. A list of open-hardware components to build a *SEnviro* node.

Figure 5. The different components used to build the *Core Power supply* and *Communication* groups.

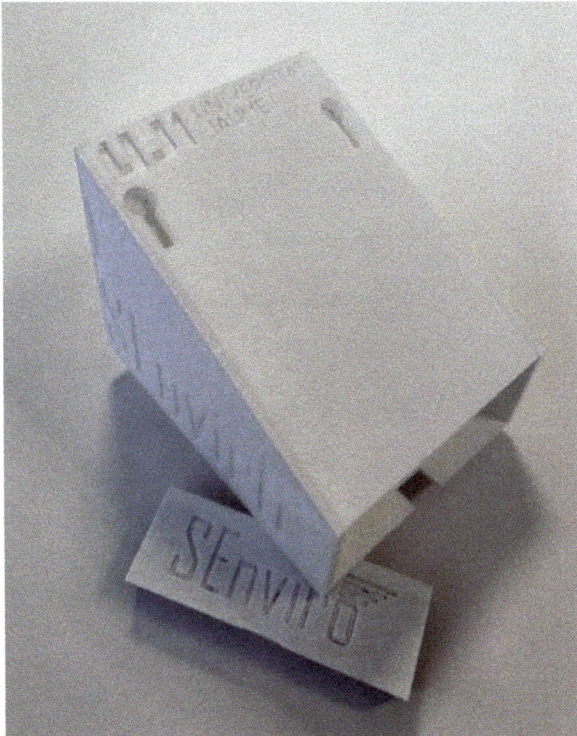

Figure 6. 3D printed enclosure.

Table 2. Components list included in the *SEnviro* node.

Category	Component	Description	Cost
Core	Microcontroller board	Particle Electron	€ 63.41
	Shield Weather	Base Shield V2	€ 27.02
	Box for sensors	3D printed box	€ 10.00
	Some connectors and cables		€ 5.00
Communication	3G module	Included in Particle Electron	€ -
Sensors	Temperature and humidity sensor		€ 12.60
	Rainfall, wind speed and direction sensors		€ 65.41
	Soil temperature		€ 8.46
	Soil humidity		€ 4.21
Power supply	Battery	Included in Lithium Battery 2200 mAh 3.7 V	€ -
	Solar panel	3W Solar Panel 138X160	€ 38.79
	Sunny Buddy	MPPT Solar Charger	€ 21.55
	Gauge counter	Included in Particle Electron	€ -

Table 3 summarizes all the information about the *Sensors* used in this proposal. All the *Sensors* that have been chosen are open hardware and low cost. Despite their low price, most of the sensors used in industrial environments obtain reliable measurements.

Table 3. Details of the included *Sensors*.

Sensor	Phenomena	Manufacturer	Model	Data Interface	Units	Range	Accuracy
Temperature and humidity sensors	Temperature	SparkFun	Si7021	Analog	Centigrade	[−10, 85]	±0.4 Degrees (C)
	Humidity				Rate	[0%, 80%]	±3 RH
Barometric pressure	Pressure	SparkFun	MPL3115A2	I2C	Hectopascal	[500, 1100]	±0.04 hPa
Temperature Sensor Waterproof	Temperature	SparkFun	DS18B20	Analog	Centigrade	[−55, 125]	±0.5 Degrees (C)
Weather meters	Wind speed	Sparkfun	SEN08942	Analog (RJ11)	km/h	Not specified	Not specified
	Wind direction				Direction (degrees)	[0, 360]	Not specified
	Rain meter				mm	Not specified	Not specified

4.2. Details of the Behavior

As indicated above, the microcontroller will be in charge of orchestrating the general operation of the node. That will be possible through a program, or sketch, that is responsible for giving functionality to each of the hardware components that make up the node. In this way, following the logic modules and state machine defined in the previous section, a sketch is developed. The Figure 7 shows at a global level how each module has been used. We will now go on to detail each of them and how they work.

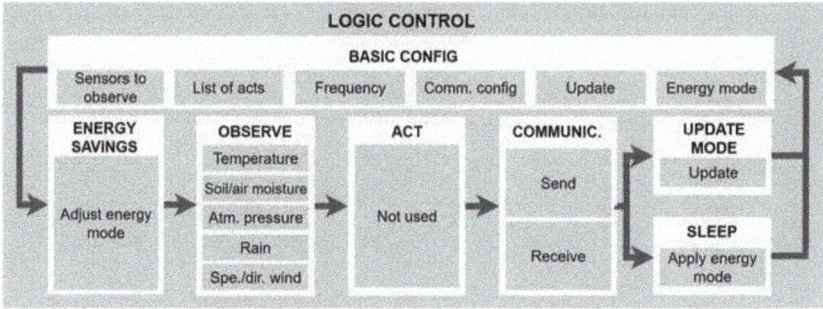

Figure 7. A workflow of the behavior modules.

The first module is *Logic control*, which does not have a specific functionality. It is responsible for joining the different modules in a coherent way. As in [7], two modes are defined, initial (INIT in states machine) and loop modes. The initial mode is executed as an initializer when the node boots. When the initial mode is complete, the loop method is executed repeatedly while the node is on.

The *Basic config* module is responsible for storing and managing the basic configuration/settings. Part of its configuration is defined in the initial mode and updated during the loop mode. The basic configuration is responsible for defining aspects such as which sensors will be consulted, the list of actions to be performed on the actuators, the observation frequency, the configuration to establish an M2M communication and checking whether the IoT node has a new update.

Within the loop mode, which is executed depending on the frequency of the *Basic config* module, the first module is the *Energy savings*. This module determines the energetic strategy depending on the battery charge. Three different modes are defined, each of which are detailed below.

- **Normal mode**: the most usual operating mode. The node works with the defined observation frequency (10 min).
- **Recovery mode**: the observation frequency is maintained, but the observations will not be sent to the server, instead they are stored in the microcontroller EEPROM memory. The stored observations are sent when the IoT node has better battery conditions.
- **Critical mode**: the IoT node does not collect new observations and goes into the most in-depth sleep mode.

The second module, *Observe*, consults from among the available sensors those that are indicated in the configuration. Among them, it includes the temperature, soil and air humidity, atmospheric pressure, rain, and wind speed/direction sensors. This module is responsible for transforming the signals from the sensors to obtain the correct values and adapt the values to the correct units for each phenomenon. For example, it treats the electric pluviometer pulsations and transforms them into the corresponding milliliters.

The next module is *Act*, which carries out actions on the available actuators. Although the current IoT node does not have an actuator due to the use case in which it has been applied, it is considered to define a general solution which is transferable to other scenarios with different requirements.

The next module is the *Communication* module. It has two differentiated functionalities: sending and receiving. To achieve this, the module establishes a connection using the 3G module included in the Electron microcontroller. Each node uses the MQTT protocol to set up M2M communication. They define an MQTT client that can send and receiving data. To do so, the node takes on the role of publisher to send sensor observations. Each observation is published on a different topic. In this way, a hierarchical organization is defined; that is, the sensor ID plus the phenomenon type name, for example */current/4e0022000251353337353037/Temperature* or */current/4e0022000251353337353037/Humidity*. The first hierarchical level (*current*) determines that the topic is in real-time. Another topic at the same level, called *lost*, is defined to send lost observations (due to in connection issues or the energy mode).

Moreover, MQTT is used to establish a dual communication between nodes. Each node is subscribed in different topics to trigger updates (OTA), or note which sensors should be consulted. The JSON format is established to code both incoming and outgoing data.

The last two modules are in parallel; this implies that they cannot be executed at the same time in the same iteration. The first one, the *Update* module, is responsible for enabling the deployment of a new update. Through the *Communication* module, when a new update message arrives in the update topic for a particular node, the update mode is activated and, to receive the update, the sleep state is not applied.

Finally, the last module, and one of the most important regarding energy consumption, is the *sleep* module. On the one hand, the Electron microcontroller offers different sleep modes, depending on which functionalities are active. The microcontroller provides a sleep operation, which is characterized by various parameters to determine which features are restricted to reduce power consumption. The sleep operation can be accompanied by seconds, which is used to define the sleep period. Also, the Electron supports the possibility to awaken using some external interruption in one of its pins.

In addition to the possibility of awakening the microcontroller, the sleep mode is also required. All the available modes are detailed below:

- Without any argument. This option does not stop the execution of the sketch. The node continues running normally, while the 3G module remains in standby mode. The energy consumption is from 30 mA to 38 mA.
- SLEEP_MODE_DEEP: this mode turns off the network module and puts the microcontroller in standby mode. When the device wakes up from the deep suspension, it restarts and executes all the user code from the beginning without maintaining the memory values. Its consumption is about 0.0032 mA.
- SLEEP_MODE_SOFT_POWER_OFF: this mode is like SLEEP_MODE_DEEP with the added benefit that the fuel gauge chip also sleeps.
- SLEEP_NETWORK_STANDBY: this mode is like SLEEP_MODE_DEEP but it does not turn off the 3G module. It reduces the amount of energy needed to reconnect with the operator when the Electron reboots from SLEEP_MODE_DEEP.

The Figure 8 shows at a generalization of how the node works at the energy level and how it applies the sleep operations previously listed. In this way, three different modes are defined depending on the battery level at any given time. These modes are listed below.

- **Normal mode**. When the battery is higher than 25% the node performs shipments depending on the observed frequency. The basic sleep mode is applied as detailed in the definition above. In addition to using the seconds to wake up, it can also use the rain gauge pin to wake up, and in this way accumulate the quantity of rainwater during the period in which the microcontroller is sleeping. The connectivity module is running while the cellular module is in standby mode.
- **Recovery Mode**. When the battery is at less than 25%, the recovery mode is activated, and no new deliveries are made until the battery exceeds 35%. The observations are saved in the EEPROM memory and are labelled with a timestamp at the moment of capture.

- **Critical Mode.** If the battery is at less than 15%, the critical mode is activated. In this mode, new observations are not produced or stored in EEPROM memory. The node goes into deep sleep and wakes up every 60 min to monitor the battery level. The critical mode is stopped when the remaining battery level is greater than 20%, and the node continues in recovery mode until it exceeds the threshold of 35%.

All the values defined above to categorize each mode have been fixed empirically. Depending on the different energy modes, we have tried to balance the behavior of the normal mode without compromising the total loss of battery, which this last would produce the loss of autonomy.

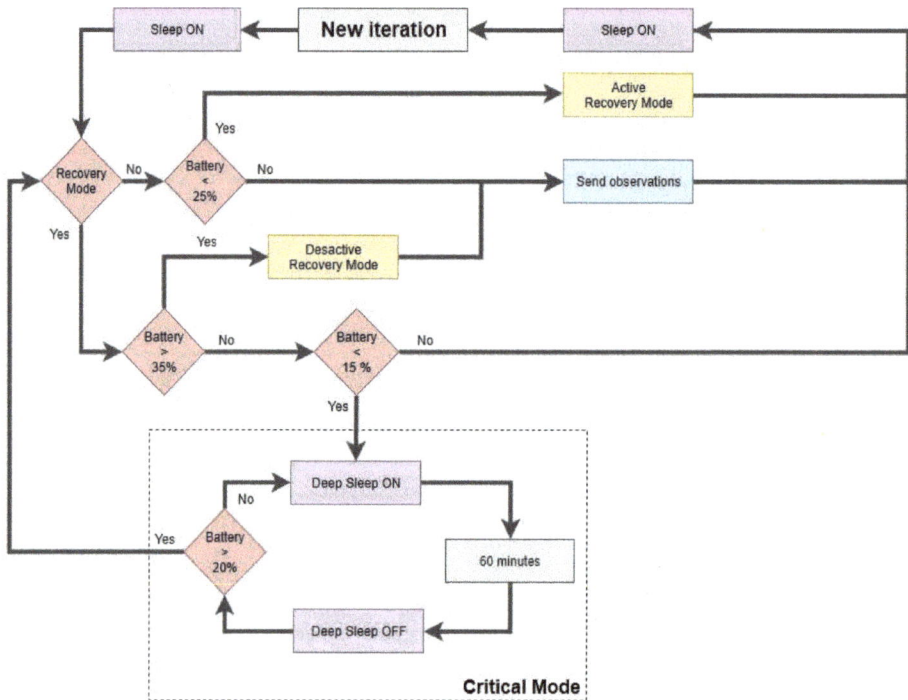

Figure 8. Energy modes diagram.

4.3. Autonomous Power Supply

This section details some aspects regarding energy consumption. The first subsection shows the theoretical energy consumption. The second subsection offers the energy consumption tests in a unit of the *SEnviro* node.

4.3.1. Theoretical Energy Consumption

Table 4 shows the energy consumption of the Electron microcontroller in the different modes (normal, sleep and deep sleep).

Table 4. Particle electron energy consumption.

Mode	Energy Consumption (mA)
Operating Current (uC on, Cellular ON)	180–250
Peak Current (uC on, Cellular ON)	800 (3 G), 1800 (2 G)
Operating Current (uC on, Cellular OFF)	47–50
Sleep Current (4.2V Li-Po, Cellular OFF)	0.8–2
Deep Sleep Current (4.2V Li-Po, Cellular OFF)	0.11–0.13

Consumption can vary considerably depending on the peripherals connected to the board. The different components along with their energy consumption are shown in Table 5.

Table 5. *Sensors* energy consumption.

Component	Energy Active (mA)	Energy Standby (mA)
Temperature and humidity sensor	0.15	0.00006
Barometer	2	0.002
Temperature (sonda)	1–1.5	0.00075–0.001
Rainfall, wind speed and direc. sensors (passive)	-	-
MPPT	2.5–3.5	0.085

To obtain the theoretical consumption, we consider the IoT node operating in the normal energy mode defined in the previous section. The other modes have a secondary role, and the normal mode is executed throughout the majority of its life assuming a proper operation.

In what follows, the Table 6 shows the seconds of each hour in which the node is applying energy-saving techniques, sending data or in the normal mode. The microcontroller consumes 19.3 mA every hour.

Table 6. Particle electron energy consumption in the normal mode.

Mode	Seconds	Energy Consumption (mA)
Peak Current	60	13.33 (3 G)
Operating Current (uC on, Cellular OFF)	300	4.17 (worst condition)
Sleep Current	3240	1.80 (worst condition)

The consumptions of all sensors included in the *SEnviro* node version are shown below. We consider that all sensors are active for 60 seconds during each iteration, which corresponds to the period in which the microcontroller is active. Table 7 shows the consumptions of the components per hour, in total it needs 2.1625 mA.

Table 7. Component energy consumption per iteration.

Component	Active Seconds	Energy Standby (mA)	Standby Seconds	Energy Standby (mA)	Total (mA)
Temperature and humidity sensor	360	0.02	3240	0.00005	0.02005
Barometer	360	0.20	3240	0.0018	0.2018
Temperature (waterproof)	360	0.15	3240	0.0009	0.1509
Rainfall, wind speed and direc. sensors	3600	-	-	-	-
MPPT	1800	1.75	1800	0.0425	1.7925

Regarding the MPPT, it has a consumption of 3.5 mA every hour when it is active (when it charges the solar panel) and 0.085 mA in standby. The number of solar hours in Spain has been taken into consideration, considering an average of 12 solar hours (minimum 9.5 solar hours and maximum

14.9 solar hours). The consumption of the rain gauge and anemometer is negligible since they are passive circuits.

In this way, the total consumption of the node (microcontroller, sensors, and components) is 21.4625 mA in the normal energy mode. Thus, with the indicated battery (2000 mA), the node can work continuously for 93.19 h, which corresponds to 3 days, 21 h and 11 min until the battery is exhausted.

As detailed above, two modes have been added (Recovery and Critical) to prolong the life of the node, with which it is expected to extend the life of the node to so that it can function in long cloudy periods with little presence of the sun.

The recovery mode starts up when the battery is at less than 25%, that is when it reaches approximately a battery load of 500 mA. Table 8 shows the energy consumption of the different states within the recovery mode. Unlike the normal mode, the recovery does not realize deliveries; otherwise, the node saves the observations in EPROM, to be sent later, when it goes into normal mode.

Table 8. Particle electron energy consumption in recovery mode.

Mode	Seconds	Energy Consumption (mA)
Operating Current (uC on, Cellular OFF)	360	5.00 (worst condition)
Sleep Current	3240	1.80 (worst condition)

The energy consumption of the microcontroller in Recovery mode is around 6.80 mA per hour, to which should be added the consumption of the components (2.1625 mA) shown in Table 7. In total on consumption of 8.9625 mA, and with the remaining battery of 500 mA, the node could extend its life to 55.788 h (2 days, 7 h, 18 min) until the entire battery was discharged.

The last and most severe mode is the critical mode, which is activated when the battery is at less than 15%, that is when it reaches approximately a load of 300 mA. Table 9 shows the consumption of the different states within the critical mode. Unlike the other modes, the critical mode keeps the node in a deep sleep and only wakes up to check the battery status, it does not consult the sensors, so any observations are lost.

Table 9. Particle electron energy consumption in critical mode.

Mode	Seconds	Energy Consumption (mA)
Operating Current (uC on, Cellular OFF)	5	0.07 (worst condition)
Deep Sleep Current (4.2V Li-Po, Cellular OFF)	3595	0.1298 (worst condition)

In this case, the sensors are not being consulted, despite this a consumption of 0.02916 mA is generated during the period in which the battery level would be consulted, since the sensors are in a standby mode. To this the consumption of the microcontroller of 0.1998 mA per hour should be added, so that the total energy consumption of the critical mode would be 0.22896 mA per hour. With this consumption and with the remaining 300 mAh, the node could extend its life to 1310.27 h (54 days, 14 h and 16 min).

Figure 9 shows the theoretical energy consumption of the node. The chart indicates when changes in energy modes occur. These changes would occur: from the normal mode to the recovery on the 3rd day; from the recovery mode to the critical the 5th day; and through the critical mode, the node would be without a battery after 37 days (864 h).

Theoretical Energy Consumption

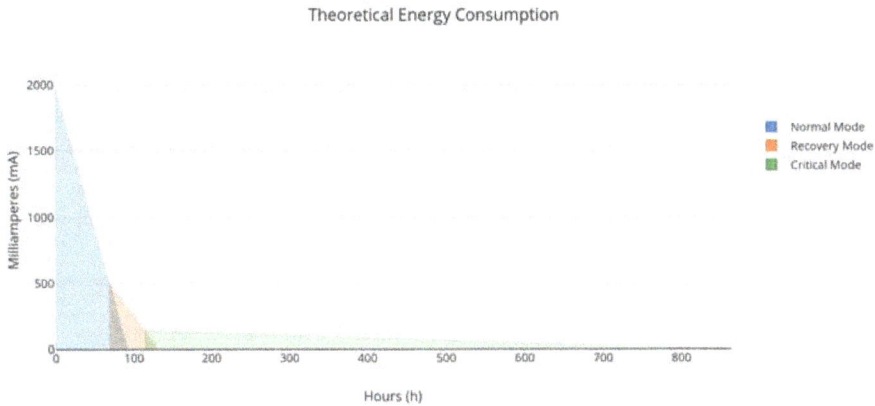

Figure 9. Theoretical energy consumption without solar panel.

4.3.2. Real Energy Consumption

The first test without a solar panel and using only the normal mode with a 2000 mAh battery reveals 73.5 h of autonomy. Nonetheless, the real test suggests that under such conditions the energy consumption is 27.25 mA per hour, instead of the 19.3 mA defined in the theoretical consumption. If we compare the results obtained with the solution presented in the previous version of *SEnviro*, we obtain a substantial improvement, tripling the energy autonomy.

This result seems satisfactory in terms of being able to deploy the nodes and keep them permanently active. The developed IoT node could keep the sensor alive for three days, one hour and 30 minutes without any charge through the solar panel. The first energy problems could appear after three cloudy days without any sun. However, even on a cloudy day, according to the tests made, the solar panel can charge with a low frequency or maintain the battery. Besides, the solar panel presents very satisfactory results; it can charge the battery with an approximate rate of 200 mA per hour.

As will be shown in the next section, five *SEnviro* nodes were deployed for 140 days, and none of them presented energy problems. During this period there were several consecutive cloudy days, which did not affect node operation. The solar panel also managed to charge during some of those days, although at a lower rate.

The chart below (Figure 10) shows the time series of the battery level of one of the nodes. It shows the recharge capacity of the solar panel, which manages to reach the highest battery charge possible in a few hours. The maximum level has been 87.4% of battery. We can assert that, on sunny days, at the end of the night, the battery never drops below 60.77% and manages to recover its charge in 3 h of sunshine. On cloudy days (like the 2nd of November) the battery can maintain the level and increase with a low rate. During the period indicated period no node entered into recovery mode.

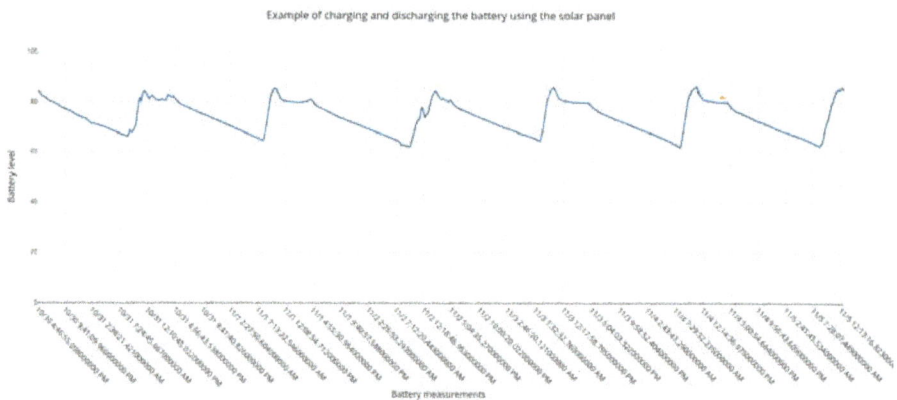

Figure 10. Example of charging and discharging the battery using the solar panel.

5. Use Case: *SEnviro* Node for Agriculture

This section presents a use case to test and validate the platform presented in the previous sections. First, the context where the *SEnviro* node has been deployed is described. In the second subsection, a sensor network deployment in vineyards is described. Finally, an IoT platform (*SEnviro* connect) developed to manage *SEnviro* nodes, and show sensor data and detected alerts is detailed.

5.1. Viticulture Context

One of the areas which has experienced the greatest adaptation of IoT platforms is viticulture. Viticulture has historically been characterized by obtaining a high-quality product. This has been made possible by many factors: the selection of suitable soil and climatic zones, as well as the varietal wines and the work of the winegrower in managing vineyards [33].

SEnviro for agriculture is used for monitoring and detecting vineyard diseases. The two components (*SEnviro* connect and Node) shown in Section 4 are applied to work together to achieve this goal. The main objective is to adapt both platforms to follow the models of diseases on vineyard crops. Based on previous studies about different models to predict vineyards diseases [34–37], we have defined four different diseases to predict. These are *Downy mildew*, *Powdery mildew*, *Black rot* or *Botrytis*. In what follows, these vineyard diseases are briefly summarized.

- **Downy mildew**: it looks like yellow to white spots on the upper surfaces of the leaves. The infection can start after rain or heavy dew. As the disease progresses, the leaves turn brown and fall off.
- **Powdery mildew**: initial symptoms appear as spots on the upper leaf surface that soon become whitish lesions. It grows in environments with high humidity and moderate temperatures.
- **Black rot**: a fungal disease that attacks grape vines during hot and humid weather. It causes complete crop loss in warm or humid climates but is unseen in regions with arid summers. This disease also attacks the shoots, leaf and fruit stems, tendrils, and fruit.
- **Botrytis**: a fungal disease that can occur anytime during the growing season, but most commonly occurs near the harvest time. At that time, birds and hail damage can encourage infection. Usually, Botrytis infects ripe berries.

All these diseases are based on meteorological conditions. For that reason, the *SEnviro* node is adapted to collect information on eight meteorological phenomena, which are directly related with these disease models. To do so, the node includes sensors to measure soil and air temperature, soil and air humidity, atmospheric pressure, rainfall, wind direction, and wind speed.

5.2. SEnviro Nodes Deployment

Five units of the *SEnviro* node have been deployed; four nodes have been installed in vineyard fields in the province of Castelló (Spain) Figure 11, and one unit has been reserved for testing in a more accessible location, closer to the laboratory (Table 10). At the time of writing this study, the nodes have run continuously and uninterruptedly for 140 days. Each node sent an observation every ten minutes during the vine season 2018.

Figure 11. Some pictures about all *SEnviro* node deployments on vineyards.

Table 10 summarizes the success of all observations during this period. The unit with the best success ratio is number five, which showed a 98.75% rate of success (this unit was only deployed for 69 days). The worst performing unit is number four, the one which had its battery changed.

Table 10. List of *SEnviro* nodes deployed.

S. Num.	Location (Lat, Lon)	Succes. Obser.	Lost Obser.	Success Rate
1	39.993934, −0.073863	18,354	1230	97.86%
2 (Figure 11a)	40.133098, −0.061000	19,055	109	98.00%
3 (Figure 11b)	40.206870, 0.015536	18,152	161	97.05%
4 (Figure 11c)	40.141384, −0.026397	18,729	440	91.96%
5 (Figure 11d)	40.167529, −0.097165	9626	68	98.75%

Some alerts were launched during this period, related to both vineyard diseases and the node itself. The objective of this study is not to evaluate or validate the different models applied to predict vineyard diseases, because there are various studies in the bibliography which have already done so. Our study only presents a validation to test the IoT architecture and how it can be used to monitor and predict disease in vineyard fields.

5.3. SEnviro Connect

The IoT node defined in the previous sections forms part of a complete IoT project called *SEnviro* for agriculture. Figure 12 summarizes all elements of the annotated project.

At the hardware level (the purple part in Figure 12), we can locate the presented IoT node. If we look at *SEnviro* at the software level (the blue part in Figure 12), we find the *SEnviro* connect [38]. It provides a technological solution to manage and analyze IoT data. There are two layers covered by *SEnviro* connect, the Middleware layer, and the Application/Business layer.

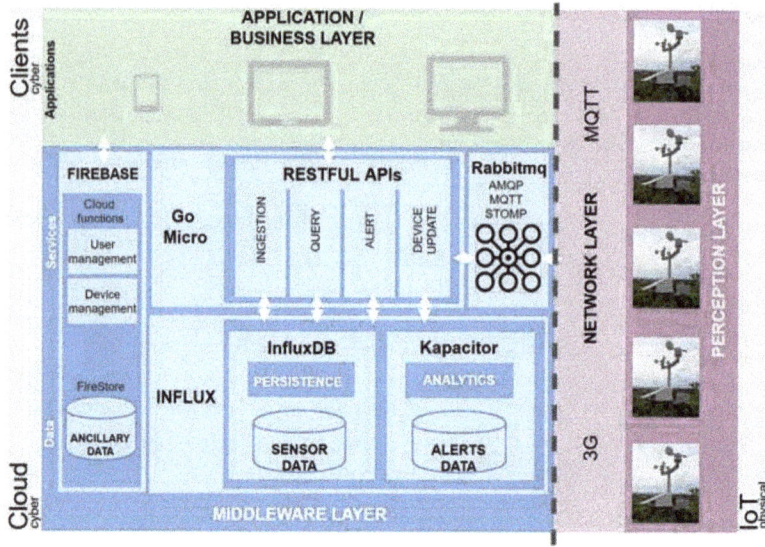

Figure 12. A general overview of the full IoT environment.

One of the most important components to establish a bridge connecting IoT nodes with *SEnviro* connect is the broker. The broker offers a publish-subscribe base messaging protocol, called MQTT. It is designed for connections with remote locations where the *SEnviro* nodes are located; usually, the network bandwidth is limited. This component is based on a RabbitMQ instance. All components listed in the Figure 12 are detailed in [38].

Moving from cloud to client, as a first prototype (Figure 13), a client based on HTML5, JavaScript and Cascading Style Sheets (CSS) has been developed. This client has been developed using Angular framework and follows a responsive design. The client is divided into two basic views, one view shows the *SEnviro* node state (battery, location, last connection, and node alerts); and the second one is focused on the vineyard use case.

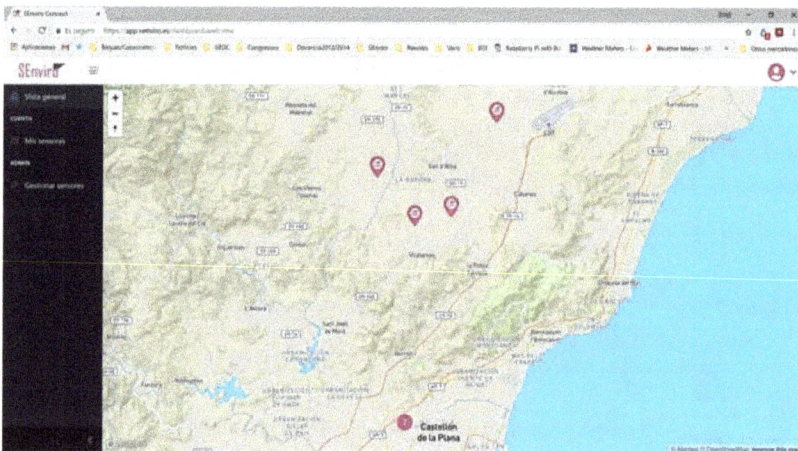

Figure 13. A screenshot of the *SEnviro* connect showing all *SEnviro* nodes deployed.

The first view (Figure 14) is more focused on managing the *SEnviro* nodes. In this way, new nodes can be claimed or edited. As stated above, each node has a QR code, the node can be claimed by recognizing the QR or adding its ID (Figure 15). A wizard has been developed to provide information about the location in which the sensor is installed, such as location (using GPS), smallholding name, and a picture of the deployment. When a *SEnviro* node is claimed it is listed and some information about it, such as battery percentage, state alerts (low battery or off-line node), and location, is shown.

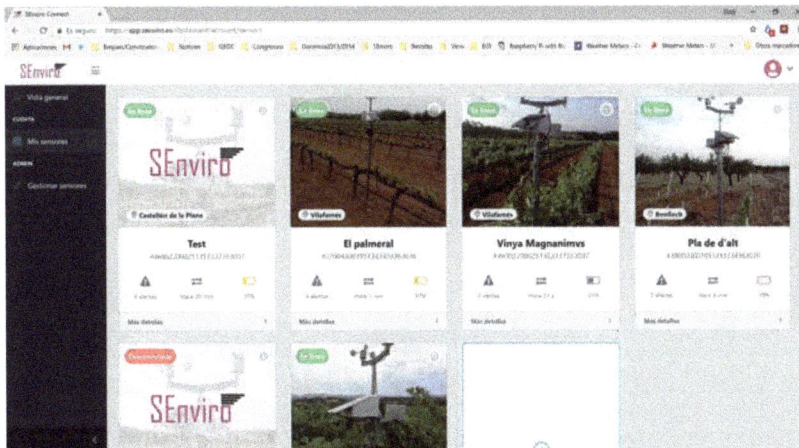

Figure 14. A screenshot of the *SEnviro* connect showing the node management view.

Figure 15. A screenshot of the *SEnviro* connect showing the wizard to claim nodes.

The second view is developed to visualize sensor observations and alerts in a way which is adapted to the stakeholder (Figure 16), in this case, vineyard farmers. The proprietary sensors are listed on a map using markers. When a user clicks on one of them, the client displays a new lateral window to show each phenomenon in a different chart (with different levels of aggregation). This view is used to show the latest detected alerts from each supported disease model.

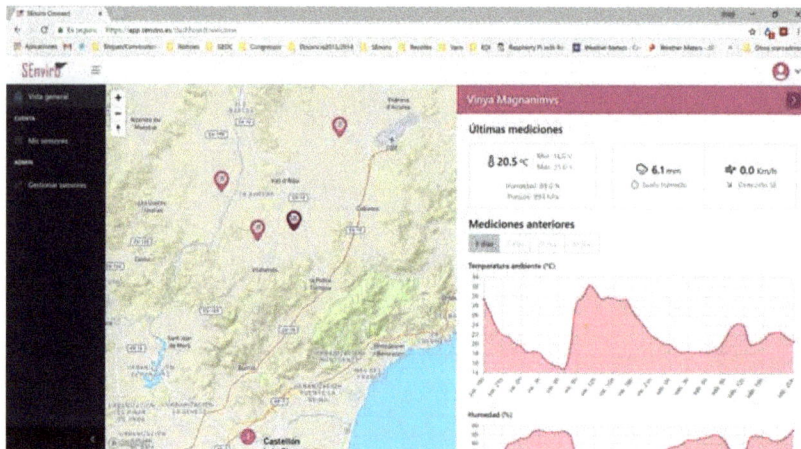

Figure 16. A screenshot of the *SEnviro* connect showing the data and alerts for each node.

Four vineyard farmers have tested the web app during the 140-day period in which the *SEnviro* nodes were deployed in their smallholdings. Their opinions have been used to reorient some design aspects relating to the visualization of the information.

6. Related Works

Related research on IoT nodes is reviewed in this section. In the literature, there are different approaches which are similar to *SEnviro*. All selected studies use open-hardware components. Some of these studies are analyzed in detail, and a comparison is provided in Table 11. To compare the previously reviewed studies, we propose using the following features to characterize each one.

- *Platform*: refers to the microcontroller model that the system uses.
- *Connection*: wireless connections available for the system. Scale: Wi-Fi, Bluetooth, ZigBee, and others.
- *Phenomena*: that the system can measure. Scale: Temp., Hum., Dust, Bar., Noise, and others.
- *Cost*: in terms of money to deploy the system. Scale: Euros.
- *Energy savings*: indicates if the IoT node applies energy strategies. Scale: Yes/No.
- *OTA*: symbolizes if the IoT node supports OTA updates. Scale: Yes/No.
- *M2M*: shows if the system follows M2M communication. Scale: Yes/No.
- *Client*: means if the system provides a client for visualizing the sensors and observations. Scale: Yes (what kind) /No.
- *Energy autonomous*: indicates if the system can be considered as a client to visualize the sensors and observations Scale: Yes (what kind) /No.

Subsequently, we describe all analyzed studies, and include only those in which open hardware has been used.

Table 11. Comparison between different sensorized platforms.

Reference	Platform	Connection	Phenomena	Cost	Energy Savings	OTA	M2M	Client	Energy Autonomous
[7]	Arduino Uno	Wi-Fi	Temp., Hum., Dust, Bar., Noise, Gases, Light, Rain and Anem.	€ 286.28	Yes	No	RESTful	Web APP	Yes
[39]	Arduino Mega	Bluetooth	Temp., Hum.,	Not specified	No	No	Serial port	Mobile APP	Yes
[40]	Lilypad	Bluetooth	Motion sensor	Not specified	No	No	Serial port	Web App	No
[41]	Stalker v3	Wi-Fi	CO, Temp., Hum, PPM2.5, Noise and UV	94US$	No	No	RESTful	Web App	Yes
[42]	Arduino UNO and Raspberry	Ethernet	Temp., Press., Rain and Anem.	165$	No	No	RESTful	Web App	No
[43]	Arduino Uno	RFID Wi-Fi (gateway)	Methan, NOx, sOx, LPG, CNG, CO and Alc.	Not specified	No	No	Not specified	No	No
[44]	Beaglebone Black	Bluetooth Wi-Fi	PIR	Not specified	No	No	RESTful	No	No
[45]	Raspberry 3	Ethernet	General events	Not specified	No	No	MQTT	No	No
Our	Particle Electron	3G	Temp., Soil and Air Hum., Barom., Rain and Anem.	€ 256.45	Yes	Yes	MQTT	Web APP	Yes

- In the previous SEnviro version [7], the authors show a full open-hardware solution to monitor environmental phenomena. An Arduino microcontroller with a Wi-Fi module is used to connect with a TCP/IP stack. They use a RESTful interface to enable M2M connectivity. The proposal is validated in a smart campus scenario, where five units were deployed.
- The paper [39] presents a smart lamp design to optimize the indoor thermal comfort and energy savings, two important workplace issues in which the comfort of the workers and the consumption of the building strongly affect the economic equilibrium of a company. An Arduino mega with a Bluetooth wireless shield is used to control different included sensors to monitor environmental conditions.
- The authors in [40] design a wearable healthcare sensors platform to monitor pervasive healthcare phenomena. The work uses an Arduino-based microcontroller called Lilypad. This platform offers a Bluetooth interface to connect with an Android app.
- Ref [41] present a low-cost environmental node to monitor Carbon Monoxide (CO), temperature, relative humidity, particulate matter 2.5, noise and UV radiation sensors. In this case, a Stalker microcontroller is selected to build the proposed solution. The solution is presented as an autonomous energy platform, it has two solar panels and a battery to support this last feature.
- The paper [42] shows a weather station with temperature, barometer, atmospheric pressure, pluviometer, and anemometer sensors. An Arduino is used as a microcontroller to collect and send data to a Raspberry. This Raspberry acts as a gateway to connect to the Internet using an Ethernet connection.
- Ref [43] proposes a node development using an Arduino microcontroller with RFID connectivity to monitor air pollution on roads and track vehicles which cause pollution over a specified limit. The proposed node has various gas sensors, such as CO, sulfur dioxide, nitrogen dioxide and methane.
- The authors in [44] present an intelligent multisensor framework based on the BeagleBone Black platform. The designed node can become an instrument for monitoring, preservation, and protection of several environments. To test the proposed framework, they conducted the prototype in an art gallery.
- Lastly, in [45], the authors combine hardware sensors with a Raspberry 3 and demonstrate an automatic telephone log device capable of capturing environmental events both automatically and through user-input. An MQTT broker is used as a data publisher.

In what follows, we will analyze the different features of the studies mentioned above. As shown in Table 11, most of the studies analyzed use the Arduino platform. Although Arduino is the most widely used platform, it is not the microcontroller that provides the best performance and features. Its popularity is based on its price and ease of purchase. Despite this, Arduino is not the best option for scenarios where connectivity (3G) and power autonomy are required, since, as we have detailed in Section 2, there are better options such as Particle Electron or Hologram Dash.

Our approach, unlike all the studies analyzed, offers a 3G connection. This kind of connectivity is the most autonomous solution among all the analyzed studies, because it does not need any installation where there is coverage.

The selected phenomena used for each study depend on the final IoT scenario in which the solution will be deployed. In our case, our validation scenario requires the monitoring of meteorological phenomena. Other projects are focused on other scenarios and require environmental or presence sensors.

Only four studies detail the final price of their solution. The total price depends on the number of sensors included and the features and functionalities that each microcontroller can perform. Our system is more expensive than [41,42], but our solution includes more sensors and adds energy components such as a solar panel, MPPT and a battery. The new *SEnviro* version is cheaper than the previous one, the difference being € 29.83.

Another analyzed feature is the IoT node energy consumption, but it is only the two *SEnviro* approaches that apply energy savings and show consumption. The improvement between both versions is substantial; the first version consumes 90 mA per hour, and the second version, despite having a 3G module, which leads to higher consumption, has a power consumption of 21.46 mA, resulting in a reduction of 68.54 mA per hour.

Only one of the analyzed projects [44], similar to our approach, follows the MQTT protocol. Five analyzed projects (including *SEnviro*) follow RESTFul communication. This kind of protocol is not appropriate for IoT solutions when considering the resource, bandwidth, and energy restrictions of IoT devices [46].

Some studies propose webApp [7,40,41] or mobile app [39] clients to manage IoT nodes in terms of visualizing the collected sensor data. Our approach offers a responsive web app client to adapt the layout depending on the final device.

Finally, only three analyzed studies [7,39,41] can be considered an autonomous energy solution. If we add the feature of autonomous connectivity (Connection column), only our approach can be considered a fully autonomous solution.

7. Conclusions

The proposed IoT node presented throughout this paper follows the DIY philosophy and is based wholly on open hardware, using low-cost components. Another main objective of the study is to detail and describe, step by step, how to form IoT nodes and present a replicable study to empower citizens to make their own creations, thus fostering citizen science developments [47,48].

First, a generic IoT node architecture is introduced, defining the physical and logical aspects intended to cover any IoT application. This architecture follows a modular design at both hardware and software levels. Section 4 proposed how to build an IoT node and fulfils all the requirements presented in Section 3, using open-hardware components. Providing a full open-hardware solution provides several benefits [7], including access to a large community of developers or makers, which increases its impact. Additionally, it facilitates the possible improvement of the platform by leaving it in the hands of the community.

Some improvements have been added concerning the preliminary *SEnviro* version; as shown in [7]. The most relevant is the 3G connectivity, which offers greater freedom when the node is set up and supports adaptation to a wide range of different IoT scenarios. Another improvement is the possibility to change the behavior of the IoT node using OTA updates, and be an energetically autonomous solution. The energy tests carried out reveal that the device can be kept alive for more than 3 days, following the normal energy mode defined. The recovery and critical modes would allow a longer duration of the life of the node without sending observations, thus providing more than 800 h of life without any power charge.

A full validation of the proposed architecture has been accomplished. It has been used in the IoT field of smart agriculture [49], more specifically to monitor small vineyards. The IoT node is integrated into an IoT management platform to analyze sensor data, called *SEnviro* connect. *SEnviro* connect can launch alerts about possible diseases in the vine.

Concerning related studies, *SEnviro* node offers a completely autonomous solution in terms of energy and connectivity. Our proposal is the only one that presents the possibility of being updated remotely. Moreover, it follows M2M connectivity (MQTT), suitable for these types of devices. *SEnviro* node applies energy saving, and the cost is adjusted to the components used.

Regarding future research, our objective is to perform a large-scale analysis of the different IoT standards and adopt one of them to increase the IoT node interoperability. Also, sensor meshes are planned to react depending on the context in which the node is located. These meshes will be used to establish strategies for energy and connection savings, and they will be deployed in the same smallholding to ascertain a more in-depth understanding of the real state of the overall area. The last improvement is to consider the particular features for each meteorological phenomenon [50] and take into account where the nodes are deployed, or if a new sensor box is designed. For example, considering a ventilated case (which we are already working on), or defining rules for when a new unit is installed (distance from the ground) to take more precise measurements, among others.

Author Contributions: All authors contributed equally to achieve the proposed work and the writing of this paper, as well. All authors read and approved the final manuscript.

Funding: This research received no external funding.

Acknowledgments: Sergio Trilles has been funded by the postdoctoral programme Vali+d (GVA) (grant number APOSTD/2016/058) and GVA doctoral stays programme (grant number BEST/2018/053). The project is funded by the Universitat Jaume I - PINV 2017 (UJI-A2017-14) and the European Commission through the GEO-C project (H2020-MSCA-ITN-2014, Grant Agreement number 642332, http://www.geo-c.eu/).

Conflicts of Interest: The authors declare no conflict of interest.

References

1. Official Arduino Website. Available online: http://www.arduino.cc (accessed on 15 October 2018).
2. Official RaspBerry Pi Website. Available online: http://www.raspberrypi.org (accessed on 15 October 2018).
3. Official BeagleBoard Website. Available online: http://beagleboard.org (accessed on 15 October 2018).
4. pcDuino. Available online: http://www.linksprite.com/linksprite-pcduino/ (accessed on 15 October 2018).
5. Fisher, R.; Ledwaba, L.; Hancke, G.; Kruger, C. Open hardware: A role to play in wireless sensor networks? *Sensors* **2015**, *15*, 6818–6844. [CrossRef] [PubMed]
6. Barroso, L.A. The price of performance. *Queue* **2005**, *3*, 48–53. [CrossRef]
7. Trilles, S.; Luján, A.; Belmonte, Ó.; Montoliu, R.; Torres-Sospedra, J.; Huerta, J. SEnviro: a sensorized platform proposal using open hardware and open standards. *Sensors* **2015**, *15*, 5555–5582. [CrossRef] [PubMed]
8. Salamone, F.; Belussi, L.; Danza, L.; Ghellere, M.; Meroni, I. Design and development of nEMoS, an all-in-one, low-cost, web-connected and 3D-printed device for environmental analysis. *Sensors* **2015**, *15*, 13012–13027. [CrossRef]
9. Fox, S. Third Wave Do-It-Yourself (DIY): Potential for prosumption, innovation, and entrepreneurship by local populations in regions without industrial manufacturing infrastructure. *Technol. Soc.* **2014**, *39*, 18–30. [CrossRef]
10. Instructables. Available online: https://www.instructables.com/ (accessed on 15 October 2018).
11. Make Magazine. Available online: https://makezine.com/ (accessed on 15 October 2018).
12. openMaterials. Available online: http://openmaterials.org/ (accessed on 15 October 2018).
13. Industries, A. Adafruit. Available online: https://www.adafruit.com/ (accessed on 15 October 2018).
14. Pearce; Sensor. SparkFun Electronics. Available online: https://www.sparkfun.com/ (accessed on 15 October 2018).
15. Gubbi, J.; Buyya, R.; Marusic, S.; Palaniswami, M. Internet of Things (IoT): A vision, architectural elements, and future directions. *Future Gener. Comput. Syst.* **2013**, *29*, 1645–1660. [CrossRef]
16. Savaglio, C.; Fortino, G.; Zhou, M. Towards interoperable, cognitive and autonomic IoT systems: An agent-based approach. In Proceedings of the 2016 IEEE 3rd World Forum on Internet of Things (WF-IoT), Reston, VA, USA, 12–14 December 2016; pp. 58–63.
17. Lasi, H.; Fettke, P.; Kemper, H.G.; Feld, T.; Hoffmann, M. Industry 4.0. *Bus. Inf. Syst. Eng.* **2014**, *6*, 239–242. [CrossRef]
18. Trilles, S.; Calia, A.; Belmonte, Ó.; Torres-Sospedra, J.; Montoliu, R.; Huerta, J. Deployment of an open sensorized platform in a smart city context. *Future Gener. Comput. Syst.* **2017**, *76*, 221–233. [CrossRef]
19. Belmonte-Fernández, Ó.; Puertas-Cabedo, A.; Torres-Sospedra, J.; Montoliu-Colás, R.; Trilles-Oliver, S. An indoor positioning system based on wearables for ambient-assisted living. *Sensors* **2016**, *17*, 36. [CrossRef] [PubMed]
20. Trilles Oliver, S.; González-Pérez, A.; Huerta Guijarro, J. An IoT proposal for monitoring vineyards called SEnviro for agriculture. In Proceedings of the 8th International Conference on the Internet of Things, Santa Barbara, CA, USA, 15–18 October 2018; pp. 20.
21. Chan, M.; Estève, D.; Escriba, C.; Campo, E. A review of smart homes—Present state and future challenges. *Comput. Methods Programs Biomed.* **2008**, *91*, 55–81. [CrossRef] [PubMed]

22. Palattella, M.R.; Dohler, M.; Grieco, A.; Rizzo, G.; Torsner, J.; Engel, T.; Ladid, L. Internet of things in the 5G era: Enablers, architecture, and business models. *IEEE J. Sel. Areas Commun.* **2016**, *34*, 510–527. [CrossRef]

23. Naik, N. Choice of effective messaging protocols for IoT systems: MQTT, CoAP, AMQP and HTTP. In Proceedings of the 2017 IEEE International Systems Engineering Symposium (ISSE), Vienna, Austria, 11–13 October 2017; pp. 1–7.

24. Hunkeler, U.; Truong, H.L.; Stanford-Clark, A. MQTT-S—A publish/subscribe protocol for Wireless Sensor Networks. In Proceedings of the 3rd International Conference on Communication Systems Software and Middleware and Workshops, Bangalore, India, 6–10 January 2008; pp. 791–798.

25. Bandyopadhyay, S.; Bhattacharyya, A. Lightweight Internet protocols for web enablement of sensors using constrained gateway devices. In Proceedings of the 2013 International Conference on Computing, Networking and Communications (ICNC), San Diego, CA, USA, 28–31 January 2013; pp. 334–340.

26. Pearce, J.M. Building Research Equipment with Free, Open-Source Hardware. *Science* **2012**, *337*, 1303–1304. [CrossRef] [PubMed]

27. Particle Community. Available online: https://community.particle.io/ (accessed on 15 October 2018).

28. Abdmeziem, M.R.; Tandjaoui, D.; Romdhani, I. Architecting the internet of things: state of the art. In *Robots and Sensor Clouds*; Springer: Berlin, Germany, 2016; pp. 55–75.

29. Fortino, G.; Savaglio, C.; Palau, C.E.; de Puga, J.S.; Ganzha, M.; Paprzycki, M.; Montesinos, M.; Liotta, A.; Llop, M. Towards multi-layer interoperability of heterogeneous IoT platforms: the INTER-IoT approach. In *Integration, Interconnection, and Interoperability of IoT Systems*; Springer: Berlin, Germany, 2018; pp. 199–232.

30. Bormann, C.; Castellani, A.P.; Shelby, Z. Coap: An application protocol for billions of tiny internet nodes. *IEEE Internet Comput.* **2012**, *16*, 62–67. [CrossRef]

31. Thangavel, D.; Ma, X.; Valera, A.; Tan, H.X.; Tan, C.K.Y. Performance evaluation of MQTT and CoAP via a common middleware. In Proceedings of the 2014 IEEE Ninth International Conference on Intelligent Sensors, Sensor Networks and Information Processing (ISSNIP), Singapore, 21–24 April 2014; pp. 1–6.

32. Cheng, K.T.; Krishnakumar, A.S. Automatic functional test generation using the extended finite state machine model. In Proceedings of the 30th Conference on Design Automation, Dallas, TX, USA, 14–18 June 1993; pp. 86–91.

33. Bramley, R. Precision Viticulture: Managing vineyard variability for improved quality outcomes. In *Managing Wine Quality: Viticulture and Wine Quality*; Elsevier: Amsterdam, The Netherlands, 2010; pp. 445–480.

34. Goidànich, G. *Manuale di Patologia Vegetale*; Edagricole: Bologna, Italy, 1964; Volume 2.

35. Carroll, J.; Wilcox, W. Effects of humidity on the development of grapevine powdery mildew. *Phytopathology* **2003**, *93*, 1137–1144. [CrossRef] [PubMed]

36. Molitor, D.; Berkelmann-Loehnertz, B. Simulating the susceptibility of clusters to grape black rot infections depending on their phenological development. *Crop Prot.* **2011**, *30*, 1649–1654. [CrossRef]

37. Broome, J.; English, J.; Marois, J.; Latorre, B.; Aviles, J. Development of an infection model for Botrytis bunch rot of grapes based on wetness duration and temperature. *Phytopathology* **1995**, *85*, 97–102. [CrossRef]

38. Trilles, S.; Gonzalez, A.; Huerta, J. An IoT middleware based on microservices and serverless paradigms. A smart farming use case to detect diseases in vineyard fields. 2018; manuscript submitted for publication.

39. Salamone, F.; Belussi, L.; Danza, L.; Ghellere, M.; Meroni, I. An open source "smart lamp" for the optimization of plant systems and thermal comfort of offices. *Sensors* **2016**, *16*, 338. [CrossRef] [PubMed]

40. Doukas, C.; Maglogiannis, I. Bringing IoT and cloud computing towards pervasive healthcare. In Proceedings of the 2012 Sixth International Conference on Innovative Mobile and Internet Services in Ubiquitous Computing (IMIS), Palermo, Italy, 4–6 July 2012; pp. 922–926.

41. Velásquez, P.; Vásquez, L.; Correa, C.; Rivera, D. A low-cost IoT based environmental monitoring system. A citizen approach to pollution awareness. In Proceedings of the 2017 CHILEAN Conference on Electrical, Electronics Engineering, Information and Communication Technologies (CHILECON), Pucon, Chile, 18–20 October 2017; pp. 1–6.

42. Brito, R.C.; Favarim, F.; Calin, G.; Todt, E. Development of a low cost weather station using free hardware and software. In Proceedings of the 2017 Latin American Robotics Symposium (LARS) and 2017 Brazilian Symposium on Robotics (SBR), Curitiba, Brazil, 8–11 November 2017; pp. 1–6.

43. Manna, S.; Bhunia, S.S.; Mukherjee, N. Vehicular pollution monitoring using IoT. In Proceedings of the Recent Advances and Innovations in Engineering (ICRAIE), Jaipur, India, 9–11 May 2014; pp. 1–5.

44. Chianese, A.; Piccialli, F.; Riccio, G. Designing a smart multisensor framework based on beaglebone black board. In *Computer Science and its Applications*; Springer: Berlin, Germany, 2015; pp. 391–397.

45. Chen, P.H.; Cross, N. IoT in Radiology: Using Raspberry Pi to Automatically Log Telephone Calls in the Reading Room. *J. Digital Imaging* **2018**, *31*, 371–378. [CrossRef] [PubMed]

46. Fysarakis, K.; Askoxylakis, I.; Soultatos, O.; Papaefstathiou, I.; Manifavas, C.; Katos, V. Which IoT protocol? Comparing standardized approaches over a common M2M application. In Proceedings of the Global Communications Conference (GLOBECOM), Washington, DC, USA, 4–8 December 2016; pp. 1–7.

47. Kera, D. Hackerspaces and DIYbio in Asia: connecting science and community with open data, kits and protocols. *J. Peer Prod.* **2012**, *2*, 1–8.

48. Fortino, G.; Rovella, A.; Russo, W.; Savaglio, C. Towards cyberphysical digital libraries: integrating IoT smart objects into digital libraries. In *Management of Cyber Physical Objects in the Future Internet of Things*; Springer: Berlin, Germany, 2016; pp. 135–156.

49. Zhao, J.C.; Zhang, J.F.; Feng, Y.; Guo, J.X. The study and application of the IOT technology in agriculture. In Proceedings of the 2010 3rd IEEE International Conference on Computer Science and Information Technology (ICCSIT), Chengdu, China, 9–11 July 2010; Volume 2, pp. 462–465.

50. Bell, S.; Cornford, D.; Bastin, L. The state of automated amateur weather observations. *Weather* **2013**, *68*, 36–41. [CrossRef]

electronics

MDPI

Article

The Platform Development of a Real-Time Momentum Data Collection System for Livestock in Wide Grazing Land

Liang Zhang [1], Jongwon Kim [2],* [ID] and Yongho LEE [3]

[1] School of Mechanical, Electrical and Information Engineering, Shandong University, Weihai 264209, China; zhangliang@wh.sdu.edu.cn
[2] Department of Electromechanical Convergence Engineering, Korea University of Technology and Education, Cheonan 31253, Chungnam, Korea
[3] Technology Research Center, INOFARM Co., Ltd. Yusung-gu, Taejon 305-701, Korea; inofarm@naver.com
* Correspondence: kamuiai@koreatech.ac.kr; Tel.: +82-41-560-1249

Received: 22 April 2018; Accepted: 13 May 2018; Published: 15 May 2018

Abstract: In the process of animal husbandry production through grazing, animals are active in large grassland or mountain areas, and it is very difficult to obtain and deal with the information on animal activity and state of life. In this paper, we propose a platform for operation of data transmission and analysis system which gathers activity and status information of livestock. The data collected in real time from integrated livestock sensor modules are anticipated to assist farmers to supervise animal activities and health. While at the same time the improvements of viable farming techniques are expected to reduce the impact of the livestock industry on the environment. For the individual management of target livestock, the data collection system needs a convergence technology with a physical active sensor, a wireless sensor network and solar power technology to cover the wide area of mountains. We implemented a momentum data collection system to collect and transfer the information of ecological and situations of livestock in grazing, which include sensor and communication modules, repeaters with solar panels to cover the problems of communications in wide grazing and a receiver connected to main server. Besides, in order to prevent data collisions and deviations on multiple transmitter operation, we renewed the format of the communication protocol and made a platform to analyze animal activities information by software. Finally, the system and platform were applied and tested in National Rural Development Administration in Republic of Korea.

Keywords: individual management of livestock; momentum data sensing; remote sensing platform; sensor networks; technology convergence

---

1. Introduction

Raw information and reliable sources of animal production are important for the economy of meat products. Meat is limited by shelf life and some meat with long shelf life is considered unhealthy due to the use of preservatives [1]. Out of consideration of food safety and economic benefits, the raw material information of meat should be controlled by the consumer, producer and distributer. Consequently, how to collect and analyze the original information of animal production has become an important research field, to which a variety of technologies have been employed [2].

Livestock management is an important issue for the green food industry, but the effective management of the target animal is very difficult because of securing of animal issues of livestock [3], increase of eco-friendly breeding methods [4], safety of foodstuffs due to the application of various technologies [5], and the famer's economics [6]. Regarding these issues, we need to be concerned with eco-friendly feeding methods and how to create and implement environments to effectively manage livestock. In particular, applying the high-level technology such as bio-chemistry and/or genetics to the food industry without appropriate treatment usually cannot guarantee safe foods. Besides, the use of these technologies is strictly protected. Thus, we have to apply the higher-level technologies carefully and selectively.

China is a big country in the production of animal husbandry, and the annual output of livestock and poultry products is considered the highest in the world. In 2015, the total output value of China's animal husbandry was 2 trillion and 978 billion 38 million Chinese Yuan, up 2.85% from the same period (Data source: China National Bureau of Statistics). However, the production efficiency is very low and the data statistics are difficult. Furthermore, China's livestock industry has led to huge carbon emissions [7]. It is a problem to improve the production animal husbandry in China. And large-scale raising of livestock cannot be maintained in places, because the amount of available arable and grazing land is decreasing due to urbanization. The livestock contamination issues are very serious social problems to maintain in terms of the green environment.

In Korea, as the agricultural economic structure changes, the total number of famers in the livestock industry is gradually decreasing. The main cause is the decrease in the income of livestock industry workers. Therefore, the increase of the livestock industry workers' income needs to be considered for the livestock environment and structure of farmers' incomes [8].

For the above-mentioned reasons, it is important to construct an effective technical path to manage the livestock according to changes in the livestock environment including livestock diseases. Health condition of livestock can be predicted with the physical activity of the target livestock. Traditional farm monitoring, such as using written notes or a simple device without data sharing capabilities, is an inaccurate method with high probability of human error. The health information of the target cannot be the useful data for management of the target and the use of Global Positioning Systems was proposed, but it required detailed field maps and was costly due to the involvement of transmission of data from satellites so, it is not suitable for real field [2].

In order to meet the requirements, the proposed system is implemented to have the ability to collect and transmit data of real time physical activities in wide range mountainous areas. After collecting raw data, the data will be analyzed to satisfy the issues. In order to monitor livestock, shed monitoring system based on a wireless sensor network [9,10] uses bio and environmental sensors such as temperature and humidity [11]. Advanced management of livestock sheds uses infrared wireless sensor nodes and unmanned surveillance cameras [12]. Studies on the monitoring of the livestock state had been carried out on the IoT-based cattle biometric information collection terminal [13], cattle health monitoring system based on biosensors [14], design and implementation of livestock disease prediction system [15] and etc. However, the application of these systems is limited to a narrow field of livestock sheds, so it is difficult to be applied in the free-range grazing system. Muhammad Fahim et al. [16–18] proposed a method of using the accelerometer sensor of the smart phone to recognize user situations (i.e., still or active) and developed a cloud-based smart phone application supported by a web-based interface to visualize the sedentary behavior. Complex and large-scale sensor communication [19] is needed under these circumstances.

In this paper, we propose a momentum data collection system to provide ecological information and situations of livestock in grazing. In order to operate an unmanned ranch, it is difficult to apply high price sensors with high performance for economic benefit. We constructed a solar power supply system for the outdoor data repeater with low power consumption using solar panel and designed a data transfer platform for the analysis and storage of physical activity information of targets. We renewed the format of the communication protocol and data analysis method. The system and platform were applied to an actual grazing field and tested in National Rural Development Administration in Republic of Korea.

2. System Design and Structure

The Data collection system is composed of three parts: data transceiver module with sensor (Sensor Network Transceiver: SNR), repeater and main system. The sensors count the physical activities of targets which can be recognized by target number by SNR. The repeaters were evenly arranged and distributed on grazing land and the main system was installed in building inside a data-base computer with data storage and analysis software.

Figure 1 shows the plane map of the grazing land. The capacity of area for a breeding stock that requires 50 kilos of forage a day is about 1.5 kg per m^2. 100 targets (Korean cattle) require an area about 4500 m^2.

Figure 1. Operation Condition of the System in Grazing Area.

When cattle stay on the grazing ground, targets have a characteristic behavior of jumping, sitting, running and rubbing for their feed. So, we have difficulties to get direct physical data by momentum sensors. In general, sensors for receiving target's Data [20] are attached directly on the leg of a target, so it is very hard to replace them by workers; it should be designed to be a low power module with a long life cycle. Unfortunately, the distance of the data transmission is related to power consumption of the SNR module which has a limited transceiver distance by module's design. Because high power consumption equipment cannot be used for signal transmission [21] in wearable devices, a low power RF-IC is the only choice for the system to minimizing power consumption. Therefore, in order to cover the wide grazing, relay stations are necessary. For long time operation, solar panels are used to power them. The conclusion is that we have to make a sensor module, a data transfer module and a solar-powered repeater station for the transmission of the data in wide grazing. Information is eventually concentrated in the main server. The information is stored in a database and analyzed by specially designed analysis software, through which the behavior patterns of livestock can be summarized and utilized.

Figure 2 illustrates the overall flowchart of the system. When a sensor is attached to a cow's neck and is ready for operation, the sensor can detect the cow's movement impact which produces electrical signals or still—no signal. In this case, the sensor recognizes the impact as an Event. The movement of livestock in general can be perceived by the sensor, but with the time interval, the validity period of information will be different. The cow might move faster or slower. If the cow moves faster, it is a general activity, signifying it is not a special activity signal. The events occur continuously within 4–6 s and when the impact signal appears once or twice within 8 s, it may be considered as a signal such as a cow's mount or a decrease in activity. If there is no more than 8 s of exercise, it can be divided into normal food or an ordinary activity. Hence, we can recognize the cow's activities. If the activity signal is divided into an event failure, the signal patterns occurring within 4~6 s and the signal patterns that occur within 8 s can be predicted to be the biological changes of the animals that have occurred. Therefore, through the Event which is an electric signal by sensor, we detect the cow's activity. The activity is within a specific time range and accumulates the number—which is count—of activities in unit time. Normally, the accumulated data (Number of events) can be transmitted to the main system for a period of time to observe the activity characteristics of the cattle by analysis with graphs or raw data.

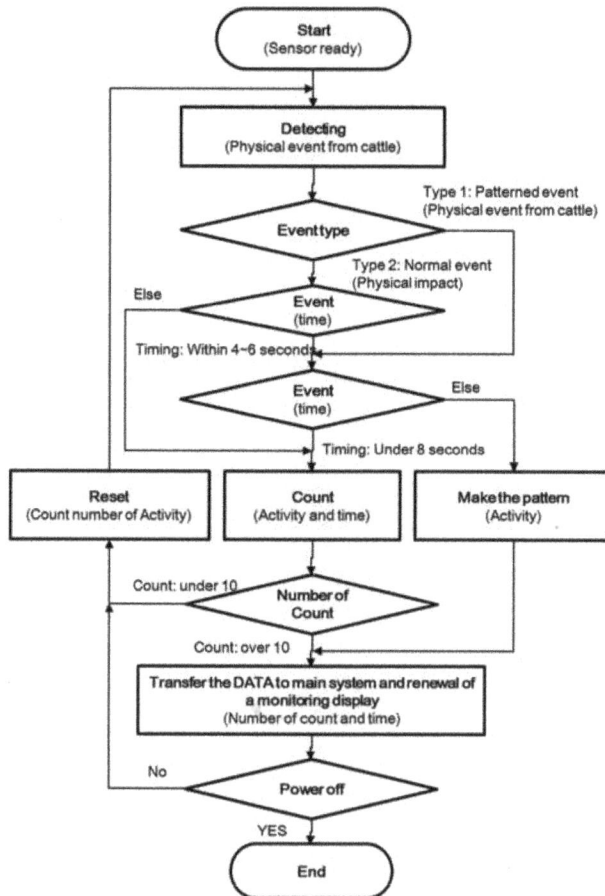

Figure 2. Operation flowchart of the system in grazing area.

3. System Implementation and Test

3.1. Data Transceiver Module with Sensor

The SNR is made up of micro-component of impact-detecting, microprocessor and RF component. The module case is waterproof and has a battery. The microprocessor operates by a call from the main system which counts the number of the target's physical activities from the impact-detecting component, and the data transfer to main the system via RF components. In the case of previous research, the sensor is attached to cattle's ears by piercing but attached/inserted type can be changed to necklace type with target's ID number [17] which enables easy recognition for each target. Figure 2 is a feature of SNR and its installation.

Figure 3 shows a feature of SNR and its installation. The SNR is made up of micro-component of impact detecting, microprocessor and RF component. In previous research, the sensor was attached on cattle's ears by perforation. The attached/inserted type was changed to a necklace type with the target's ID number [22]. In this way, each target could be easily distinguished.

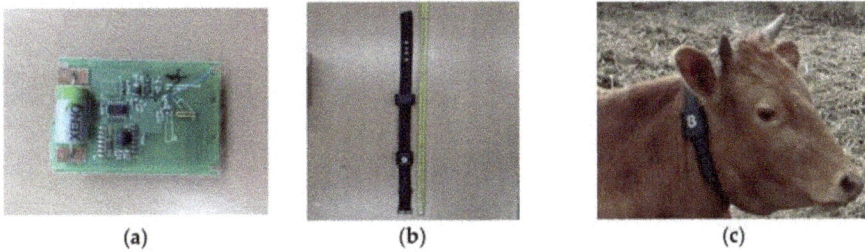

Figure 3. Feature and installation of the SNR on a target (**a**) SNR, (**b**) Feature of the module, (**c**) Installation module on the target.

Figure 4 shows the IC structure of the SNR. Data are transmitted to main system through the RF module—RFM69HCW [23] which is optimized for low power consumption while offering high RF output power and channelized operation. The module case is waterproof and has a battery. The microprocessor operates by a call from the main system which counts the number of the target's physical activities from the impact-detecting sensor.

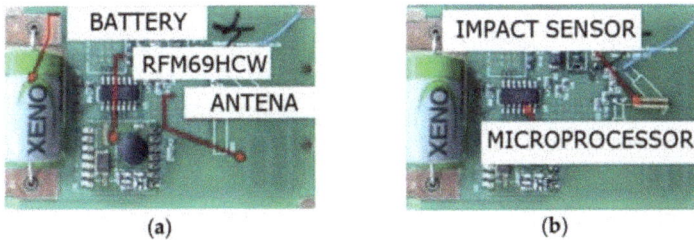

Figure 4. The PCB circuit of SNR (**a**) Battery and transceiver, (**b**) Microprocessor and sensor.

3.2. DATA Transmission and Repeater

To manage the wide area of grazing, the system needs repeaters shown in Figure 5. The grazing has no power line, so wired electrical devices cannot be used. The repeater gets energy from solar panels and transmits the data from SNR to the main system receiver [24], and transfers the call signal from the main system to each sensor which can be recognized by the target ID. In that case, the wireless signals get errors for several reasons such as target movements, weather and obstacles. So considering these reasons, the data transmitter needs a special platform.

3.3. Format of Communication Protocol and Data Analysis

Detour path and traffic concentration problems are the fundamental problems of the general tree routing protocols [25], which cause overall network performance degradation. The forwarding scheme is prone to cause uneven load distribution and further lead to network congestion [26]. Nodes mounted on targets move frequently, and constant end-to-end paths are rarely available [27]. It is a challenging task to maintain quality of service with respect to parameters such as high throughput, and minimum end-to-end delay [28,29].

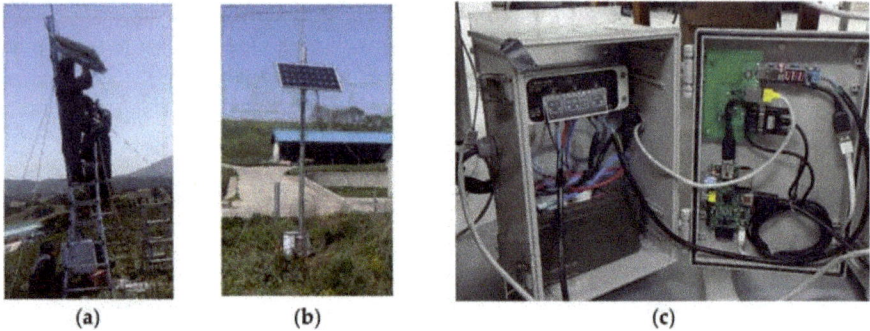

(a) (b) (c)

Figure 5. Installation of the repeater with solar panel, (**a**) Installation, (**b**) Installation result, (**c**) Repeater.

As shown in Figure 6, the receiver and each SNR use a program to measure the time to synchronize the data transmission cycle. However, if the data transmits and receives for a long time, data signal collisions and deviations occur at the same time in each transmitter. In order to solve this problem, each transmitter is synchronized by measuring time on each transmitter, and each transmitting time is revised by the receiver's origin time, which is the correct method.

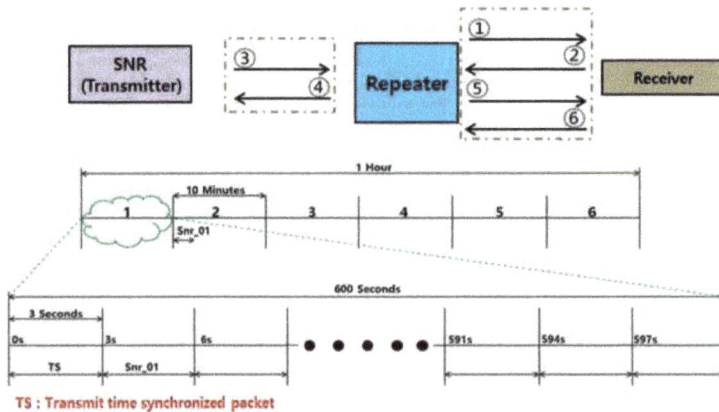

Figure 6. DATA Transmission in single repeater with single transmitter.

In Figure 7, each transmitter is performed by the program in its own time, but each transmitter's data packaged by the algorithm with original time is allocated a data packet. The final data packet can prevent data collisions and deviations on multiple transmitters operations. The data collects in the data base in the main system from each sensor which resulted from a packet of single transmitter or multi transmitter format. In some cases of the data transmission, the repeater makes a multi transmission format in a 600-s time limit and the raw data format can cover one target cattle every three seconds, so one packet can cover 200 target cattle every ten minutes [30].

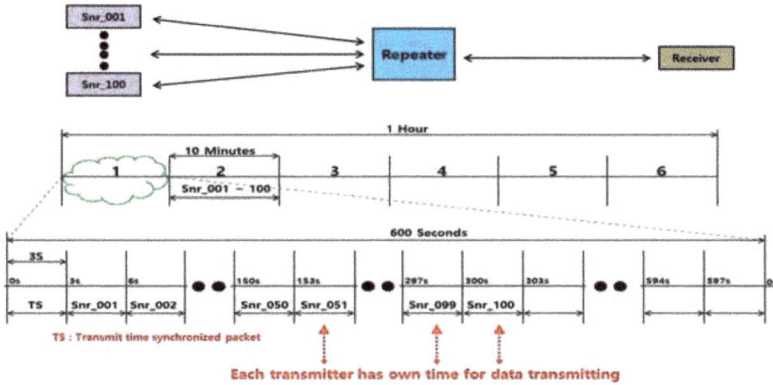

Figure 7. DATA Transmission in single repeater with multi transmitter.

The main system (receiver) has a packet communication protocol as shown in Tables 1 and 2. The data protocol has 15 units of packet. In this study, one kind of data is used, which is count number of physical activities of the targets. The packets could be extended, to such as temperature and humidity. The sensors of temperature and humidity need a lot of electric power for a long time operation. Thus, those sensors are not used in this study.

Table 1. The Sample of raw data received from packet.

Repeater No. [01]	Receive Data
Origin Time	Received Data Packet
13:02:28	FF FB FC F3 02 01 8C 9C 1B 39 00 07 FD FE FF
13:02:31	FF FB FC F3 03 01 01 0B E2 E5 00 0A FD FE FF
13:02:34	FF FB FC F3 04 01 0D 06 E4 EF 00 0C FD FE FF
13:02:36	FF FB FC F3 05 01 71 33 C5 62 00 0F FD FE FF
13:02:39	FF FB FC F3 06 01 55 68 1D D4 00 12 FD FE FF
13:02:42	FF FB FC F3 07 01 16 35 D8 1E 00 15 FD FE FF
13:02:45	FF FB FC F3 08 01 01 06 DE E1 00 18 FD FE FF
13:02:49	FF FB FC F3 09 01 0D 14 D5 F3 00 1B FD FE FF
13:02:52	FF FB FC F3 0A 01 03 27 CB F3 00 1E FD FE FF
13:02:55	FF FB FC F3 0B 01 06 1B 19 39 00 21 FD FE FF

The communication code is shown in the Table 1, the first column is the original time, and the second column is the received data packet. There are 15 units in every data packet shown as Figure 8.

Data unit of Packet No.0~14(index)

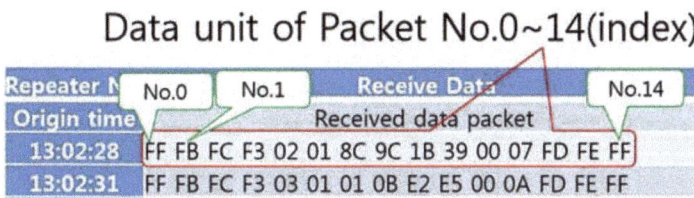

Figure 8. Structure of DATA unit of packet.

Table 2 interprets the unit formation by typical examples. The no. 0–2 units are for the Sync. from sensor, repeater and main system by wireless communication. The no. 3 is Sensor ID. The no. 4 is flag data for dividing from communication and activity data. The no. 5 is activity data which is accumulated from

41

sensing unit time (NOW). The no. 6 is activity data which is accumulated from sensing unit time (Before 10 min). The no. 7 is activity data which is accumulated from sensing time (Before 20 min). The no. 8 is packet sequence number which is divide the packets from get a lot of sensors (at a same time the receiver can get a signal from many sensors at a same time, so the packet sequence information using for filtering of not-useful packets). The no. 9 is the check sum bit. The no. 10 and 11 are real time activities data (MSB and LSB). The no. 12 to 14 are end of communication Sync. Bit. After receiving the data, the main system performs the formalization the data with the data base software shown as Figure 9.

After receiving the data, the main system performs formalization of the data with the data base software as Figure 9.

Normally in livestock industrial field, the farmer needs an estrus [31] time of targets. The activity graph shows the most important information about the state of the target, so it could be possible to estimate insemination time [32].

Table 2. Definition of the packet units.

Packet Buffer	Index No.	Meaning	Value Min.	Value Max.
	0	0xFF (Start-Sync.-First check bit)	252	252
	1	0xFB (Start-Sync.-Mid. check bit)	1	250
	2	0xFC (Start-Sync.-Final check bit)	252	252
	3	Sensor ID	1	250
	4	Flag Bits (DATA)	0	127
	5	Number of activities (present)	0	250
	6	Number of activities (before 10 Min.)	0	250
RF TX Buffer	7	Number of activities (before 20 Min.)	0	250
	8	Packet Sequence Number	0	250
	9	Check Sum	0	250
	10	Time data bit—MSB (counting the time)		
	11	Time data bit—LSB (counting the time)		
	12	0xFD (End-Sync. check bit)		
	13	0xFE (End-Sync. check bit)	It can be extension.	
	14	0xFE (End-Sync. check bit)		

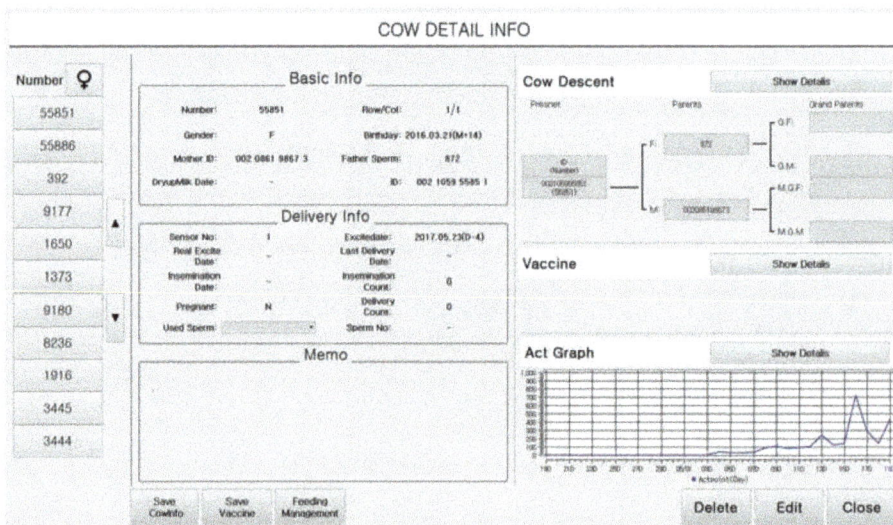

Figure 9. Analyzed information from the data in the Main system.

Figure 10 is an activity graph about one target. The red line in the graph represents the average activity every seven days and the blue line is activity number within one day. Farmers can estimate the insemination time about the target with graph analysis. After formalizing the data, it is processed into information and analyzed by the software that can provide a customized service by the needs from the famers, researchers, government officers, supporters and etc. [33,34]. On a specific basis, the continuous collection of data regarding the amount of activity of the animals of the object has a role in distinguishing the animal from the object, such as fertilization, estrus, disease, fertility and other biological changes (water medicine). Therefore, one months of activity (band curve) and average activity (fast curve) Change has the most important meaning. For more specific judgment, one day of dramatic changes in the amount of activity, the change of activity (the Bohr icon) and the change of the average activity (red chart) are used as important information to determine the biological state of the animal, if the results of the analysis are used to isolate livestock by isolation, germination, disease and so on. Measures of protection and so on to enlargement the effect of animal husbandry.

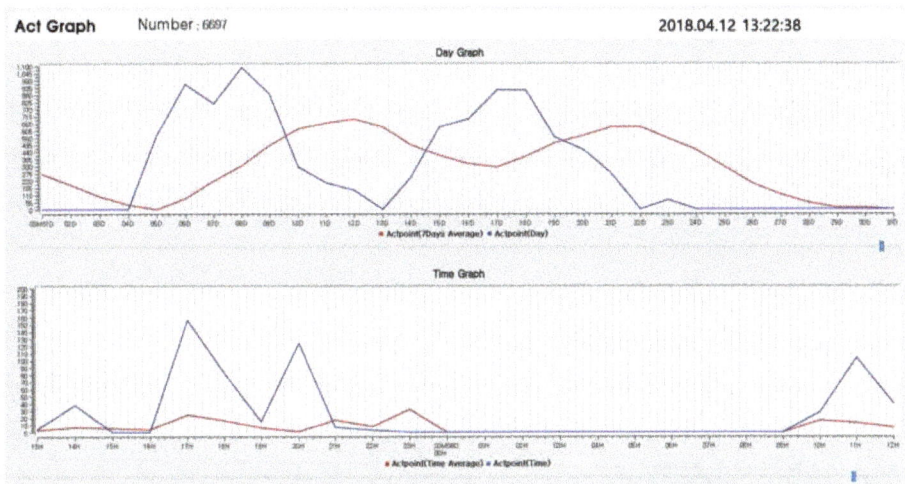

Figure 10. Activities information analyzing of one target.

4. Conclusions and Discussion

The platform and systems developed in this study provide a complete implementation of a momentum data collection system using solar power type of repeaters and sensors that collect data of physical activities in the process of feeding and breeding in grazing. The transmission platform was designed and its effectiveness was tested through experiments. The platform and system was verified in the Korea-cattle Experiment Station of National Rural Development Administration in Republic of Korea. This study may provide solutions to prevent the expected problems for changing environment of the livestock industry using different kinds of useful technologies in this field. The data of physical activities can be used for prevention of diseases and breeding managements. In this study, the expert system of veterinary judgment can be based on future technical research. The current system is the processing of the collected data, as veterinary experts and other experts can determine the basic data of the biological status of livestock. In order to monitor the activity of livestock, the information collection system suitable for animal husbandry environment is set up, and the information collected by quantitative analysis is used to reflect the combination of different technologies and use it for the purpose of industry. The proposed system is the first step considering several multiple issues to human and animals by different methods with the convergence of technology.

The platform of the system solved the problems about sensing the target's active-moment, transferring the data to a main system and energy supply with solar panels. The data analysis will be defined by animal experts or famers, so the system is partly automated to management of the target's state. In the future, the system should be fully automatic system with other physical and bio sensors and expert's knowledge by IT technology. The convergence technology should be popularized more quickly to solve problems such as power consumption, collection of the different kind of data, increment number of targets, etc. Once the system is stabilized, the data mapping algorithm with artificial intelligence, IoT, Big-DATA technology and expert-knowledge can be added and applied on the system for monitoring several events such as diseases, insemination and delivery of the target.

Author Contributions: Conceptualization, J.K. and L.Z.; Methodology, J.K.; Software, J.K.; Validation, L.Z., J.K. and Y.L.; Formal Analysis, L.Z.; Investigation, J.K.; Resources, J.K.; Data Curation, Y.L.; Writing-Original Draft Preparation, L.Z.; Writing-Review & Editing, J.K. and L.Z.; Visualization, Y.L.; Supervision, J.K.; Project Administration, J.K.

Funding: This research received no external funding.

Conflicts of Interest: The authors declare no conflict of interest.

References

1. Nakade, K.; Ikeuchi, K. Optimal ordering and pricing on clearance goods. *Int. J. Ind. Eng.* **2016**, *23*, 155–165.
2. Neethirajan, S. Recent advances in wearable sensors for animal health management. *Sens. Bio-Sens. Res.* **2017**, *12*, 15–29. [CrossRef]
3. Hahm, T.S. A Legal study on trends and issues of animal law in the US—Focusing on criminal issues. *Study Am. Const.* **2015**, *26*, 337–378.
4. Jung, K.S. The efficient policy programs of the livestock pollution abatement. *Korean J. Agric. Manag. Policy* **2001**, *28*, 167–185.
5. Jang, W.K. Improvement of livestock environment for the livestock's epidemic. In Proceedings of the Spring Conference of the Korean Journal of Environment Agriculture, Daejeon, Korea, 17 May 2011; pp. 3–23.
6. Kim, G.N. *The National Guide for Raising of Korean Cow*; Ministry of Agriculture, Food and Rural Affairs: Guelph, ON, Canada, 2002; pp. 86–90.
7. Chen, Y.; Shang, J. Disconnect analysis and influence factors of animal husbandry in China. *China Popul. Resour. Environ.* **2014**, *24*, 101–107.
8. Park, M.S. Structural change in agriculture—Raw data analysis of 2005 agricultural census report 2005. *Coop. Manag. Rev.* **2005**, *37*, 1–28.
9. Keshtgari, M.; Deljoo, A. A wireless sensor network solution for precision agriculture based on Zigbee technology. *Wirel. Sens. Netw.* **2012**, *4*. [CrossRef]
10. Othman, M.F.; Shazali, K. Wireless sensor network applications: A study in environment monitoring system. *Procedia Eng.* **2012**, *41*, 1204–1210. [CrossRef]
11. Kwong, K.H.; Wu, T.T.; Goh, H.G.; Sasloglou, K.; Stephen, B.; Glover, I.; Shen, C.; Du, W.; Michiel, C.; Andonovic, I. Implementation of herd management systems with wireless sensor networks. *IET Wirel. Sens. Syst.* **2011**, *1*, 55–65. [CrossRef]
12. Yoon, M.; Chang, J.W. Design and implementation of an advanced cattle shed management system using an infrared wireless sensor nodes and surveillance camera. *J. Korea Contents Assoc.* **2010**, *12*, 22–34. [CrossRef]
13. Kim, Y.B.; Choi, D.W. Design of business management system for livestock pens based of IoT. *J. Korean Entertain. Ind. Assoc.* **2014**, *8*, 207–216. [CrossRef]
14. Park, M.C.; Ha, O.K. Development of effective cattle health monitoring system based on biosensors. *Adv. Sci. Technol.* **2015**, *117*, 180–185.
15. Kim, H.G.; Yang, C.J.; Yoe, H. Design and implementation of livestock disease forecasting system. *J. Korean Inst. Commun. Inf. Sci.* **2012**, *37*, 1263–1270. [CrossRef]
16. Muhammad, F.; Thar, B.; Masood, K.A.; Babar, S.; Saiqa, A.; Francis, C. Context mining of sedentary behaviour for promoting self-awareness using a smartphone. *Sensors* **2018**, *18*, 874. [CrossRef]

17. Muhammad, F. Alert Me: Enhancing active lifestyle via observing sedentary behavior using mobile sensing systems. In Proceedings of the 2017 IEEE 19th International Conference on e-Health Networking, Applications and Services (Healthcom), Dalian, China, 12–15 October 2017.

18. Fahim, M.; Khattak, A.M.; Baker, T.; Chow, F.; Shah, B. Micro-context recognition of sedentary behaviour using smartphone. In Proceedings of the 2016 Sixth International Conference on Digital Information and Communication Technology and its Applications (DICTAP), Konya, Turkey, 21–23 July 2016.

19. Senthilnath, J.; Harikumar, K.; Suresh, S. Dynamic area coverage for multi-UAV using distributed UGVs: A two-stage density estimation approach. In Proceedings of the Conference: IEEE International Conference on Robotic Computing, Laguna Hills, CA, USA, 31 January–2 February 2018.

20. Gong, D.; Yang, Y. Low-latency sinr-based data gathering in wireless sensor networks. *IEEE Trans. Wirel. Commun.* **2014**, *13*, 3207–3221. [CrossRef]

21. Chang, Y.S.; Lin, Y.S.; Wu, N.C.; Shin, C.H.; Cheng, C.H. Scenario planning and implementing of a dairy cattle UHF RFID management system. In *Proceedings of 2013 4th International Asia Conference on Industrial Engineering and Management Innovation (IEMI2013)*; Springer: Berlin/Heidelberg, Germany, 2014; pp. 643–654.

22. Gutiérrez, A.; Dopico, N.I.; González, C.; Zazo, S.; Jiménez-Leube, J.; Raos, I. Cattle-powered node experience in a heterogeneous network for localization of herds. *IEEE Trans. Ind. Electron.* **2013**, *60*, 3176–3184. [CrossRef]

23. RFM69HCW Datasheet, HOPERF. Available online: http://www.hoperf.com/upload/rf/RFM69HCW-V1.1.pdf (accessed on 22 April 2018).

24. Senthilnath, J.; Kandukuri, M.; Dokania, A.; Ramesh, K.N. Application of UAV imaging platform for vegetation analysis based on spectral-spatial methods. *Comput. Electron. Agric.* **2017**, *140*, 8–24. [CrossRef]

25. Wadhwa, L.K.; Deshpande, R.S.; Priye, V. Extended shortcut tree routing for ZigBee based wireless sensor network. *Ad Hoc Netw.* **2016**, *37*, 295–300. [CrossRef]

26. Wei, K.M.; Dong, M.X.; Weng, J.; Shi, G.Z.; Ota, K.R.; Xu, K. Congestion-aware message forwarding in delay tolerant networks: A community perspective. *Concurr. Comput. Pract. Exp.* **2015**, *27*, 5722–5734. [CrossRef]

27. Mavromoustakis, C.X.; Karatza, H.D. Real-time performance evaluation of asynchronous time division traffic-aware and delay-tolerant scheme in ad hoc sensor networks. *Int. J. Commun. Syst.* **2010**, *23*, 167–186. [CrossRef]

28. Bali, R.S.; Kumar, N.; Rodrigues, J.J.P.C. An efficient energy-aware predictive clustering approach for vehicular ad hoc networks. *Int. J. Commun. Syst.* **2017**, *30*, e2924. [CrossRef]

29. Wei, K.; Guo, S.; Zeng, D.; Xu, K. A multi-attribute decision making approach to congestion control in delay tolerant networks. In Proceedings of the IEEE International Conference on Communications (ICC), Sydney, Australia, 10–14 June 2014; pp. 2748–2753.

30. Kim, S.J.; Jee, S.H.; Cho, H.C.; Kim, C.S.; Kim, H.S. Implementation of unmanned cow estrus detection system for improving impregnation rate. *J. Korean Acad. Ind. Coop. Soc.* **2015**, *16*, 1–11. [CrossRef]

31. Andersson, L.M.; Okada, H.; Miura, R.; Zhang, Y.; Yoshioka, K.; Aso, H.; Itoh, T. Wearable wireless estrus detection sensor for cows. *Comput. Electron. Agric.* **2016**, *127*, 101–108. [CrossRef]

32. Bauckhage, C.; Kersting, K. Data mining and pattern recognition in agriculture. *KI-Künstliche Intell.* **2013**, *27*, 313–324. [CrossRef]

33. Guo, Y.; Corke, P.; Poulton, G.; Wark, T.; Bishop-Hurley, G.; Swain, D. Animal behaviour understanding using wireless sensor networks. In Proceedings of the 31st IEEE Conference on Local Computer Networks, Tampa, FL, USA, 14–16 November 2006; pp. 607–614.

34. Gutierrez-Galan, D.; Dominguez-Morales, J.P.; Cerezuela-Escudero, E.; Rios-Navarro, A.; Tapiador-Morales, R.; Rivas-Perez, M.; Dominguez-Morales, M.; Jimenez-Fernandez, A.; Linares-Barranco, A. Embedded neural network for real-time animal behavior classification. *Neurocomputing* **2018**, *272*, 17–26. [CrossRef]

electronics

MDPI

Article

Design and Implementation of a Sensor-Cloud Platform for Physical Sensor Management on CoT Environments

Lei Hang [1] , Wenquan Jin [1] , HyeonSik Yoon [2], Yong Geun Hong [2] and Do Hyeun Kim [1,*]

1   Department of Computer Engineering, Jeju National University, Jeju 63243, Korea;
    hanglei@jejunu.ac.kr (L.H.); wenquan.jin@jejunu.ac.kr (W.J.)
2   KSB Convergence Research Department, Electronics and Telecommunications Research Institute,
    Daejeon 34129, Korea; x7hyhs@etri.re.kr (H.Y.); yghong@etri.re.kr (Y.G.H.)
*   Correspondence: kimdh@jejunu.ac.kr; Tel.: +82-64-7543658

Received: 6 July 2018; Accepted: 24 July 2018; Published: 7 August 2018

Abstract: The development of the Internet of Things (IoT) has increased the ubiquity of the Internet by integrating all objects for interaction via embedded systems, leading to a highly distributed network of devices communicating with human beings as well as other devices. In recent years, cloud computing has attracted a lot of attention from specialists and experts around the world. With the increasing number of distributed sensor nodes in wireless sensor networks, new models for interacting with wireless sensors using the cloud are intended to overcome restricted resources and efficiency. In this paper, we propose a novel sensor-cloud based platform which is able to virtualize physical sensors as virtual sensors in the CoT (Cloud of Things) environment. Virtual sensors, which are the essentials of this sensor-cloud architecture, simplify the process of generating a multiuser environment over resource-constrained physical wireless sensors and can help in implementing applications across different domains. Virtual sensors are dynamically provided in a group which advantages capability of the management the designed platform. An auto-detection approach on the basis of virtual sensors is additionally proposed to identify the accessible physical sensors nodes even if the status of these sensors are offline. In order to assess the usability of the designed platform, a smart-space-based IoT case study was implemented, and a series of experiments were carried out to evaluate the proposed system performance. Furthermore, a comparison analysis was made and the results indicate that the proposed platform outperforms the existing platforms in numerous respects.

Keywords: Internet of Things; wireless sensor networks; Cloud of Things; virtual sensor; sensor detection

1. Introduction

The Internet of Things (IoT) paradigm will be the next wave in the era of computing [1]. The Internet of Things is the fundamental idea of essentially empowering a worldwide connected network of any devices with an on and off switch to the Internet between the real world and a virtual world. The Internet of Things guarantees to enable accomplishing imaginative applications in application areas such as building and home control, smart environment, framing, intelligent transportation, and healthcare services. The embodiment of IoT is to ensure secure connection among heterogeneous physical sensors and actuators with the Internet. Depending upon the Gartner report, the IoT market is predicted to incorporate 20.8 billion IoT devices by 2020 [2]. Huge measures of the data generated by all these devices together should be gathered, processed, stored, and retrieved [3]. Consequently, how to deal with the expanding number of devices has consistently been a vital issue in the area of IoT.

Cloud computing [4] is a much more appropriate technology which provides magnificent elastic computation and data management abilities for IoT. Indeed, it is diffusely acknowledged that cloud computing can be invested in utility services in the immediate future [5]. Cloud computing has been involved and comprised of miscellaneous technologies like smart grids, networking, computational virtualization, and software services [6]. Normally, the IoT consists of a large variety of objects which are limited in storage and computing power. Cloud management techniques are progressively employed to manage IoT components, as IoT systems are growing complex [7]. The goal of cloud computing is to give a better utilization of distributed resources to accomplish better services for the end users. This model is furthermore transforming the way software is made, as an ever increasing number of applications nowadays are intended to be carried out in the cloud. In this manner, there are particular programming models in light of the idea of "XaaS" where X is hardware, software, data, and so forth [8]. Software-as-a-service (SaaS), platform-as-a-service (PaaS), and infraconfiguration-as-a-service (IaaS) are three fundamental categories of cloud computing services.

In consideration of how the cloud is produced, the virtualization of objects is the accompanying normal step in this field. Virtualization [9] is a rising technique that creates virtual representations of physical resources and enables usage of resources as shared resources and on an on-demand basis. Virtual sensors [10] are required for taking care of complex tasks which cannot be handled by physical sensors. They can be utilized to remotely interface two regions of interest with a single function. Researchers have recently focused on sensor virtualization to advance applications in business, health, and academic domains, and they have made several proposals [11] which expand the paradigm of physical sensor networks in which client applications can access virtualized sensors that are capable of being operated from anyone else in software control [12]. For example, mobile crowdsensing could be one of the areas that can benefit from these virtualized sensors through participatory sensing [13]. A crowdsensing platform is presented in Reference [14] where thousands of sensors are required to collect context data from users in the city. Vehicular cloud computing [15] is another typical application, which merges mobile cloud computing and vehicular networking to integrated communication-computing platforms. A distributed and adaptive resource management controller [16] is proposed which aims at allowing resource-limited car smartphones to perform traffic offloading towards the cloud in vehicular access networks.

However, existing studies of physical sensors center around sensor data processing [17], power consumption [18], sensor localization [19,20], and sensor node programming [21–23]. There are few studies that concentrate on the management of physical sensors since these sensors are linked to their own application directly. This means that these sensors would reside on the local system, and hence have no other way to be accessed from external servers beyond the local domain. Furthermore, the requirements for sensor management have not been clarified, and thus, users may feel discontent if they cannot use the sensors they need. Moreover, users should be informed of the status of sensor nodes if the sensor is disconnected or faults occur so that they can select other functioning sensors. However, most of these systems have poor capability and infraconfiguration to detect fault nodes in sensor networks.

Our contributions in this paper are as follows: First, we propose a new sensor-cloud-based platform with the main objective of empowering more advanced capability to cope with physical sensors, by means of virtual sensors. With the progression of web front-end technologies, for instance, JavaScript and HTML5, the growing trend of IoT platforms moves in a web-driven direction using Representational State Transfer Application Programming Interfaces (REST APIs) in combination with product-specific services and web-based dashboards to help users to rapidly configure and monitor the connected objects through the platform. This study concentrates on the virtualization of wireless sensor networks (WSNs) to the cloud by utilizing these web technologies, proposing a platform which is intended to achieve a consistent combination with physical sensor networks. This could be applicable by giving software agents running in the cloud, likewise alluded to as virtual sensors that act simply like physical sensors dispersed in the cloud, giving the environmental circumstances from the zone where they are deployed. Along these lines, this is the so-called sensor-data-as-a-service (SDaaS) paradigm [24]. This paradigm lessens the excess data capture as data reusability in WSNs that is straightforward to the cloud users. The end users can directly control and utilize the WSNs through standard functions with an assortment of parameters like sampling frequency, latency, and security. Moreover, multiple users from different regions can share the data captured by WSNs with a specific end goal to reduce the burden of data collection for both the system and the users. The designed platform also provides different interfaces for enrolling physical sensors, for provisioning virtual sensors, for monitoring and controlling virtual sensors, and for creating and supervising end users. Besides, the users do not need to concern the low-level points of specification of sensors, for example, the types of sensors utilized and how to set up the sensors.

The most prominent aspect about this platform, which makes it a stage forward in reinforcing coordination of WSNs with the cloud, is the novel approach to recognize faulty sensors via virtual sensors. Since the status of virtual sensors have a relationship with their corresponding physical counterpart, the virtual sensors will be able to inform of sensor errors if the associated physical sensors are faulty. Lastly, we demonstrate the practicality of our designed platform by implementing a real-life case study in smart space. The Raspberry Pi, which provides the computing power and wireless communication capabilities, has been used to integrate with various low-resource sensors. The IoTivity [25] is utilized for the communication between physical sensors and the cloud, but for communication between the cloud and application client, HTTP is used. The IoTivity is a standard and open source framework to permit adaptability making solutions which can address a wide range of IoT devices from various application environments. In general, the IoTivity conforms to the User Datagram Protocol (UDP)/IP stack, hence the Constrained Application Protocol (CoAP) [26] is chosen as the mandatory protocol. It originally characterizes an HTTP-like transfer protocol which also entirely complies with the representative architecture of state transfer (REST) architecture style.

The rest of this paper is organized as follows: Section 2 gives a concise introduction on cloud infra configurations towards IoT and discusses a portion of the similar related research. Section 3 explains the designed platform for integrating physical sensors in the cloud. Section 4 gives some insight into the implementation of the case study over the designed platform. Also, we describe an overview of the implementation of the smart space case study using various snapshots. Section 5 reports the evaluation results of the proposed platform. Section 6 outlines the significance of the proposed work through a comparative analysis with existing work. Finally, Section 7 summarizes the main conclusions of the paper and discusses some directions for future research.

2. Related Work

The Internet of Things offers potentialities which make it possible for the development of a huge number of applications. Some of the mentioned domains are transportation, healthcare, smart environment, personal and social domains. Each of the domains include its own unique characteristics in terms of real-time processing, volume of data collection and storage, identification and authentication, and security considerations. For example, real-time processing is of utmost importance in the transportation industry, while identification and authentication are important aspects in healthcare. Cloud computing [27], with its virtually unlimited resources of computing power, storage space, and networking capabilities, is well appropriate for scaling in the IoT world.

As of late, an extensive measure of research in the field of the probability of combining cloud computing with WSNs has been explored [28]. This paradigm has been proposed as a feasible mechanism to accomplish the best use of a wireless sensor infraconfiguration and allows data sharing among multiple applications. Recently, the REST architecture style appeared, leading to the development of the Web of Things [29]. Uniform resource identifiers (URIs) are used to identify web things, and the HTTP protocol is used for stateless reciprocation between clients and servers. Uniform resource identifiers which contain both name and locators are put to use in resources in the real world to identify web things. It describes web services with a uniform interface (HTTP method) which provide the pathways for consumers to obtain possible representations from servers for interactions [30]. This makes it an ideal way to construct application programming interfaces (APIs) for allowing mashups to be created that allow end users to associate data from physical data sources to virtual sources on the web [31]. The resulting approach significantly improves the integration of service deployment for resource constrained IoT devices, while reducing the burden on both the devices and the service developers.

SenseWeb [32] is one of the essential architectures being presented on merging WSNs with the Internet for sensor information sharing. This system provides diverse web APIs which capacitate users to enroll and distribute their own sensor data. In the idea of the Web of Things, smart things and their services are completely organized in the web, and the REST architecture style is associated with the resources in the physical world. In Reference [33], the authors propose a resource-oriented architecture for the IoT, where distinctive web technologies can be used to configuration applications on smart things. The interfaces of things have turned out to be similar to those found on the web, and this principle can be applied in various prototypes, for instance, environmental sensor nodes, energy monitoring system, and Radio-Frequency Identification (RFID)-labeled things. The utilization of an organized Extensible Markup Language (XML) document or a JavaScript Object Notation (JSON) object energizes the compatibility of a large amount of sensors and permits describing services offered by these sensors. sMAP [34] has been expected to represent the data from sensors and actuators over HTTP in JSON schemas. The readings themselves are sent as JSON objects with strict formats and data semantics that are characterized in a number of sets of JSON schemas. The architecture supports resource-constrained devices through proxies that translate the data between JSON and binary JSON. SensorML [35], proposed by the OGC (Open Geospatial Consortium), is an XML encoding intended to absolutely model physical sensors' description and measurement processes, in addition to context like geolocation data and legal data. This approach depicts the metadata of physical sensors and the mapping between physical sensors and virtual sensors, enabling the requests interpreted from end users to virtual sensors for the related physical sensors. A comprehensive work on the cloud-based IoT paradigm is introduced in Reference [36], as it specifies the inspiration, applications, research challenges, related works, and platforms for this paradigm. One perceived research challenge is the coordination of colossal measures of exceptionally heterogeneous things into the cloud. To address this issue, Reference [37] presents a service-oriented infraconfiguration and toolset called LEONORE for provisioning application components on edge devices in substantial-scale IoT deployments. This solution supports pull-based provisioning which enables devices to autonomously schedule provisioning runs to off-peak times, while push-based provisioning takes into account the greater control over the deployed application landscape. Madria et al. [38] propose a new paradigm for interacting with wireless sensors and the sensor-cloud in order to overcome restricted resources and efficiency. The designed infraconfiguration spans over a wide geographical area, bringing together multiple WSNs composed of different physical sensors. Misra et al. [39] make one of the first attempts in the direction of the sensor-cloud by proposing a novel theoretical modeling. A mathematical formulation of the sensor-cloud is presented, and a paradigm shift of technologies from traditional WSNs to a sensor-cloud architecture is suggested as well. Existing cloud pricing models are limited in terms of the homogeneity in service-types, and in order to address this issue Chatterjee et al. [40] propose a new pricing model for the heterogeneous service-oriented architecture of Sensors-as-a-Service (Se-aaS) with the sensor-cloud infraconfiguration. The proposed pricing model comprises two components: pricing attributed to hardware (pH) that focuses on pricing the physical sensor nodes, and pricing attributed to infraconfiguration (pI) that deals with pricing incurred due to the virtualization of resources. An interactive model is proposed in Reference [41] to enable the sensor-cloud to provide on-demand sensing services for multiple applications, and this model is designed for both the cloud and sensor nodes to optimize the resource consumption of physical sensors as well as the bandwidth consumption of sensing traffic. Abdelwahab et al. [42] further expand in this direction by proposing a virtualization algorithm to deploy virtual sensor networks on top of a subset of selected physical devices, as well as a distributed consensus approach to provide high-quality services from unreliable sensors. In order to improve the lifetime of conventional WSNs, Dinh et al. [43] propose a new decoupling model for physical sensors and information providers toward a semantic sensor-cloud integration. This model takes advantage of data prediction to minimize the number of networked sensors as well as the traffic load from these sensors. In order to make the sensor-cloud be able to satisfy multiple applications with different latency requirements, the authors in Reference [44] propose an automatic scheduling method to meet the requirements of all applications. A request aggregator is designed to aggregate latency requests from applications to minimize the workloads for energy saving, and a feedback-based control theory is designed to handle the sensing packet delivery latency. Sen et al. [45] propose a risk assessment framework for a WSN-integrated sensor-cloud using attack graphs to measure the potential threats. The Bayesian-network-based approach analyzes attacks on WSNs and predicts the time frames of security degradation on the grounds of integrity, availability, and confidentiality. Mils-Cloud [46] is a sensor-cloud architecture utilizing the networks-as-a-service paradigm for the integration of military tri-services in a battlefield area. Furthermore, users are assigned different levels of priority in order to boost the system performance. A location-based interactive model [47] specified for mobile cloud computing applications is proposed to render sensing services on the demand of a user's interest and location in order to save energy. The cloud controls the sensing scheduling of sensors, for example, sensors are put into inactive mode when there is no demand. Zhu et al. [48] propose a multi-method data delivery scheme for sensor-cloud users, which comprises four kinds of delivery. The proposed approach could achieve lower delivery times and delivery times according to the evaluation results under different scenarios. Cloud4sens [49] is a new architecture which combines both the data-centric and device-centric models, enabling the end users to choose on-demand cloud services. IoTCloud [50] is an open source platform with a view to incorporate distinctive terminals (e.g., smart phones, tablets, robots, etc.) with backend services for controlling sensors and their messages; it gets RESTful-based APIs to share information with applications. ThingSpeak [51] is another open source platform for putting away and retrieving data from physical things through a local area network. With the numeric APIs given by the ThingSpeak, users can build sensor-logging applications, location tracking applications, and a social-network-of-things with announcements. The DIGI [52] enables users to interface a physical device to the cloud and utilize an online web application for remote access. This platform is a machine-to-machine (M2M) platform as a service. It is outfitted for dealing with the correspondence between enterprise applications and remote device resources, regardless of location or network. The platform incorporates the device connector software that promotes remote device connectivity and combination. The application additionally provides cache and permanent storage options available for generation-based storage and on-demand access to historical device samples. As the desire for low-resource IoT devices is raised, some researchers have put forth efforts to enhance the incorporation in the field of constrained devices. In Reference [53], the authors exhibit a flexible framework for IoT services in light of both The Constrained Application Protocol (CoAP) and HTTP. The designed architecture incorporates three phases (i.e., network, protocol, and logic) which shape a processing pipeline where each phase has its own particular separate thread pool. The evaluation result represents that CoAP is an appropriate protocol for IoT service. Other platforms like Heroku [54], Kinvey [55], Parse [56], CloudFoundry [57], and Bluemix [58], as illustrated in Reference [59], are also used broadly, but they only offer abilities at the platform level, thus creating PaaS solutions which are not adaptable to the general public.

A sensor network is made out of an expansive number of sensor nodes which involve sensing, data processing, and communication capabilities. Faults in sensor information can occur for some reasons. Factors such as extraordinary temperatures or precipitation can influence sensor performance. A novel strategy to distinguish physical sensors with data faults is provided in FIND [60], by ranking the physical sensors on their readings, in addition to their physical distances from an event. It considers a physical sensors faulty if there is a noteworthy mismatch between the sensor data rank and the distance rank. The expansion in the quantity of sensor nodes genuinely improves the level of difficulty in identifying the inside status of sensor nodes. As a result, many researchers focus on the utilization of machine learning to find faulty nodes. In Reference [61], the authors propose a distributed Bayesian network to detect faulty nodes in WSNs. Border nodes placed in the network are used for adjusting the fault probability. This approach improves the correctness of fault probability by reducing the negative effect of considerable number of faulty nodes. However, the conventional fault detection algorithms face low detection accuracy. A fuzzy-logic-based system [62] is proposed to enhance the detection accuracy by using a three input fuzzy inference system, which identifies hardware faults including transceiver and battery condition. However, these systems mainly center on detecting the faults on physical sensors, while we identify faulty sensors by monitoring the virtual sensors.

As mentioned above, these systems are either intended for a very limited application domain or are closed source which are not flexible to the generic public. It is also crucial to inform the end users whether the physical sensors are accessible and whether sensor faults occur so as to maintain the data quality from physical sensors. However, most of the existing approaches only focus on detecting faulty physical sensors. To the best of our knowledge, the proposed sensor detection approach in this paper is the first ever step towards faulty node detection in WSNs via virtual sensors. In one word, it is essential for the realization of the IoT to build up a generic sensor-cloud based architecture, which can easily be adapted to multiple domains while providing smart device management, monitoring, processing, and detection functionalities.

3. Designed Architecture for Sensor-Cloud Platforms

3.1. Basic Architecture of the Cloud-Based Platform

Figure 1 presents the basic architecture of the Cloud-based platform. The IoT nodes consist of various physical sensors capable of communicating with the Internet. The cloud extracts profile data from the sensors, thus representing them as virtual sensors via the web interface. The cloud also provides RESTful APIs to offer functionalities such as discovering physical sensors and reading sensing data from them.

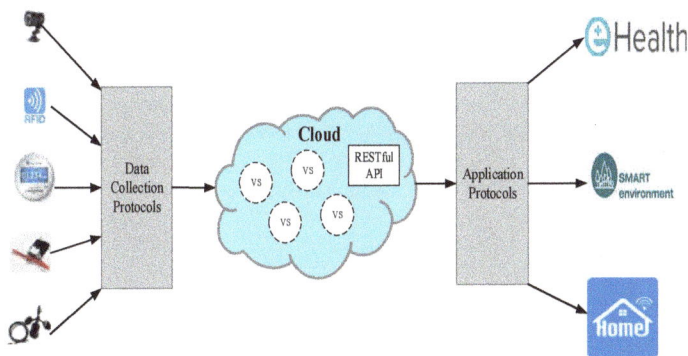

Figure 1. Basic architecture of Cloud-based platform.

The mapping physical sensors to virtual sensors on the Cloud-based platform is described in Figure 2. We describe virtual sensors in order to enable the users to use sensors without any concern about considerations like specifications or location of physical sensors. The virtual sensors are used to encapsulate the basic attributes and behaviors of their corresponding physical counterparts in the physical space. The physical space consists of various IoT devices across different application domains such as smart home, healthcare, etc. These devices have the ability to perform actions according to commands from the virtual space. For each device, a device profile is preserved, which contains the properties (ID, URI, location, etc.) of the device and the basic action offered by the service. The profiles are stored by the cloud platform for further processing, as in other previous platforms. This platform provides various visual interfaces that provide better understanding of virtual sensors and enable the end users to manipulate IoT devices associated with the platform in an intuitive way, thereby reducing the burden under the premise of performing some actions on the physical sensors.

Figure 2. Mapping physical sensors to virtual sensors.

3.2. Data Representation using Virtualization and Grouping

Figure 3 shows the data production and representation for the proposed platform. The data are presented as a single instance illustrating the metadata used for that layer. The physical sensor layer comprises various physical sensors, which are the IoT resources responsible for the production of data. These IoT resources are represented as virtual sensors in the system. The metadata for a virtual sensor is depicted in the second layer of the figure. In order to register an IoT resource with the system, information must be provided corresponding to the metadata in the virtual sensor layer. This data includes the target URI of the resource, which describes the protocol and the network address of the resource so that it can be accessed from anywhere. Other metadata includes the ID of the resource so it can be identified from other resources. The type metadata is a representation associated with the resource referenced by the target URI. The interface metadata defines the instance supported by a resource that can be either a sensor or an actuator. The endpoint metadata represents the endpoint information of the target resource.

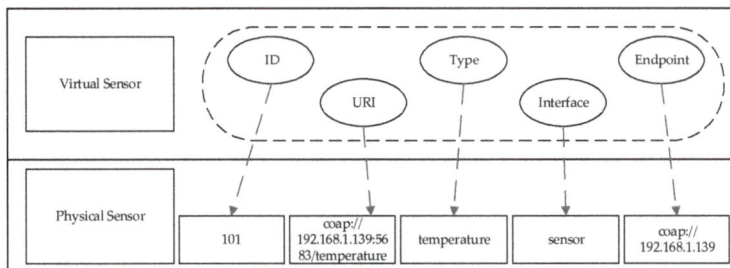

Figure 3. Metadata representation between physical sensors and virtual sensors.

The relationships among physical sensors, virtual sensors, and virtual sensor groups are discussed in Figure 4. There is a one-to-one correspondence between a virtual sensor and the physical sensor from which it was created. Although physical sensors are varied in type, size, and use, a specific application does not need to use all of them. An application uses some types of physical sensors according to the scenario and requirement. The proposed sensor-cloud system divides virtual sensors into virtual sensor groups depending on the corresponding application scenario. A virtual sensor group consists of one or more virtual sensors. Users can freely use the virtual sensors included in the group as if they are on their own. For example, they can control the behaviors of virtual sensors and check their status. The proposed platform prepares typical virtual sensor groups for the smart space, and this use case will be discussed in detail in the use case Section. Users can also create new virtual sensor groups by selecting virtual sensors according to their custom requirements.

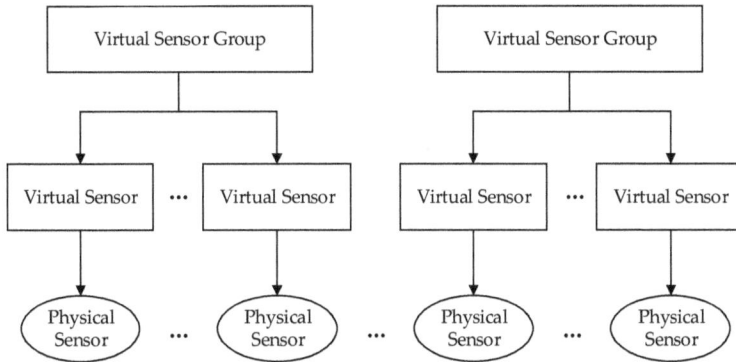

Figure 4. Relationship among virtual sensor groups, virtual sensors, and physical sensors.

3.3. Proposed Layer-Based Platform Architecture

Figure 5 represents the layer-based architecture of the Cloud-based platform, where each layer is decoupled from other layers so that they can evolve independently. The physical layer comprises various network connected devices with the abilities of sensing, computer processing, and storage. The major function provided by the network layer is routing, as physical sensors do not have a global ID and have to be self-organized. It also contains other modules for providing services including network management, security management, and message broker. The virtualization layer represents the physical sensors as virtual sensors in cyber space. These objects contain various behaviors, including the services or functions that can be utilized by the system to interact with the corresponding physical sensors, and information such as URI and location of these objects are the attributes with them. The service layer contains all modules that organize common services for providing various features including user authentication, device monitoring, service provisioning, and the device access. The data storage provides storage space for the profile and sensing data provided by physical sensors, and contains the instructions to process such information. The data processing receives and converts the data from the sensor networks to the file format understood by the platform. The resource allocation assigns and manages the available resources to various uses such as determining whether requests for services can be responded to. The device discovery is used to detect all the available devices connected to the platform or reject access when the devices are disconnected. Finally, the message offers the required components for enabling communication among the elements of the architecture. The top layer is the application layer, this layer visualizes the physical sensor network in either desktops or mobile devices and can mash content with data or apps from third parties.

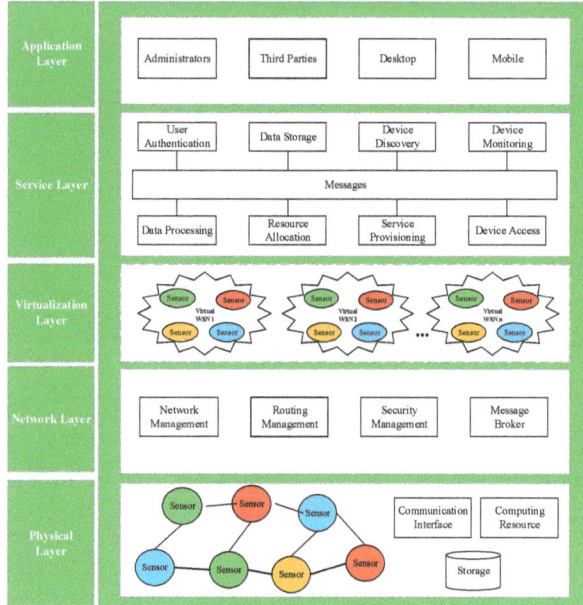

Figure 5. Proposed layer-based platform architecture.

3.4. Designed Sensor-Cloud Application

Figure 6 represents the detailed functional configuration of the cloud-based web application for preserving the profiles of physical sensors. This application is deployed on the Amazon Elastic Compute Cloud (Amazon EC2) in the Amazon Web Services (AWS) cloud. The cloud plays a simple role in the proposed platform by providing a repository of profile information associated with the linked physical sensors through a more user-friendly way. Thus, exploiting a fully developed cloud platform such as AWS or Azure will be seen as overqualified [63]. The user authentication allows the system to verify the identity of all users accessing to the network resources. The resource manager hosts the resource descriptions of physical sensors and provides the functionalities to register, maintain, lookup and remove these descriptions. The DB connector provides interfaces to connect to the database and allows to run Create, Read, Update, and Delete (CRUD) procedures on the database. The resource information which contains web links to all resources is stored in the MySQL database. The abstraction manager prepares the abstract for presenting virtual resource information in a template, and this template will be released to the app client whenever the user builds the request. The access manager is responsible for network communication with physical sensors. This module is implemented in RESTful style, which is flexible enough to adapt to other communication protocols, and is not limited to a single protocol such as HTTP. The device registration manager processes the incoming device information while the sensing data manager handles the sensing data from the physical sensors. The request manager is capable of receiving and handling the request from the client. It decouples the attributes from the request in order to discover the appropriate action based on the URI requested. The response manager generates the response based on the request and converts the data in the form which the client can understand. The client provides various interfaces such as user login and registration, device configuration, and device visualization. Physical sensors can be visualized whenever a new device is connected to the system. In addition, the cloud provides a unique interface for every virtual sensor, through which the end user can monitor the state of the sensor, and consequently the physical sensor can be easily controlled through the visual interface even though the user is outside the local space.

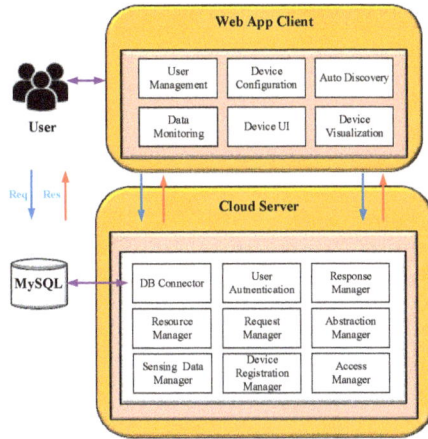

Figure 6. Functional configuration of the cloud-based web application.

3.5. RESTful API-Based Architecture

The business logic of the cloud-based web application as shown in Figure 7 is basically built on heterogeneous web APIs, which are synonymous with online web services that client applications can use to retrieve and update data. A web API, as the name suggests, is an API over the web which can be accessed using HTTP. Processing tasks can be executed by requesting these APIs in the cloud. The business logic contains various web APIs such as user management, device management, resource management, and sensing data management. Each client application utilizes the same API to get, update, and manipulate data in the database. For example, users can perform read and write operations to a virtual sensor by calling the read and write APIs. In addition, user can retrieve the details of the virtual sensor or send requests for getting sensing data without requiring much knowledge about the specification of physical sensors, which significantly lowers the training requirements for the general public. As an open source architecture, the developer can extend his own module according to the requirements of applications and runtime environments.

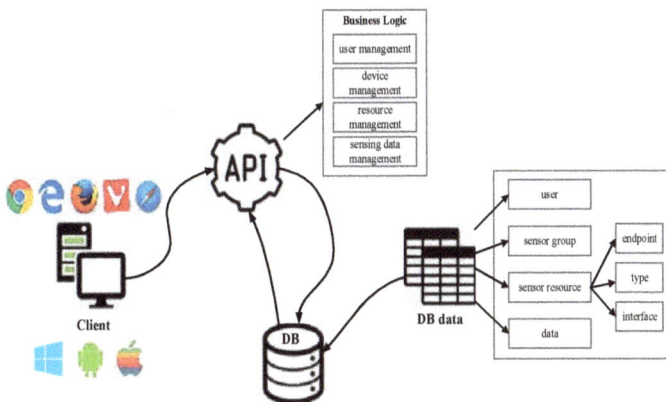

Figure 7. Web Application Program Interface (API) of the cloud-based web application.

3.6. Sensor Detection Approach Using Virtual Sensors for Physical Sensor Management

A number of issues arise when directly detecting the physical sensors, not least when selecting suitable diagnostic systems and encountering difficulties of placing a sensor in the required position for a given task. These issues can be overcome by using virtual sensors which provide the knowledge obtained from the behaviors of physical sensors. Virtual sensors can collect and even replace data from physical sensors. The central premise of this approach is that the physical sensors need to be registered to the cloud before accessing through virtual sensors.

The execution sequence between the IoT device and a cloud-based application is described in Figure 8. The IoT device connects to WiFi so that the local router assigns an IP address to the server, and meanwhile configures the resources of the physical sensors as part of the server. The server inserts the device profile (name, ID, status) into the request payload and requests the cloud to upload the profile using the POST method. This information is stored in the device registry of the cloud application, and a response acknowledge message is generated by the cloud to inform that the device information was successfully registered. After registering to the cloud, the IoT device reads sensing data from physical sensors and uploads the data to the cloud. The cloud sends back a response acknowledge message to inform the device that the sensing data was uploaded. However, if the device goes wrong, such as losing the network connection, the device status value will turn into inactive, therefore, the device stops uploading the sensing data until it reconnects to the cloud. Hence, this process is implemented as a cycle process as the device status might change with time.

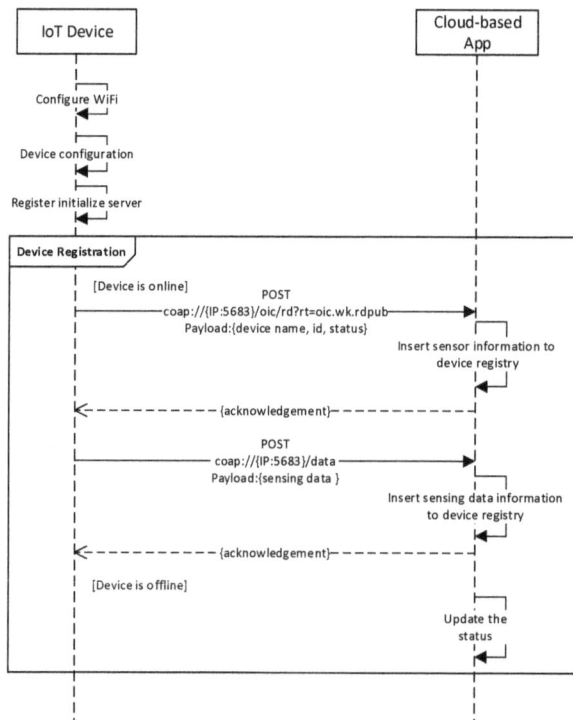

Figure 8. Sequence diagram of operations between Internet of Things (IoT) device and the cloud-based application.

Figure 9 proposes the virtual sensor-based detection mechanism with the architecture. Our concept is to monitor the status of virtual sensors and decide whether to continue showing updates and keep communications open between the client and the cloud-based application or to postpone them when the virtual sensors are not available. To implement this solution, we utilize one of the real-time techniques, so-called polling, to apply our concept, as it benefits from JavaScript features for providing advanced front-end user experience. Polling [64] is a time-driven technique that sends the client request to the cloud application at regular and frequent intervals (e.g., every second) to check whether an update is available to be sent back to the browser. The client initializes a request to the cloud-based application within the specified interval time that should be defined in terms of different scenarios. The application receives the request and retrieves the information from the device registry to check whether the device status is active or not. If the status is active, the application pushes the sensor information to the client, whereas it calls the postpone function when devices are inactive, to suspend sending requests to the cloud application provisionally.

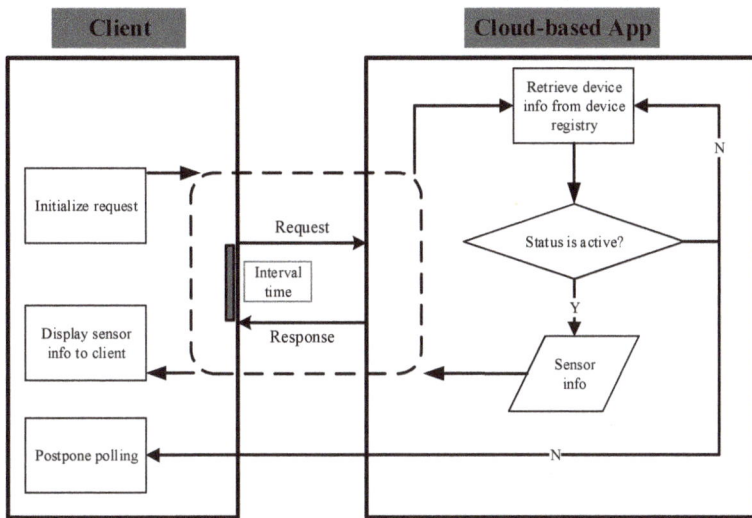

Figure 9. Virtual sensor-based auto detection mechanism.

3.7. Interaction Model between Client and Cloud-Based Application

Figure 10 illustrates the various operation processes between the client and the cloud-based application. The user profile including username and password is submitted to the cloud for identity authentication. If authentication succeeds, the cloud returns the success message to the client and allows the user to access the system. The client can request the cloud for information from a sensor group, and the cloud decides whether the request can be responded to depending upon the status of the sensor group. If the sensor group status is active, the cloud provides the sensor group information to the client. The user can retrieve detailed information from the sensors from the group, and the cloud provides the sensor information which is visible to the client. Furthermore, the user can get the sensing values of a specific sensor by a single click, and then the cloud continuously feedbacks the behaviors and readings of the selected sensor.

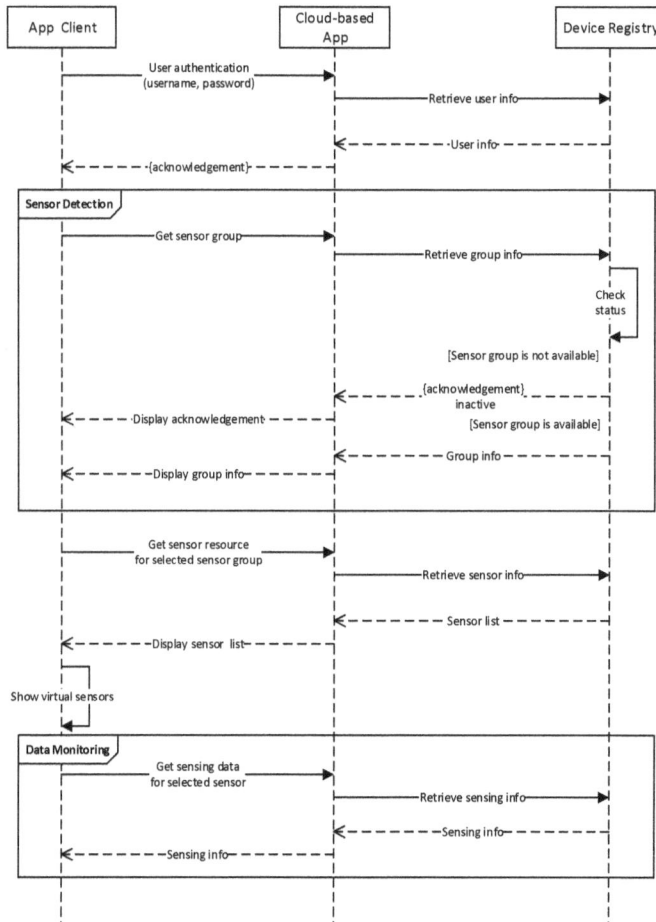

Figure 10. Sequence diagram of various operations within the cloud-based application.

4. Prototype Implementation of Sensor-Cloud Platform in Smart Space

This section describes the topology of virtual sensors and introduces a review of the implementation Integrated Development Environments (IDEs), hardware, and technologies used to develop the smart space use case. Figure 11a presents the hierarchical topology of virtual sensors, where the parent node is the network itself, grouping any number of virtual sensor nodes. Given that the end user requires data from the parent node (7), which is a virtual sensor group, all intermediate nodes (4, 5, and 6) in the topology are virtual sensors, and all leaf nodes (1, 2, and 3) represent the physical sensors in the WSN. The data objects used to form virtual sensors consist of various data tables as illustrated in Figure 11b. Table t_parent summaries the information of a virtual sensor group, where the di is used as the identifier so that each group can be distinguished from each other. Furthermore, the isactive attribute represents the status value (active or inactive) of the group, the ttl attribute stands for the live time of the group, and the name attribute represents the group name which can also be used as the identifier. Table t_node, t_if, t_rt, and t_ep represent the properties referred to in a virtual sensor, while table t_data records the data and the timestamp of the sensing.

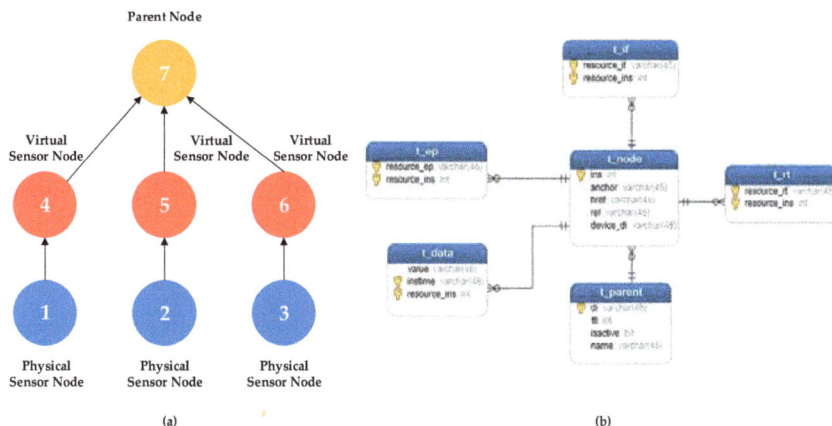

Figure 11. (**a**) Hierarchical virtual sensor topology; (**b**) data objects of virtual sensors.

The project consists of three main modules so that the development environments have been categorized into three tables to describe the development environment for each module separately.

Table 1 depicts the development tools and technologies utilized for implementing the IoT server hosted on the Raspberry Pi. The operating system on the Raspberry Pi is flashed into Android Things so that the applications can be developed using the Java programming language. To support the communication between the IoT server and the cloud, the IoTivity framework for Advanced RISC Machine (ARM) architecture is used. Physical sensors measuring temperature, humidity, and pressure have been implemented into resources as part of the IoTivity server. Each resource is assigned with a unique URI so that it can be identified by the server.

Table 1. Development environment of the IoT server.

Component	Characterization
Hardware	Raspberry Pi3 Model B
Operating System	Android Things 0.4
Memory	1 GB
Server	IoTivity Server
Resources	Temperature, Humidity, Pressure
IDE	Android Studio 2.3.1
Library and Framework	IoTivity (ARM), bmx280
Programming Language	Java

Table 2 describes the development tools and technologies used for the cloud-based application. The application is deployed to the AWS EWC2 and is developed in the Java programming language. Apache Tomcat was used as the container to host the web content and applications. A MySQL database was used as the repository to hold resources from the IoT server. The cloud listens for the request on the appointed port and performs the operation on the corresponding resources.

Table 2. Development environment of cloud-based application.

Component	Characterization
Operating System	Linux AWS EC2 Compute Node
IDE	Eclipse Luna (4.4.2)
Server	Apache Tomcat
Database Management System (DBMS)	MySQL
Library and Framework	IoTivity (Linux), dbutils, gson, mysql-connector
Programming Language	Java

Table 3 presents the technology stack for developing the web client. The client is implemented using various web techniques, including HTML, Cascading Style Sheets (CSS), and JavaScript. In order to present the information in a more user-friendly way, we utilized Bootstrap and JQuery, which are two open-source toolkits for quick web application prototyping. The client can interact with the cloud-based application to operate on the resources and perform their functionalities by HTTP methods like GET and POST.

Table 3. Development environment of web client application.

Component	Characterization
Operating System	Linux AWS EC2 Compute Node
IDE	WebStorm (2017.1.4)
Browser	Google Chrome, Firefox, Safari, IE
Library and Framework	Bootstrap, JQuery
Programming Language	HTML, CSS, JavaScript

For the purpose of assessing the usability of the designed system, we have realized a smart space case study as part of the experimental work. Figure 12 illustrates the implementation environment for the case study, and also presents the means of interconnection between the IoT nodes, the cloud application, and the web client. The HTTP is used for the communication between the cloud and client application, while IoTivity is used between the cloud and physical resources. The Raspberry Pi serves on the IoT server, which is integrated with some physical sensors. The temperature, humidity, and pressure sensors were used for the smart space case study, which is apparent in the figure.

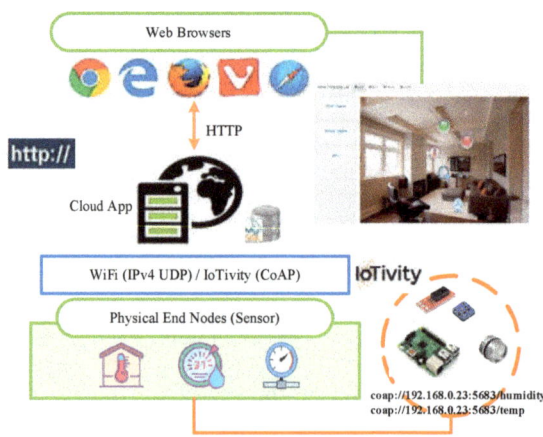

Figure 12. Implementation environment and use case deployment.

A set number of endpoints for communication used between the web client and cloud application are summarized in Table 4. The resource typically represents the data entity and the verb sent along with the request informs the API what to do with the resource, for instance, a request with GET is used to get data about an entity, while a POST request creates a new entity on the resource. There is an observance in place such that a GET request to an entity URI such as /getDevice returns a list of devices, probably matching some criteria that are sent with the request. Query string parameters are also allowed to be used in an API, for example, /getResource&id=? returns all the resources with the specific ID.

Table 4. HTTP requests in RESTful API.

Resource	Verb	Action	Response Code
/UserLogin?username=?&password=?	GET	User Sign In	200/OK
/UserReg?username=?&password=?	POST	User Sign Up	201/Created
/getDevice	GET	Device Info	200/OK
/updateDevice&name=?&id=?	PUT	Update Device	201/Created
/getResource&id=?	GET	Resource Info	200/OK
/getEp&ins=?	GET	Endpoint Info	200/OK
/getRt&ins=?	GET	Resource Type Info	200/OK
/getIf&ins=?	GET	Interface Info	200/OK
/getData&ins=?	GET	Data Info	200/OK

The following section describes some snapshots of various interfaces depending upon the communications endpoints shown in Table 4, along with their respective responses displayed through the web interface. When developing the cloud-based application, developers should take potential security issues into account in order to ensure reliable resources and data transferring [65]. The main issues existing in the cloud can be categorized as follows: identity authenticity, availability, and data confidentiality. Identity authenticity is a core security requirement in a sensor-cloud architecture, which is to verify the identities of all the participants involved in the cloud. Data confidentiality ensures the data secrecy, as the data is only accessible to authorized end users. Availability makes sure that resources and data are available from anyplace and anytime. This paper focus on identity authentication to prevent non-authenticated users from accessing the system. Users' credentials are stored centrally and the main advantage is that the user can just authenticate once to the server and start using the resources across the system. Figure 13a represents the user sign-up interface which requires the input of an email address, a username, and a password. The user sign-in interface shown in Figure 13b is used to submit the user information (username, password) to the cloud for identity authentication.

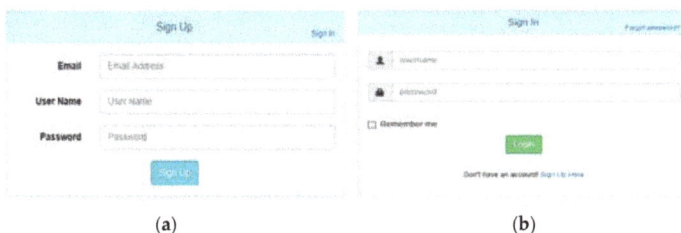

(a)   (b)

Figure 13. Snapshot of end user management in web. (**a**) User Sign Up; (**b**) User Sign In.

The client initializes a request to the cloud application for obtaining the information of available sensor groups after the end user is authenticated by the system. In this system, the status of the virtual sensor group is synchronous with the related physical counterpart, through which the status of the physical sensors can be remotely informed to the end user without requiring any on-site inspections.

The cloud application checks the status of the virtual sensor group from the device registry at each minute, and returns the information of the sensor group if the status value is active. Figure 14a represents the snapshot of the sensor group dashboard which includes the sensor group ID, sensor list, status, and name. The dashboard provides an entry to access all the sensors belonging to the selected group and an editor which enables users to update, create, and delete the sensor group. For instance, the user can modify the properties of the sensor group as shown in Figure 14b, and after confirming the operation, the request is sent to the cloud and the dashboard will be updated accordingly.

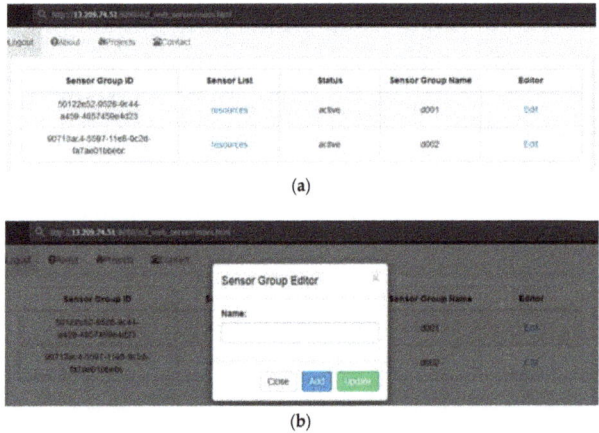

(a)

(b)

Figure 14. Snapshot of sensor group dashboard. (**a**) Sensor Group Dashboard; (**b**) Sensor Group Setting.

The detailed sensor dashboard shown in Figure 15 depicts the descriptions of each virtual sensor from the selected sensor group along with properties such as sensor ID, resource URI, endpoint, type, interface, and status. Each line in the dashboard represents a resource that encapsulates the complete information regarding a virtual sensor.

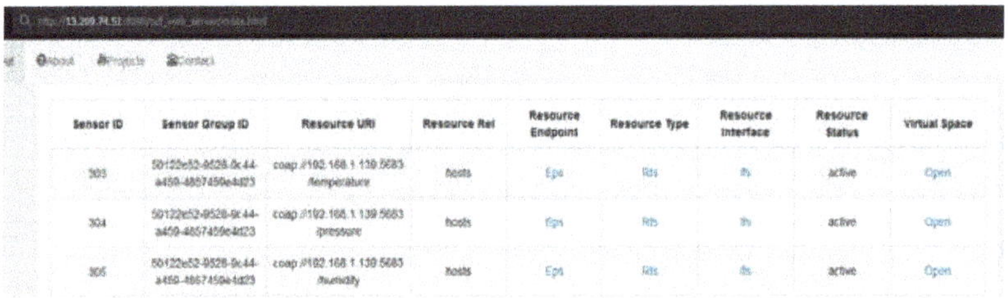

Figure 15. Snapshot of detailed sensor dashboard.

Figure 16a represents the overview of the smart space in web, where a graphical representation is presented for each virtual sensor located in different zones of the home. Once the virtual sensor obtains a request from the end user, the virtual sensor forwards the request to the cloud-based application. A data object mentioned in the implementation details section is created at the back-end, and then the cloud application retrieves the request URI of the virtual sensor from the data object so that the application can communicate with the corresponding physical sensor in the network. The physical sensor provides the sensing data, which is then parsed to the virtual sensor accordingly. The process of reading sensing data from the temperature sensor is illustrated in Figure 16b, where the temperature sensor is clicked to display a pop-up displaying the current readings of the indoor temperature.

(a)   (b)

Figure 16. Snapshot of the web smart space. (**a**) Virtual Sensor Group; (**b**) Data Monitoring.

This work presents a real-life case study for smart space, which is implemented as part of the experiment to evaluate the scalability of the proposed system. We believe this system has the potential to be extended to larger-scaled scenarios like smart hospitals or smart cities, which can easily benefit from the significance of the work. For example, the proposed system can be expanded in the smart farm to facilitate the management of all kinds of wireless sensors and to monitor remote devices on the farm. Our designed system can frequently measure the framing sensors and store the sensing data in the cloud where the data can be additionally visible in the front-end interface which is accessible from anywhere, anytime. Furthermore, the designed sensor detection approach uses virtual sensors, which is useful, especially when farmers have to leave the farm for a long time.

5. Performance Evaluation of Sensor-Cloud Platform

This section illustrates the evaluation results to comprehensively assess the performance of the proposed sensor-cloud platform. We evaluated the service execution time for a different number of virtual sensors under three cases. The first case corresponding to the costing time to create virtual sensors on the cloud-based web application, the second one to the time spent on detecting virtual sensors, and the last one to the time required to initialize the request to the virtual sensors. The first case was analyzed for performance and the results are displayed in Figure 17. For this analysis, three sets of 50, 250, and 500 virtual sensor information was provided to the proposed platform, and each set of virtual sensor information was allowed to be ingested by the platform ten times at randomly selected system resource utilization levels.

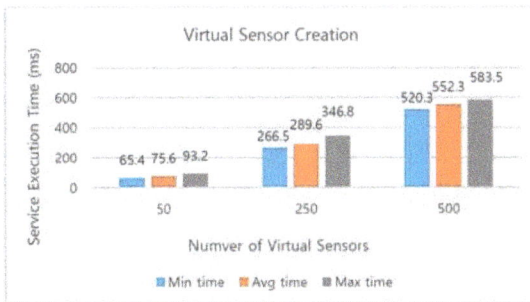

Figure 17. Performance analysis graph of virtual sensor creation.

The graph in Figure 17 presents the minimum, average, and maximum time in ms taken by the proposed platform to parse and instantiate the corresponding visual representations of the physical sensors. For the 50 virtual sensor information set, the minimum time taken in the ten iterations was recorded to be 65.4 ms, averaging at 75.6 ms, and the maximum delay was recorded to be 93.2 ms. For the 250 virtual sensor information set, the minimum time taken in the ten iterations was recorded to be 266.5 ms, averaging at 289.6 ms, and the maximum delay was recorded to be 346.8 ms. For the 500 virtual sensor information set, the minimum time taken in the ten iterations was recorded to be 520.3 ms, averaging at 552.3 ms, and the maximum delay was recorded to be 583.5 ms.

Similarly, the evaluation results for the second and third case are reported in Figures 18 and 19, respectively. It can be seen from these above figures that the service execution time increased in accordance with the number growth of virtual sensors. However, the time increase was at such a low level that it can even be disregarded, in other words, it will not intuitively influence the user experience.

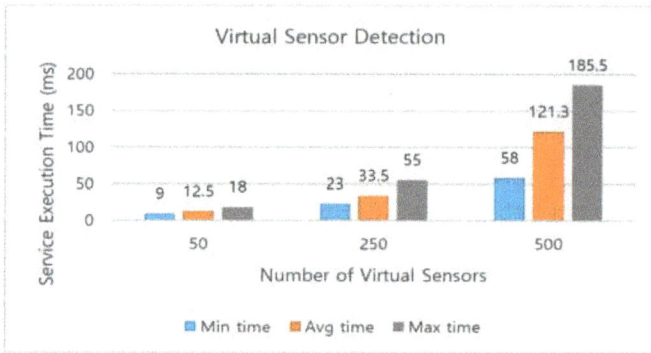

Figure 18. Performance analysis graph of virtual sensor detection.

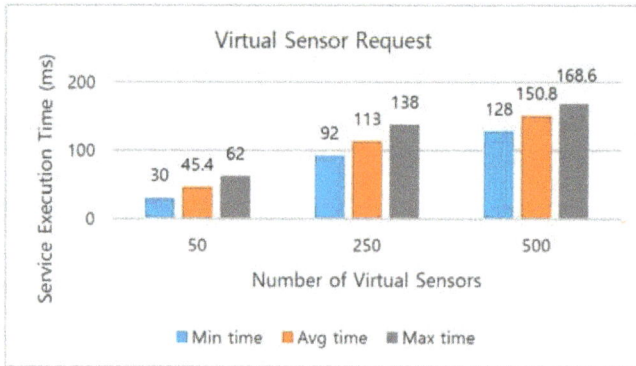

Figure 19. Performance analysis graph of virtual sensor request.

Figure 20 measures the end-to-end access latency performance with variations of increasing numbers of virtual sensors. The end-to-end access latency means the time needed for a request to be transferred from the virtual sensor to the target physical sensor in the network. It is obvious to see from the graph that the access latency raises as the number of virtual sensors grows. The minimum, average, and maximum delay time in ms taken by the proposed platform to access the physical sensors were recorded twenty times at randomly selected system resource utilization levels. For the worst-case performance with 500 virtual sensor information set, the minimum delay taken in the twenty iterations was recorded to be 124.3 ms, averaging at 246.4 ms, and the maximum delay was recorded to be 337.2 ms. The measured end-to-end latency of the dataflow was controlled in an acceptable range and maintained under the 400 ms, which denotes that the proposed system has the ability to assure high efficiency of the end-to-end access latency.

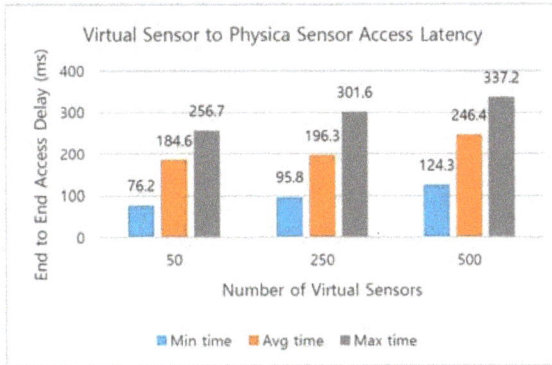

Figure 20. Performance analysis graph of end-to-end latency with different numbers of virtual sensors.

6. Comparison and Significance

This section presents a comparative analysis of the proposed platform with some of the related projects in the related work section. A benchmark study has been executed for the purpose of demonstrating the effectiveness and feasibility of the system, and the evaluation results are depicted in Table 5. The following properties that play a pivotal role to compare the overviewed platforms are considered for this study. It is obvious to see from the table that Bluemix is a somewhat similar approach with the highest variety of IoT-related services. Message Queuing Telemetry Transport (MQTT) is utilized as the basic protocol to the Bluemix, but many other platforms just support REST-based interface. However, the Bluemix is closed source, which is not flexible for the general public, and has a limited hosting environment so that end users cannot decide the place of deployment. These two limitations are common issues existing in most of the overviewed systems. Another main problem is that some of the systems have no support for visualizing the data coming from the IoT environment, which is one of the most important features in the cloud as described in the preceding sections. Moreover, the platforms, like Parse, Heroku, Kinkey, CloudFoundry, IoTCloud, ThingSpeak, Mils-Cloud, and Cloud4sens lack the capability to detect device faults, which results in sensor network partitioning so as to reduce the WSN availability. The demand for an open source application that offers sensor detection, different protocols, and visualization is growing rapidly, and this paper aims to find out the potential to solve all these issues mentioned above.

Table 5. Comparative analysis of the proposed platform with the related platforms.

Name	Open-Source	Hosting	Protocols	Remote Access	Data Store	Programming Language	Sensor Detection	Visualization
Bluemix	No	Closed	MQTT	Yes	Yes	Many	Yes	Yes
Parse	Yes	Open	REST	Yes	Yes	JS	No	No
Senshare	No	Closed	REST	Yes	Yes	Java	No	Yes
CoAP-based Cloud	Yes	Open	CoAP	Yes	Yes	Java	No	No
LEONORE	No	Closed	REST	Yes	Yes	Java, C	No	Yes
Heroku	No	Closed	MQTT	Yes	Yes	Many	No	No
Kinvey	No	Closed	REST	Yes	Yes	JS	No	No
CloudFoundry	Yes	Open	REST	Yes	Yes	Many	No	No
IoTCloud	Yes	Closed	REST	Yes	Yes	Many	No	Yes
ThingSpeak	Yes	Closed	REST	Yes	Yes	Many	No	Yes
DIGI	No	Closed	REST, Zigbee	Yes	Yes	Many	No	Yes
Mfis-Cloud	No	Closed	REST	Yes	Yes	Many	No	Yes
Cloud4sens	No	Open	REST, Zigbee	Yes	Yes	Many	No	Yes
Proposed platform	Yes	Open	REST	Yes	Yes	Java, JS	Yes	Yes

7. Conclusions and Future Direction

This paper outlines the procedures for the design and implementation of a sensor-cloud platform for providing virtualization environments to efficiently manage heterogeneous wireless sensors. The resources of the physical sensors are ingested by the cloud-based web application which provides a graphical interface for better user experience and remote availability. Raspberry Pi is used for implementing the IoT server, and an IoTivity framework was utilized for communication between the IoT sever and cloud. The cloud synchronizes the resource descriptions of the physical sensors and represents them in the form of virtual sensors. End users can either manipulate or read sensing data from the physical sensors by means of virtual sensors. The novelty of the proposed work is the sensor detection approach, which can voluntarily detect faulty sensor nodes through virtual sensors. A smart space has been implemented as the proof of concept, and a series of experiments were performed, which indicated a very steady level, allowing effective and powerful access and control of the virtual sensors. This designed system is scalable enough to be deployed across various applications domains including smart cities and other industrial fields according to the use case study. The significance of this work has been highlighted by a comparison analysis of the designed system with some existing works, and it has been shown that the proposed system performs better than the existing works. The future direction of this work is to enable users to design and deploy the application logic via virtual sensors. The virtual sensors shall be available in a widget that can be easily dragged and dropped to compose IoT services. Nowadays, the growing trend of the IoT have moved in a global ecosystem of connected devices providing services to the general public, and we believe this work has great potential for inexperienced users to prototype IoT applications according to their own requirements.

Author Contributions: L.H. conceived the idea for this paper, designed the experiments and wrote the paper; W.J. implemented the physical sensor network part of the use case; H.Y. and Y.G.H. supported the development of OCF IoTivity; D.H.K. conceived the overall idea of Sensor-Cloud platform for physical sensor management, and proof-read the manuscript, and was correspondence related to this paper.

Acknowledgments: This work was supported by the National Research Council of Science & Technology (NST) grant by the Korea government (MSIT) (No. CRC-15-05-ETRI), and this research was supported by Basic Science Research Program through the National Research Foundation of Korea (NRF) funded by the Ministry of Education, Science and Technology (2015R1D1A1A01060493).

Conflicts of Interest: The authors declare no conflict of interest.

References

1. Gubbi, J.; Buyya, R.; Marusic, S.; Palaniswami, M. Internet of things (IoT): A vision, architectural elements, and future directions. *Future Gener. Comput. Syst.* **2013**, *29*, 1645–1660. [CrossRef]
2. Eddy, N. Gartner: 21 Billion IoT Devices to Invade by 2020—InformationWeek. Available online: http://www.informationweek.com/mobile/mobile-devices/gartner-21-billion-iot-devices-to-invade-by-2020/d/d-id/1323081 (accessed on 8 January 2016).
3. Kelly, S.D.T.; Suryadevara, N.K.; Mukhopadhyay, S.C. Towards the implementation of IoT for environmental condition monitoring in homes. *IEEE Sens. J.* **2013**, *13*, 3846–3853. [CrossRef]
4. Shyam, S.M.; Prasad, G.V. Framework for IoT applications in the Cloud, is it needed? A study. In Proceedings of the 2017 International Conference on Computing Methodologies and Communication (ICCMC), Erode, India, 18–19 July 2017.
5. Buyya, R.; Shin, C.; Venugopal, S.; Broberg, J.; Brandic, I. Cloud computing and emerging IT platforms: Vision hype and reality for delivering computing as the 5th utility. *Future Gener. Comput. Syst.* **2009**, *25*, 599–616. [CrossRef]
6. Gai, K.; Li, S. Towards cloud computing: A literature review on cloud computing and its development trends. In Proceedings of the 2012 Fourth International Conference on Multimedia Information Networking and Security, Nanjing, China, 2–4 November 2012.
7. Babu, S.M.; Lakshmi, A.J.; Rao, B.T. A study on cloud based internet of things: CloudIoT. In Proceedings of the 2015 Global Conference on Communication Technologies (GCCT), Thuckalay, India, 23–24 April 2015.

8. Rimal, B.P.; Choi, E.; Lumb, I. A taxonomy and survey of cloud computing systems. In Proceedings of the 2009 Fifth International Joint Conference on INC, IMS and IDC, Seoul, Korea, 25–27 August 2009.

9. Zeng, D.; Miyazaki, T.; Guo, S.; Tsukahara, T.; Kitamichi, J.; Hayashi, T. Evolution of software-defined sensor networks. In Proceedings of the 2013 IEEE 9th International Conference on Mobile Ad-Hoc and Sensor Networks, Dalian, China, 11–13 December 2013.

10. Kabadayi, S.; Pridgen, A.; Julien, C. Virtual sensors: Abstracting data from physical sensors. In Proceedings of the 2006 International Symposium on a World of Wireless, Mobile and Multimedia Networks (WoWMoM'06), Buffalo-Niagara Falls, NY, USA, 26–29 June 2006.

11. Khan, I.; Belqasmi, F.; Glitho, R.; Crespi, N. A multi-layer architecture for wireless sensor network virtualization. In Proceedings of the 6th Joint IFIP Wireless and Mobile Networking Conference (WMNC), Dubai, UAE, 23–25 April 2013.

12. Kortuem, G.; Kawsar, F.; Sundramoorthy, V.; Fitton, D. Smart objects as building blocks for the internet of things. *IEEE Internet Comput.* **2010**, *14*, 44–51. [CrossRef]

13. Cardone, G.; Cirri, A.; Corradi, A.; Foschini, L. The participact mobile crowd sensing living lab: The testbed for smart cities. *IEEE Commun. Mag.* **2014**, *52*, 78–85. [CrossRef]

14. Cardone, G.; Foschini, L.; Bellavista, P.; Corradi, A.; Borcea, C.; Talasila, M.; Curtmola, R. Fostering participaction in smart cities: A geo-social crowdsensing platform. *IEEE Commun. Mag.* **2013**, *51*, 112–119. [CrossRef]

15. Shojafar, M.; Cordeschi, N.; Baccarelli, E. Energy-efficient adaptive resource management for real-time vehicular cloud services. *IEEE Trans. Cloud Comput.* **2013**, *51*, 112–119. [CrossRef]

16. Cordeschi, N.; Amendola, D.; Shojafar, M.; Baccarelli, E. Distributed and adaptive resource management in cloud-assisted cognitive radio vehicular networks with hard reliability guarantees. *Veh. Commun.* **2015**, *2*, 1–12. [CrossRef]

17. Ma, Y.; Wang, L.; Liu, P.; Ranjan, R. Towards building a data-intensive index for big data computing—A case study of remote sensing data processing. *Inf. Sci.* **2015**, *319*, 171–188. [CrossRef]

18. Biswas, S.; Das, R.; Chatterjee, P. Energy-efficient connected target coverage in multi-hop wireless sensor networks. In *Industry Interactive Innovations in Science, Engineering and Technology*; Bhattacharyya, S., Sen, S., Dutta, M., Biswas, P., Chattopadhyay, H., Eds.; Springer: Berlin, Germany, 2018.

19. Jiang, M.; Luo, J.; Zou, X. Research on algorithm of three-dimensional wireless sensor networks node localization. *J. Sens.* **2016**, *2016*, 2745109. [CrossRef]

20. Khelifi, M.; Benyahia, I.; Moussaoui, S.; Naït-Abdesselam, F. An overview of localization algorithms in mobile wireless sensor networks. In Proceedings of the 2015 International Conference on Protocol Engineering (ICPE) and International Conference on New Technologies of Distributed Systems (NTDS), Paris, France, 22–24 July 2015.

21. Miller, J.S.; Dinda, P.; Dick, R. Evaluating a basic approach to sensor network node programming. In Proceedings of the 7th ACM Conference on Embedded Networked Sensor Systems (SenSys 2009), Berkeley, CA, USA, 4–6 November 2009.

22. Fortino, G.; Giannantonio, R.; Gravina, R.; Kuryloski, P.; Jafari, R. Enabling effective programming and flexible management of efficient body sensor network applications. *IEEE Trans. Hum.-Mach. Syst.* **2013**, *43*, 115–133. [CrossRef]

23. Mottola, L.; Picco, G.P. Programming wireless sensor networks: Fundamental concepts and state of the art. *ACM Comput. Surv.* **2011**, *43*, 19. [CrossRef]

24. Manujakshi, B.C.; Ramesh, K.B. SDaaS: Framework of sensor data as a service for leveraging service in internet of things. In Proceedings of the International Conference on Emerging Research in Computing, Information, Communication and Applications, Bangalore, India, 29–30 July 2016.

25. IoTivity. Available online: https://openconnectivity.org/developer/reference-implementation/iotivity (accessed on 5 April 2018).

26. Shelby, Z.; Hartke, K.; Bormann, C. *The Constrained Application Protocol (CoAP)*; Internet Engineering Task Force (IETF): Fremont, CA, USA, 2014.

27. Botta, A.; De Donato, W.; Persico, V.; Pescapé, A. On the integration of cloud computing and internet of things. In Proceedings of the 2014 International Conference on Future Internet of Things and Cloud, Barcelona, Spain, 27–29 August 2014.

28. Atzori, L.; Iera, A.; Morabito, G. The internet of things: A survey. *Comput. Netw.* **2010**, *54*, 2787–2805. [CrossRef]

29. Richardson, L.; Ruby, S. *RESTful Web Services*; O'Reilly Media: Seville, CA, USA, 2007.

30. Duquennoy, S.; Grimaud, G.; Vandewalle, J. The web of things: Interconnecting devices with high usability and performance. In Proceedings of the International Conference on Embedded Software and Systems, Hangzhou, China, 25–27 May 2009.

31. Du, Z.; Yu, N.; Cheng, B.; Chen, J. Data mashup in the internet of things. In Proceedings of the International Conference on Computer Science and Network Technology (ICCSNT), Harbin, China, 24–26 December 2011.

32. Kansal, A.; Nath, S.; Liu, J.; Zhao, F. SenseWeb: An infrastructure for shared sensing. *IEEE MultiMedia* **2007**, *14*, 8–13. [CrossRef]

33. Guinard, D.; Trifa, V.; Mattern, F.; Wilde, E. From the internet of things to the web of things: Resource-oriented architecture and best practices architecting the internet of things. In *Architecting the Internet of Things*; Uckelmann, D., Harrison, M., Michahelles, F., Eds.; Springer: Berlin, Germany, 2011.

34. Dawson-Haggerty, S.; Jiang, X.; Tolle, G.; Ortiz, J. sMAP: A simple measurement and actuation profile for physical information. In Proceedings of the 8th International Conference on Embedded Networked Sensor Systems (SenSys 2010), Zurich, Switzerland, 3–5 November 2010.

35. SensorML. Available online: http://vast.uah.edu/SensorML/ (accessed on 10 May 2018).

36. Leontiadis, I.; Efstratiou, C.; Mascolo, C.; Crowcroft, J. Senshare: Transforming sensor networks into multi-application sensing infrastructure. In Proceedings of the 9th European Conference on Wireless Sensor Networks, Trento, Italy, 15–17 February 2012.

37. Vögler, M.; Schleicher, J.; Inzinger, C.; Nastic, S.; Sehic, S.; Dustdar, S. LEONORE—Large-scale provisioning of resource-constrained IoT deployments. In Proceedings of the 2015 IEEE Symposium on Service-Oriented System Engineering, San Francisco Bay, CA, USA, 30 March–3 April 2015.

38. Madria, S.; Kumar, V.; Dalvi, R. Sensor cloud: A cloud of virtual sensors. *IEEE Softw.* **2014**, *31*, 70–77. [CrossRef]

39. Misra, S.; Chatterjee, S.; Obaidat, M.S. On Theoretical modeling of sensor cloud: A paradigm shift from wireless sensor network. *IEEE Syst. J.* **2017**, *11*, 1084–1093. [CrossRef]

40. Chatterjee, S.; Ladia, R.; Misra, S. Dynamic optimal pricing for heterogeneous service-oriented architecture of sensor-cloud infrastructure. *IEEE Trans. Serv. Comput.* **2017**, *10*, 203–216. [CrossRef]

41. Dinh, T.; Kim, Y. An efficient interactive model for on-demand sensing-as-a-servicesof sensor-cloud. *Sensors* **2016**, *16*, 992. [CrossRef] [PubMed]

42. Abdelwahab, S.; Hamdaoui, B.; Guizani, M.; Znati, T. Cloud of things for sensing-as-a-service: Architecture, algorithms, and use case. *IEEE Internet Things J.* **2016**, *3*, 1099–1112. [CrossRef]

43. Dinh, T.; Kim, Y. Information centric sensor-cloud integration: An efficient model to improve wireless sensor networks' lifetime. In Proceedings of the 2017 IEEE International Conference on Communications (ICC), Paris, France, 21–25 May 2017.

44. Dinh, T.; Kim, Y. An efficient sensor-cloud interactive model for on-demand latency requirement guarantee. In Proceedings of the 2017 IEEE International Conference on Communications (ICC), Paris, France, 21–25 May 2017.

45. Sen, A.; Madria, S. Risk assessment in a sensor cloud framework using attack graphs. *IEEE Trans. Serv. Comput.* **2017**, *10*, 942–955. [CrossRef]

46. Misra, S.; Singh, A.; Chatterjee, S.; Obaidat, M.S. Mils-cloud: A sensor-cloud-based architecture for the integration of military tri-services operations and decision making. *IEEE Syst. J.* **2016**, *10*, 628–636. [CrossRef]

47. Dinh, T.; Kim, Y.; Lee, H. A location-based interactive model of internet of things and cloud (IoT-Cloud) for mobile cloud computing applications. *Sensors* **2017**, *17*, 489. [CrossRef] [PubMed]

48. Zhu, C.; Leung, V.C.M.; Wang, K.; Yang, L.T.; Zhang, Y. Multi-method data delivery for green sensor-cloud. *IEEE Commun. Mag.* **2017**, *55*, 176–182. [CrossRef]

49. Fazio, M.; Puliafito, A. Cloud4sens: A cloud-based architecture for sensor controlling and monitoring. *IEEE Commun. Mag.* **2015**, *53*, 41–47. [CrossRef]

50. Open Sourense IoT Cloud. Available online: https://sites.google.com/site/opensourceiotCloud/ (accessed on 9 May 2018).

51. Thingspeak. Available online: https://www.thingspeak.com/ (accessed on 10 March 2018).

52. DIGI. Available online: http://www.digi.com/ (accessed on 6 March 2018).

53. Kovatsch, M.; Lanter, M.; Shelby, Z. Californium: Scalable cloud services for the internet of things with CoAP. In Proceedings of the International Conference on the Internet of Things (IOT'14), Cambridge, MA, USA, 6–8 October 2014.

54. Available online: https://www.heroku.com/ (accessed on 10 April 2018).

55. Available online: http://www.kinvey.com/ (accessed on 16 May 2018).

56. Available online: http://parseplatform.org/ (accessed on 16 May 2018).

57. Available online: http://Cloudinary.com/ (accessed on 14 May 2018).

58. Available online: https://console.ng.bluemix.net/ (accessed on 14 May 2018).

59. Pflanzner, T.; Kertesz, A. A survey of IoT cloud providers. In Proceedings of the 2016 39th International Convention on Information and Communication Technology, Electronics and Microelectronics (MIPRO), Opatija, Croatia, 30 May–3 June 2016.

60. Guo, S.; Zhong, Z.; He, T. FIND: Faulty node detection for wireless sensor networks. In Proceedings of the 7th ACM Conference on Embedded Networked Sensor Systems (SenSys 2009), Berkeley, CA, USA, 4–6 November 2009.

61. Krishnamachari, B.; Sitharama, I. Distributed Bayesian algorithms for fault-tolerant event region detection in wireless sensor networks. *IEEE Trans. Comput.* **2004**, *53*, 241–250. [CrossRef]

62. Jadav, P.; Babu, V.K. Fuzzy logic based faulty node detection in wireless sensor network. In Proceedings of the 2017 International Conference on Communication and Signal Processing (ICCSP), Chennai, India, 6–8 April 2017.

63. Ahmad, S.; Hang, L.; Kim, D.H. Design and implementation of cloud-centric configuration repository for DIY IoT applications. *Sensors* **2018**, *18*, 474. [CrossRef] [PubMed]

64. Aziz, H.; Ridley, M. Real-time web applications driven by active browsing. In Proceedings of the 2017 Internet Technologies and Applications (ITA), Wrexham, UK, 12–15 September 2017.

65. Rasslan, M.; Nasreldin, M.; Elkabbany, G.; Elshobaky, A. On the security of the sensor cloud security library (SCSlib). *J. Comput. Sci.* **2018**, *14*, 793–803. [CrossRef]

electronics

MDPI

Article

Blockchain-Oriented Coalition Formation by CPS Resources: Ontological Approach and Case Study

Alexey Kashevnik [1,*] and **Nikolay Teslya** [2]

[1] International research center "Computer Technologies", ITMO University, 49 Kronverksky Pr., 197101 St. Petersburg, Russia

[2] Laboratory of Computer Aided Integrated Systems, St. Petersburg Institute for Informatics and Automation of the Russian Academy of Sciences (SPIIRAS), 39, 14th Line, 199178 St. Petersburg, Russia; teslya@iias.spb.su

* Correspondence: alexey@iias.spb.su; Tel.: +7-911-211-9880

Received: 15 February 2018; Accepted: 7 May 2018; Published: 8 May 2018

Abstract: Cyber-physical systems (CPS), robotics, Internet of Things, information and communication technologies have become more and more popular over the last several years. These topics open new perspectives and scenarios that can automate processes in human life. CPS are aimed at interaction support in information space for physical entities communicated in physical space in real time. At the same time the blockchain technology that becomes popular last years allows to organize immutable distributed database that store all significant information and provide access for CPS participants. The paper proposes an approach that is based on ontology-based context management, publish/subscribe semantic interoperability support, and blockchain techniques. Utilization of these techniques provide possibilities to develop CPS that supports dynamic, distributed, and stable coalition formation of the resources. The case study presented has been implemented for the scenario of heterogeneous mobile robots' collaboration for the overcoming of obstacles. There are two types of robots and an information service participating in the scenario. Evaluation shows that the proposed approach is applicable for the presented class of scenarios.

Keywords: blockchain; ontology; context; cyber-physical systems; robotics; interaction; coalition

1. Introduction

The second generation of cyber-physical systems (CPS) consider different aspects of artificial intelligence aimed at self-awareness and/or self-adaptation between CPS resources. There is a lot of research in the areas of mobile robotics [1,2], unmanned vehicles [3,4], wearable electronics [5,6] and others that are required by second generation CPS for effective operation. This paper considers the problem of dynamical decentralized coalition formation by heterogeneous CPS resources for joint task performing. For this purpose, resources should not only exchange information with each other but also have to understand the semantics in CPS, take into account the current situation for self-adaptive behavior, and recognize compromised situations.

Tasks distribution and assignment among the coalition participants requires a special method. An example of such a method is the sensor assignment to missions approach, based on knowledge representation using ontologies and reasoning techniques to distribute tasks among coalition participants [7]. However, the participants of the coalition do not have a full picture of how the tasks were distributed. They can get an approximate idea of this, only by making full reasoning among the knowledge of all coalition members. This behavior is more typical for swarm interaction models and reduces overall efficiency of the coalition. Usually the existing method is characterized by the absence of a central or distributed processing log, in which results or coalition participants' work are displayed. Such a log allows us to reduce the number of calculations on the coalition formation stage and to distribute tasks taking into account the capabilities and characteristics of all the participants in the coalition. At the same time, with such a log each participant at any time will have a view of task distribution as well as when it is expected to receive the result from other participants and what resources are involved in solving the overall problem. The overall view makes it possible to reconfigure the coalition on the fly by adding or replacing participants who will know what they should do from the very beginning.

The paper presents a blockchain-oriented approach to coalition formation of CPS resources that is based on context management methodology and publish/subscribe-based semantic interoperability mechanism as well as utilization of a distributed, decentralized, immutable transaction log technology, also known as a blockchain for creation of the coalition working log. These technologies allow us to implement self-adaptive behavior of CPS resources for coalition formation. Its distinctive feature is the combination of private transactions in a block of a given length (defined by the developer of a particular block) using hash functions, as well as the connection of each new block with the last block in the chain that is also provided by using the hash function. It is proposed to use two types of transactions. The first one contains task assignment for coalition participants. It can be used to control the overall progress of the coalition participants and adjust CPS resource's ratings, as well as adjusting coalition structure by replacing the resource that cannot complete the assigned task. The second type of transaction contains the facts of the system's consumables spending. Each CPS resource can use consumables to process the assigned task. The fact of spending is written to the distributed ledger and can be checked by the CPS manager. Each transaction in the block is signed by the private key of the transaction owner. Such a mechanism makes it possible to unequivocally determine who entered the information in the block, track its sequence of changes, and ensure the immutability of information contained in the log.

The main contribution of the paper is an extension of the classical multi-agent approach to resource interaction by the ontology-based smart space and blockchain technology. The main features of the approach are usage of context management techniques for determination of the resources knowledge as well as its competencies that are related to the current task. The smart space technique used supports ontology-based publish/subscribe mechanism that allows to create the actual for the task model of physical space in the information space. This model provides interoperability support for CPS resources. Based on resource interaction in information space, the coalitions are formed to perform the task. Every resource is described by the rating, as well as its competencies. Ratings are taken into account for decision about the resource participation in a coalition. To support the immutability of the resource ratings the blockchain technology is used for keeping the history of resources interaction, as well as consumables spending, and preventing compromised situations.

For the approach evaluation, a case study has been presented that considers the scenario of heterogeneous mobile robots collaboration work for obstacles overcoming. The scenario is based on the proposed approach and Smart-M3 semantic interoperability platform that implement concept of smart spaces and provides possibilities of ontology-based mobile robots interaction organization for their semantic interoperability. The platform makes possible to significantly simplify further development of CPS, include new information sources and services, and to make the system highly scalable. Smart-M3 assumes that devices and software entities can publish their embedded information for other devices and software entities through simple, shared information brokers. The platform is open source and accessible for download at Sourceforge (Smart-M3 at Sourceforge, URL: http://sourceforge.net/projects/smart-m3). There are two types of mobile robots that have been developed for participation in the case study: manipulating robots and measuring robots. The task is aimed at cargo delivery by the manipulating robot with increased passability. During the delivery, several obstacles should be overcome. For obstacle overcoming, the manipulating robot should form a coalition with the measuring robot and knowledge base service to get the appropriate overcoming algorithm.

The prototype of the manipulating robot has been assembled for this scenario based on Lego Mindstorms EV3 kit. EV3 control block has been re-flashed for LeJOS operating system that allows to connect it to Wi-Fi network and Smart-M3 information sharing platform. The robot consists of four motors, four pair of wheels, and an ultrasonic sensor. The measuring robot has been assembled based on the Makeblock robotics kit. The Raspberry Pi 3 Model B single-board microcomputer was connected to the control board to expand its computation power. The main task of the Raspberry Pi is to connect the robot to the Smart-M3 platform, as well as to process data from sensors and calculate the further robot actions.

The rest of the paper is organized as follows. The next section describes related work in the area of mobile robot interaction and blockchain technology utilization for next generation CPS. Then the authors present a blockchain-oriented approach to coalition formation in CPS. After that the CPS ontology is described. Next, the authors propose a case study that is aimed at heterogeneous mobile robots collaboration work for the overcoming of obstacles. Finally, the conclusion summarizes the paper.

2. Related Work

The section presents the related work analysis in the area of interaction between CPS resources on the example of mobile robots and blockchain technology utilization for next generation CPS. The following main technologies are actively used for CPS resources coalition formation in the considered research papers: smart space, ontology management, context management. The usage of blockchain technologies for CPS resources coalition formation is a rather new research field despite the great potential of the technology in distributed trustless systems. However, there are several research works aiming to utilize blockchain technology in the field of coalition operation and Internet of Things (IoT).

2.1. Related Work Analysis in the Area of Interaction between CPS Resources on the Example of Mobile Robots

The aim of the RoboEarth project is building the Internet of robots. Authors determine the Internet of robots as a web community that autonomously shares descriptions of tasks they have learned, object models they have created, and environments they have explored [8]. Authors developed a formal language for encoding this information and propose methods for solving the inference problems related to finding information, to determining if information is usable by a robot, and to grounding it on the robot platform; the infrastructure for using this representation to reason about the applicability of information in a given context and to check if all required robot capabilities are available, and mechanisms for creating and uploading shared knowledge.

The authors of the paper [9] deal with an intelligent system of robot and human interaction. This system uses adaptive learning mechanisms to account for the behavior and preferences of a human. It includes algorithms for speech and face recognition, natural language understanding, and other ones that provide the robot possibility of recognition the human actions and commands. A robot used in this paper (Pioneer 3-AT) is equipped with multiple sensors, scanners and a camera. The paper focuses on the process of teaching the robot and its mechanisms of interaction with a human.

In the paper [10] authors describe a coordinated process for the agents (in particular, mobile robots) jointly perform a task, using the knowledge about the environment, their abilities and possibilities for communication with other agents. The effective use of knowledge about the agents' capabilities is implemented by using suitability rates which allow agent to choose the most appropriate action. This approach was tested during a football match between mobile robots. The interaction in this article is considered in terms of the choice of an agent to perform an action.

The authors of the paper [11] propose a methodology for coordination of autonomous mobile robots in jams in congested systems. The methodology is based on reducing a robot's speed in jams and congested areas, as well as in front of them. Thus, the robots slow down if detecting other robots in front of themselves. The effectiveness of the methodology was tested in a simulation of a congested system with a jam.

The paper [12] discusses the combining of wireless sensor networks and mobile robots. Authors suggest layered swarm framework for the interaction of decentralized self-organizing complex adaptive systems with mobile robots as their members. This framework consists of two layers (robots and wireless LAN) and communication channels within and between layers. During the experimental verification of framework mobile robots independently reached the destination.

The paper [13] is devoted to the problem of robots group control in non-deterministic, dynamically changing situations. The authors propose a method for solving formation task in a group of quadrotors. The method makes it possible to ensure accurate compliance with distances between quadrotors in the formation, as well as featuring low computational complexity.

The authors of the paper [14] shows a system of robots interaction based on BML—Battle Management Language. This language is intended for expressing concise, unambiguous orders read by both humans and robots. The orders are transmitted from the system of multiple robots and processed for further distribution specific commands between them.

In the paper [15], design of the human-robot interaction, improving human situational awareness on the basis of an agent-oriented approach is described. The approach is applied when mobile agents move (autonomous robots and astronauts in protective gear) on the Moon's surface in a limited geographically area.

The paper [16] presents the concept of the learning factory for cyber-physical production systems (LVP) and presents robotics and human-robot collaboration. Authors use ontologies in the context of flexible manufacturing scenarios to equip autonomous robots with the necessary knowledge about their environment, opportunities and constraints. They discuss the use of semantic technologies together with cyberphysical systems for integrating decision making into smart production machinery.

The paper [17] presents extensions to a core ontology for the robotics and automation field. Authors of the paper proposed ontological approach aims at specifying the main notions across robotics subdomains.

The paper [18] defines the current issues and solutions possible with ontology development for human-robot interaction. It describes the role of ontologies in robotics at large, provide a thorough review of service robot ontologies, describe the existing standards for robots, along with the future trends in the domain, and defines the current issues and solutions possible with ontology development for human-robot interaction.

The paper [19] evaluates the proposed ontology through a use case scenario involving both heterogeneous robots and human-robot interactions, showing how to define new spatial notions using an ontology. It discusses experimental results, presenting the ontology strengths. The paper is concerned with the development of model-based systems engineering procedures for the behavior modeling and design of CPS. Three independent but integrated modules compose the system: CPS, ontology and time-reasoning modules. This approach is shown to be mostly appropriate for CPS or which safety and performance are dependent on the correct time based prediction of the future state of the system. Consequently, the ontological approach applicability for modeling the problem area is justified.

The approaches described by [9,11] require a well-defined functionality of robots, so the set of tasks undertaken is restricted. The paper [10] describes the knowledge of the capabilities of robots that is generated dynamically during the execution of tasks what allows to manage the robots effectively. Thus, no limitation to robots' functionality and dynamic determination of the options available can provide the widest range of tasks to execute. Authors of the paper [12] show the effectiveness of the indirect interaction of robots, that is, the interaction through an intermediary that is Smart-M3 platform in the paper. The paper [13] discusses a formal model that can be used for interaction. In the paper [14] the effectiveness of using specialized language for robots control is shown. Authors of the paper [15] demonstrate the applicability of the multi-agent systems approach for robot and human interactions in various fields. Many researches inherit the idea of ontologies usage for modelling context in social-cyber physical systems. An ontology should include description of physical components, agents, model of problem area. It consists of classes, subclasses, properties, and relationships between classes.

2.2. Related Work Analysis in the Area of Blockchain Technology Utilization for Next Generation CPS

The paper [7] considers the use of blockchain technology to create a logically centralized system for managing the distribution of tasks among coalition members. The blockchain technology itself is considered as a mechanism for recording and tracking transactions between participants performing control over the distribution of tasks. At the same time, it is possible to create a virtual log that contains all distributions of the assets available among the coalition participants.

The paper [20] provides an example of the E-Business model for decentralized autonomous corporations based on IoT and blockchain. The blockchain network is used here to create smart contracts and perform payment transactions using Bitcoin and the system's own currency—IoTcoin. The approach presented provides the following advantages: internal resources of the system are presented in the form of payment system tokens; distributed ledger for transactions, built-in mechanism of digital signature; and protection of transactions by linking blocks with a hash function of a given complexity. Interaction between elements is carried out by transactions passing through four phases: preparing a transaction, negotiating, signing a contract, and fulfilling the terms of the contract.

A similar method is considered in relation to a Smart Home scenario in [21]. Also, there is an unchanged transaction log as an advantage of integrating IoT and Blockchain technologies. The proposed framework provides the following transaction processing functions: transaction storage, access to the transactions history and real-time monitoring. Also, due to the adaptation of the hierarchical structure of the proposed framework (which includes three levels: the smart home, overlay network, and cloud storage), optimization of resource consumption is achieved, and scalability of the network is increased. The functions of the blockchain are focused on the level of interaction of elements in the overlay network.

It is also noted that the combination of the peer-to-peer network and the cryptographic algorithms that are underlying the blockchain technology allow for a negotiation process and consensus building without the presence of any controlling authorities. The distributed nature of the blockchain can be used in swarm robotics to store global knowledge about swarm actions [22]. At the same time, due to blockchain, the security of the transmitted data is ensured (garbage data can affect the achievement of a common goal), distributed decision making (creating a distributed voting system for the solution and use of the multi-signature), separation of robots behavior (switching between behavior patterns depending on the role in the swarm), the emergence of new business models using the swarm. In addition, the availability of a distributed transaction ledger allows new robots to join the swarm and gain all the knowledge they have gained prior to the moment of inclusion by downloading and analyzing the transaction history.

The results of the research presented can be summarized in the following way. The blockchain is mostly used as immutable storage for information exchange between robots. Information stored in the blockchain could contain records about task and assets distribution [7,21], smart contracts and payment transactions [20], or global knowledge about coalition actions [22]. In combination with IoT, concept blockchain technology can provide more trust for transaction through IoT, due to the storing information about transactions in immutable log that can be verified by every IoT participant. In contrary to existing approaches, blockchain does not require a central authority that provides trust for all nodes. All nodes negotiate with each other and come to consensus using one of the possible mechanisms: Proof of Work, Proof of Stake, or practical Byzantine fault tolerance [23]. Assuming that robots are forming a coalition without any control center, the blockchain can be used for safe and trustiness logging of robots' task distribution and reward for task solving.

3. A Blockchain-Oriented Ontological Approach to Coalition Formation

For coalition formation in CPS it is required to support the semantic interoperability between resources, as well as providing infrastructure for information & knowledge sharing between them and keeping agreements between resources made in scope of interaction in the system. The CPS concept is aimed at tight integration of the physical and cyber spaces based on interactions between these spaces in real time. Resources are exchanging the information with each other in information space while their physical interaction occurs in physical space. For interoperability support, it is needed to create the model of CPS and support the interaction of resources based on this model. One of the possible approaches to problem domain modelling is ontology management. The ontology formally represents knowledge as a set of concepts within a domain, using a shared vocabulary to denote the types, properties, and interrelationships of those concepts. For current situation modelling and reducing the search space of potential coalition members, the utilization of context management technology is proposed. The aim of this technology is a context model creation that is based on the ontology of problem domain and information about current situation in physical space. Context is defined as any information that can be used to characterize the situation of an entity. An entity is a person, place or object that is considered relevant to the interaction between a user and an application, including the user and application themselves [24]. Context is suggested to be modeled at two levels: abstract and operational. These levels are represented by abstract and operational contexts, respectively. The process of coalition formation based on abstract and operational contexts is shown in Figure 1 (enhanced based on [25]).

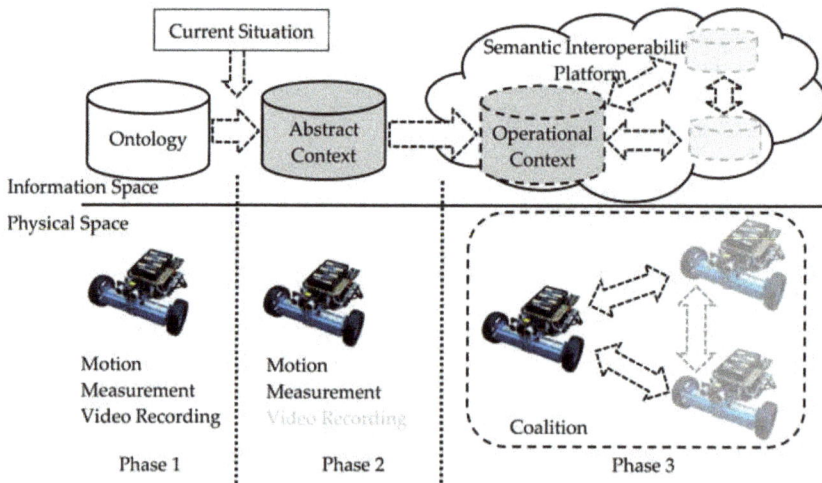

Figure 1. Abstract and operational contexts for coalition formation.

Abstract context is an ontology-based model of a potential coalition participant related to the current task. Abstract context is built based on integrating information and knowledge relevant to the current problem situation. Operational context is an instantiation of the domain constituent of the abstract context with data provided by the contextual resources. Thereby, coalition formation is implemented with the following phases. In the first phase the resources are waiting to participate in the process of task performing. Every resource is described by an ontology. For the second phase the abstract context is created based on the resource ontology, a task appeared in CPS, and current situation. Abstract context includes all accessible knowledge relevant to the task. For the third phase, the process of concretization of these knowledge by information accessible in information space for operational context formation is implemented. The operational context is published in information space and becomes accessible for other potential coalition participants. The coalitions of resources are created based on their operational context intersections (see

For information space organization, a semantic interoperability platform is used that is based on blackboard architecture that provides possibilities for potential coalition participants to implement indirect interaction with each other. Thereby, a virtual coalition is created in information space and then a physical coalition appears in physical space (CPS resources implement the joint task). Interaction of potential coalition participants in information space is implemented using the ontology-based publish/subscribe mechanism that provides possibilities for resources publishing their information and knowledge, and subscribing to interesting information using ontologies. When a resource registers to be a potential coalition participant it uploads its own ontology to the semantic interoperability platform. This resource ontology formalizes the main resource capabilities and constraints that have to be satisfied to use the capabilities. Thereby, the interoperability is supported based on open information space and ontology-based publish/subscribe mechanism. A potential coalition participant can participate in a joint task if the operational context in information space is matched with its abstract context. More details about the ontology matching can be found in the paper [26].

Simultaneously with the publishing of the operational context to the semantic interoperability platform the signed transaction is writing to the blockchain network. The transaction consists of the information published to the platform by a particular resource on each stage of the joint task performance, signed by a private key of the resource. The transaction is created on the coalition participant hardware and includes the new block that writes to the local copy of the blockchain. Then the blockchain spreads the new block among the other coalition participants that can verify block's source as well as transactions inside it using the public key. Figure 2).

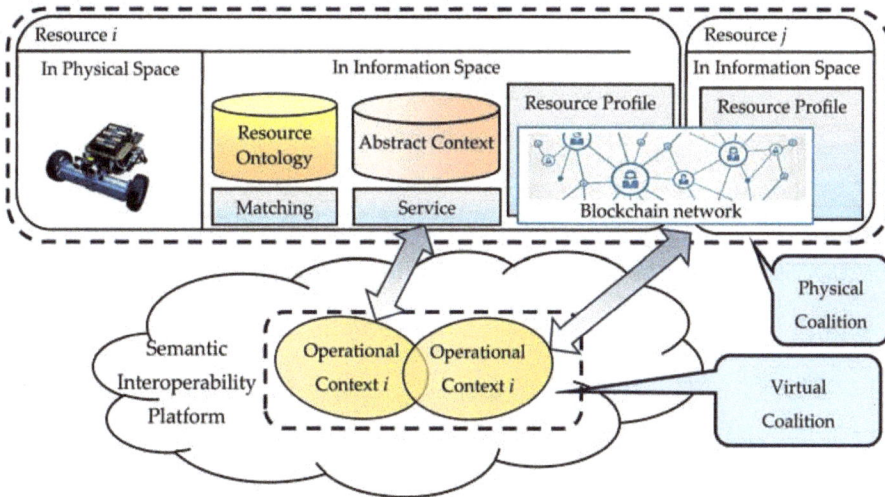

Figure 2. Blockchain-Oriented Conceptual model for resource interaction for coalition formation.

Figure 3 shows the resource profile model that is aimed at automating the process of coalition creation and takes into account the ratings of the resources considered for coalition formation. Every time the resource participated in the coalition, other resources and users have the possibility to estimate their competencies' ratings and the overall rating of the resource. The model consists of the following components: basic resource information, its competence & constraints, and resource interaction history. Basic resource information includes the following information about the resource: identifier, name, description, and rating. Competence and constraints of resource is aimed at description of the tasks the resource can perform and constraints that have to be satisfied to perform these tasks. Competence and constraints of resource includes: basic competencies, extended competencies, and constraints. Basic competencies and extended competencies can be estimated with ratings by another resource, and users. The blockchain network is used for the following purposes. First is a coalition formation. During the coalition formation process the potential resources share their competencies and get subtasks that are parts of the task. Each subtask assignment is stored in the blockchain network. In this case for other resources it is easy to check who is responsible for a certain subtask. If some resource fails in the subtask, its competence rating is decreased, and the resource can be replaced by another one with a higher competence rating. In the case of a subtask performed successfully, the ratings of competencies related to this subtask are increasing. All rating changes are stored in the blockchain network and can be tracked for preventing the unauthorized changes. The second purpose is aimed at information exchange and consumables tracking. During the subtask performance, each resource can spend some of the system's consumables (e.g., power, time or task-specific resource, like seeds for planting). The details of consumables spending is stored in the blockchain network. Every resource can check at any time the value of consumables that was spent by another resource for subtask performance. It is helpful to track the task performance as well as for in-time consumables supplying and increasing the efficiency of joint work by adjusting the coalition. In these cases the blockchain network is used to prevent the submitting and use of malicious information. Only signed information from coalition members can be used in joint task performance. The immutability property of the blockchain allows us to track actions performed by each resource and detect the guilty resource that causes the failure and decrease its competence rate. Unlike existing approaches, the blockchain can provide an immutable log with distributed trustiness between robots without any central point. In cases when robots are acting only with each other, without any central node, the blockchain can simplify trustiness providing.

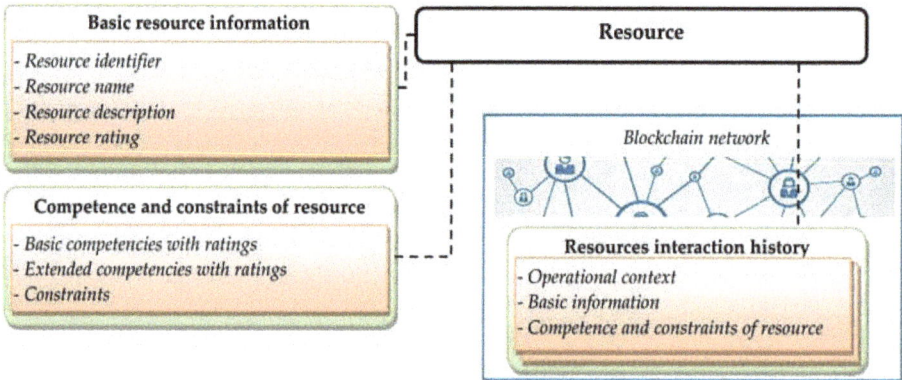

Figure 3. Resource Profile Model.

4. CPS Ontology

Developed CPS ontology is based on definitions and abbreviations from Suggested Upper Merged Ontology (SUMO), proposed in [27]. The model is built for CPS, where the mobile robots are interacting with each other for joint task performance. The ontology includes description of the main processes that occur when coalitions of robots are created to solve problems in cooperation. The protégé editor has been used to create the ontology. The ontology classification consists of two main classes: "Physical Space" and "Information Space".

Class "Physical Space" described all physical entities which participated in a coalition (devices, sensors, motors, end etc.). "Information Space" refers to computational resources and processes for coalition formation. Considering the classification of "Physical Space" in more detail (see Figure 4), it contains classes "Object", "Environment", and "Process". Class "Object" describes the physical facilities and collections. Class "Device" describes the physical entities and their parts participated in the task performing: "Battery", "Hull", "Motor", "Sensor", "Switch", "Whell". Class "Motor" is divided into subclass of motor drives—«ServoMotor». «Sensor» consists of subclasses specifying sensors installed on the device: "DistanceSensor", «HeatSensor», «LightSensor». Class "Wheel" includes two subclasses "LeftWheel" and "RightWheel".

The class "Object" specifies physical characteristics of devices depending on the scenario. For example, in the case study considered it is important to formalize what types of hull, motor, and wheels are used, since it is necessary to overcome obstacles. For all scenarios the class "Collection" joins together several devices to making the complex device.

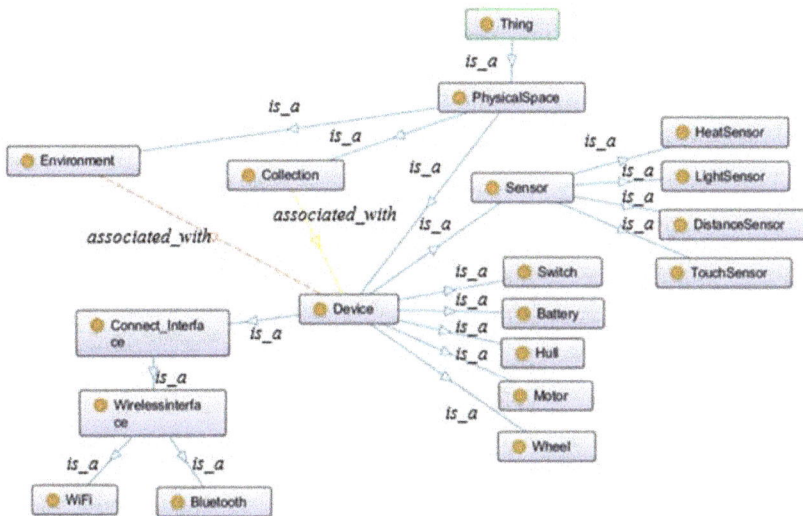

Figure 4. An ontology of the physical space in cyber-physical systems (CPS).

Class "Process" consists of subclasses "Action" and "Interaction". "Action" consists of "Movement", "Interaction", "Photo", "Grip", "Fly", "Clean", "Check" and includes a route options "Route" ("IndoorRoute"; "OutdoorRoute"). It means that the class «Action» represents the description of a physical object. The class "Interaction" consists of the interaction types "HumanInteraction" and "RobotInteraction" ("InterGroupInteraction", "IntraGroupInteraction").

Now consider the information space (see Figure 5). Class "InformationSpace" consists of "CompetenceProfile", "Configuration", "Context", "Policy", and "Process_Model". Class "CompetenceProfile" includes classes "BasicInformation" ("User_account", "Robot_description"), class "Competency" ("Basic_competencies", "Special_competencies"), and classes "History" and "Constraints". The robot competency determines the robot's ability to perform certain tasks. The basic information contains a description of the robot and information about its characteristics. Basic competencies include motion in space, turns, and the special ones are the ability to overcome obstacles, take photos, to do cleaning of the space and to execute the gripping of components in the process of object assembly. Class "History" is aimed at resources interaction history keeping using the blockchain network. Restrictions allow to provide options, when a task cannot be implemented due to initial or emerging condition while fulfilling the process. Class "Context" includes "Device Context" and "Environment Context" that can be "Spatial" ("Left", "Off", "On", "Right"), "Temporal" ("Distance": "Far", "Near"), and "Interface" ("Wireless interface": "Bluetooth", "WiFi").

Class "Interface" demonstrates the way of interaction between the robots through a wireless link. Class "Policy" defines the access rights, including authorization, standard definitions, opportunities and application limits. Class "Configuration" defines settings and components that have to be taken into account in the physical space. Class "Process_model" describe scenarios that need to be implemented.

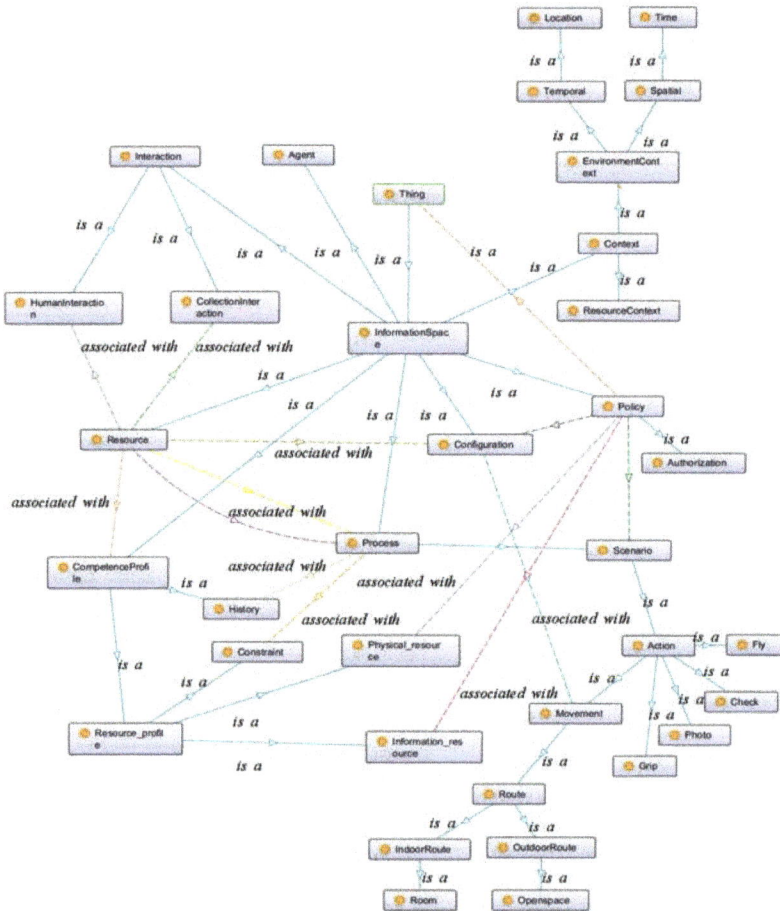

Figure 5. An ontology of the information space in CPS.

5. Case Study

To evaluate to approach to CPS resource coalition formation presented, the scenario of heterogeneous mobile robots' collaboration work for the overcoming of obstacles is proposed in the paper. The scenario presented is a good example of next generation CPS that includes two heterogeneous automated robots (manipulating robot and measuring robot) that are set to perform a task together, taking into account the current situation. The task is aimed at cargo delivery by the manipulating robot, which is a 6WD robot with lifting front and back chassis. During the delivery, obstacles can be met and the manipulating robot should overcome them. Before overcoming, the measuring robot should scan and measure the obstacle to create the obstacle model. Based on this model the manipulating robot should decide if it can overcome this obstacle and which overcoming algorithm should be used for these purposes.

For the interaction between robots, a Smart-M3 platform [28] is used for providing possibilities of CPS participants interact in the information space. The Smart-M3 is an open source semantic interoperability platform, it is accessible for downloading and testing, supported by a development community, and supports modern mobile platforms (Android, Symbian, Windows Phone). The key idea of the platform is that robots can publish their information for each other through simple, shared information brokers. Information exchange in smart space is implemented via HTTP using Uniform Resource Identifier (URI). Semantic Web technologies have been applied for interoperation purposes.

The prototype of the manipulating robot has been developed for this scenario based on the 6WD robot with increased passability developed by the Science and Technology Center "ROCAD" (http://rocad.ru/en/gallery/). The Lego Mindstorms EV3 kit has been used for assembling the robot prototype. This kit allows us to easily design robots with the required functionality for education purposes. At the same time, there is the possibility to use electronic units, motors and sensors and program them in Java. The Lego Mindstorms EV3 kit supports connection Bluetooth wireless network. For Wi-Fi network connection the USB module Netgear WNA1100 has been used. The authors installed the LeJOS (http://www.lejos.org/) operation system to EV3 control block that allows us to connect the manipulating robot Smart-M3 platform. The manipulating robot prototype is shown in Figure 6.

In real life, the measuring robot is a helicopter that flies to the obstacle and implements scanning and measuring. For the research prototype the measuring robot has been assembled based the Makeblock robotics kit. The control board of this set is based on a scheme compatible with Arduino, which allows equipping the robot by a wide range of sensors and motors, as well as to develop and connect own equipment for interacting with the physical world. For the control board it is possible to connect to Wi-Fi and Bluetooth wireless networks. However, the microcontroller provides low computation power as well as highly limited memory. The Raspberry Pi 3 Model B single-board microcomputer was connected to the control board to expand it computation power. The main task of the Raspberry PI is to connect the robot to the Smart-M3 platform, as well as to process data from sensors and calculate further robot actions. The interaction between Raspberry Pi and the robots' control board is implemented through the serial port on both devices. The microcontroller of the robot has official firmware, which provides the functions to control sensors and motors. For Raspberry Pi, the MegaPi library is used to develop code for controlling robot functions via a serial port using Python programming language. The measuring robot prototype is presented in Figure 7.

During the scenario (see Figure 8) the manipulating robot should deliver cargo to a destination. If the robot finds an obstacle that has to be overcame to complete the task, it publishes the operational context to Smart-M3 (e.g., location, available time to wait). The measuring robots get a notification about the obstacle found and decides who can create a coalition with the manipulating robot to perform the obstacle overcoming task together. Decision making robots find who can perform the task, as well as checking their ratings stored in blockchain. The robots with the highest ratings and appropriate hardware are selected to perform the task. The ID of the selected robot is assigning with the task ID and writing into the blockchain network. Then the measuring robot participating in the coalition moves to the obstacle, implements scanning and measuring, and publishes the operational context (obstacle parameters) to Smart-M3. Based on the operational context from the measuring robot, the knowledge base service finds an algorithm for overcoming the obstacle and publishes it in Smart-M3. The manipulating robot is notified by the Smart-M3 platform that the algorithm for obstacle overcoming is found and implements it in physical space. If the measuring robot has provided correct measurements and the coalition has performed a task, the ratings of the measurement and manipulating robots increase. These changes are stored in the blockchain network. If the robots should spend the consumables to perform the task, it requests the value of this consumables from the blockchain network. If the robot has these consumables enough it reserve them by creating a new transaction with the consumable identifier and the needed consumables' value. Each transaction in the blockchain network is stored to the block that is linked to the last existing one. Transactions are verified by other robots using Byzantine fault tolerance [29] consensus mechanism and stored into blockchain in case of reaching consensus.

Providing transaction information to the blockchain allows for all coalition participants at any time to check the information authorship in Smart-M3 as well as prevent the malicious information to be used in joint task performance. Due to participants being able to check the source of the information, it becomes possible to ignore information from unauthorized sources that are not coalition participants.

Figure 6. The manipulating robot prototype.

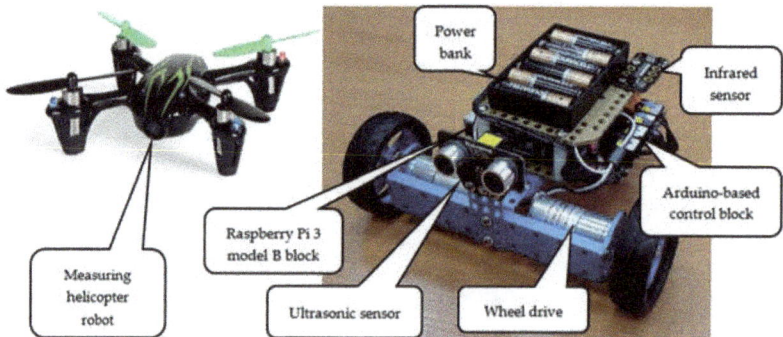

Figure 7. The measuring robot prototype.

An interaction between participants on the Smart-M3 platform is implemented on top of their operational contexts represented by RDF (Resource Description Framework) triples. A simple example of a triple published in Smart-M3 platform is <"robot", "task", "goToLocation">. The subject "robot" describes a robot, which the information in the triple relates to, the predicate "task" defines what this information is, and the object "goToLocation" is a particular "task" for the subject "robot". At the same time "goToLocation" is also a subject in the triple <"goToLocation", "coordinates", "40; 200">. Devices connected to the information space based on Smart-M3 platform are able to subscribe to a triple to get a notification when the needed information is published in the information space. In the case of a new triple satisfying pattern, all devices that are subscribed to triples of this pattern are notified.

The proposed case study has been evaluated for two types of obstacles: the first one has emulated the stairs; and the second one has a higher stair first and then a smaller one (see Figure 9). During the evaluation, the measuring robot recognizes these stairs and the manipulating robot implements the algorithm for the overcoming of obstacles, based on this recognition. For the stairs recognition the ultrasonic sensor has been used. The results of measurements for two different obstacles (the first is one higher stair first and then the smaller one, and the second one is regular stairs) are shown in Figure 10. The measurement robot publishes the information determined about the obstacle to the smart space, the knowledge base services find the appropriate algorithm to overcome the obstacle, and the manipulating robot overcomes the obstacle using the algorithm found. The knowledge base service uses the comparison function that is based on the standard deviation calculation. Based on the measurement accuracy the successfulness of the obstacle identification the manipulating robot adjusts the rating of the measurement robot as well as rating of the competencies related to the obstacle searching and measurement.

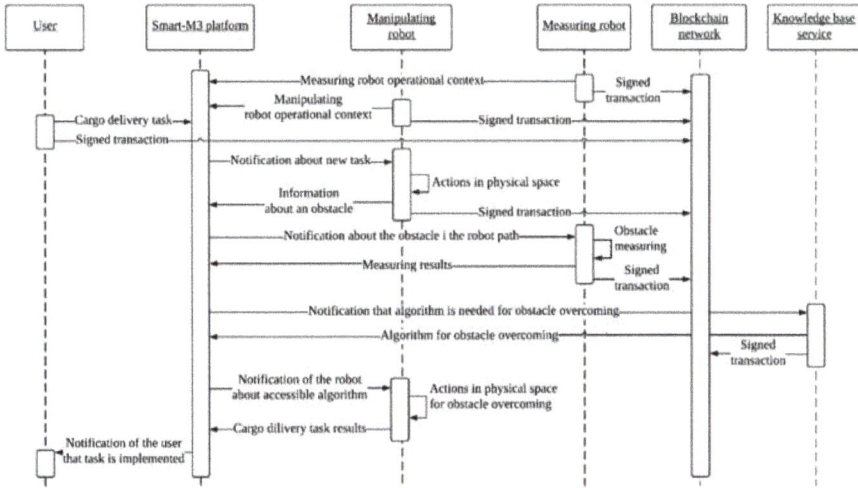

Figure 8. The scenario of heterogeneous robots collaboration work for obstacles overcoming.

Figure 9. The case study evaluation: the manipulating robot overcomes two obstacles.

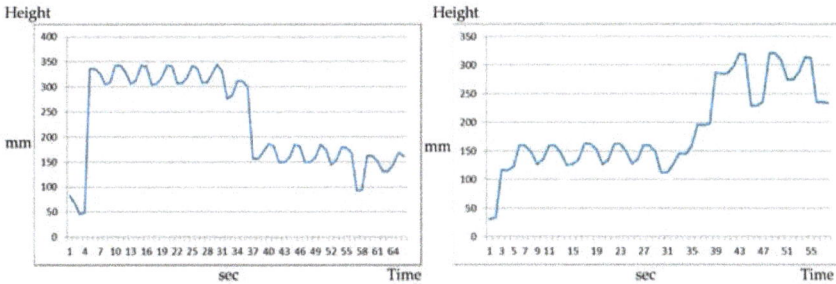

Figure 10. The case study evaluation: measuring robot determines higher stair first and then the smaller one at the left picture, and stairs at the right picture.

The blockchain network for the case study has been implemented based on the Hyperledger Fabric platform that is provided by community of software and hardware companies leading by IBM [30]. The platform provides possibilities of wide range configurations: changing of a core database for transactions and block storing, changing of consensus mechanisms, and changing signature algorithms for peers' interaction with the blockchain. For the case study presented in this paper, the default configuration has been used that includes Byzantine Fault Tolerate consensus mechanism based on BFT-SMaRT core [31], Apache CouchDB as a database and an internal solution for peer certification. This configuration provides the ability to process more than 3500 transactions per second with latency of a hundred ms. Also, the platform provides the possibility to create smart-contracts called chaincodes (program code that describes interaction between resources) using Go or Java programming languages. The chaincodes are running in isolated containers of core peers of Hyperledger based on the Docker technology stack. Figure 11 shows a system log snapshot from the blockchain during the task performance process.

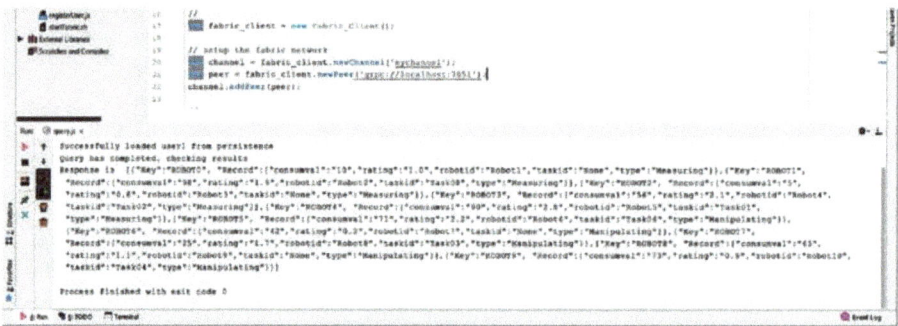

Figure 11. System log snapshot from the Hyperledger Fabric platform.

6. Conclusions

The paper presents an ontological approach and case study for blockchain-oriented coalition formation by CPS resources. For these purposes, the model of abstract and operational contexts for CPS resources coalition formation, the blockchain-oriented conceptual model for CPS resources interaction, CPS ontology that is used for interoperability support between CPS resources, and the case study for evaluating the proposed approach, have been presented. The approach presented develops existing multi-agent-based approaches by providing a two-level interaction model based on abstract and operational contexts, as well as providing immutable transaction logs based on the blockchain technology that allows us to track task assignation, task performance process, resource's rating changes, and consumables allocation between resources. Utilization of the semantic interoperability platform (smart space) for ontology-based information & knowledge exchange between CPS resources provides possibilities of ontology-based information and knowledge exchange between participants for joint task performance. Utilization of the blockchain technology for CPS participants allows us to confirm authorship for information such as ratings, and support its immutability. At any time, every resource has the possibility to check the trustiness of the information and decide if it should use it or not. The information such as ratings is stored at the blockchain database. In this case it is possible to simplify checking that a resource rating, accessible at the semantic interoperability platform, corresponds to this resource or not. The blockchain network does not provide the possibility to edit such information without agreement between all resources. The case study presented has been implemented for the scenario of heterogeneous robots' collaboration work for the overcoming of obstacles. Two types of robots and information services participate in the scenario. During the task performance, robots adjust ratings of each other and store this information in the blockchain network.

The experiments show that the proposed approach is applicable for the presented class of scenarios. The blockchain network for the proposed case study is based on the Hyperledger Fabric platform that provides high possibilities of configuration as well as a high speed of transaction processing, that is more than 3500 transactions per second with latency of a few hundred ms.

Author Contributions: A.K. implements the model for mobile robots coalition formation based on abstract and operational contexts. A.K. and N.T. developed the blockchain-oriented model for mobile robots interaction for coalition formation. A.K. has developed the cyber-physical system ontology. A.K. and N.T. have developed the case study.

Funding: The presented results are part of the research carried out within the project funded by grants ## 16-29-04349, 17-29-03284, 17-29-07073, 16-07-00462 of the Russian Foundation for Basic Research, program I.29 of the Russian Academy of Sciences. The work was partially supported by the Russian State Research # 0073-2018-0002 and by Government of Russian Federation (Grant 08-08).

Conflicts of Interest: The authors declare no conflict of interest.

References

1. Cebollada, S.; Payá, L.; Juliá, M.; Holloway, M.; Reinoso, O. Mapping and localization module in a mobile robot for insulating building crawl spaces. *Autom. Constr.* **2018**, *87*, 248–262. [CrossRef]
2. Sun, T.; Xiang, X.; Su, W.; Wu, H.; Song, Y. A transformable wheel-legged mobile robot: Design, analysis and experiment. *Robot. Auton. Syst.* **2017**, *98*, 30–41. [CrossRef]
3. Shah, N.; Czarkowski, D. Supercapacitors in Tandem with Batteries to Prolong the Range of UGV Systems. *Electronics* **2018**, *7*, 6. [CrossRef]
4. Manrique, P.; Johnson, D.; Johnson, N. Using Competition to Control Congestion in Autonomous Drone Systems. *Electronics* **2017**, *6*, 31. [CrossRef]
5. Scilingo, E.; Valenza, G. Recent Advances on Wearable Electronics and Embedded Computing Systems for Biomedical Applications. *Electronics* **2017**, *6*, 12. [CrossRef]
6. Kekade, S.; Hseieh, C.-H.; Islam, M.M.; Atique, S.; Mohammed Khalfan, A.; Li, Y.-C.; Abdul, S.S. The usefulness and actual use of wearable devices among the elderly population. *Comput. Methods Programs Biomed.* **2018**, *153*, 137–159. [CrossRef] [PubMed]
7. Verma, D.; Desai, N.; Preece, A.; Taylor, I. A block chain based architecture for asset management in coalition operations. In Proceedings of the SPIE 10190, Ground/Air Multisensor Interoperability, Integration, and Networking for Persistent ISR VIII, Anaheim, CA, USA, 10–13 April 2017; Pham, T., Kolodny, M.A., Eds.; p. 101900Y.
8. Tenorth, M.; Clifford Perzylo, A.; Lafrenz, R.; Beetz, M. The RoboEarth language: Representing and exchanging knowledge about actions, objects, and environments. In Proceedings of the 2012 IEEE International Conference on Robotics and Automation, Saint Paul, MN, USA, 14–18 May 2012; pp. 1284–1289.
9. Sekmen, A.; Challa, P. Assessment of adaptive human–robot interactions. *Knowl.-Based Syst.* **2013**, *42*, 49–59. [CrossRef]
10. Ibarra-Martínez, S.; Castán-Rocha, J.A.; Laria-Menchaca, J.; Guzmán-Obando, J.; Castán-Rocha, E. Reaching High Interactive Levels with Situated Agents. *Ing. Investig. Tecnol.* **2013**, *14*, 37–42. [CrossRef]
11. Hoshino, S.; Seki, H. Multi-robot coordination for jams in congested systems. *Robot. Auton. Syst.* **2013**, *61*, 808–820. [CrossRef]
12. Li, W.; Shen, W. Swarm behavior control of mobile multi-robots with wireless sensor networks. *J. Netw. Comput. Appl.* **2011**, *34*, 1398–1407. [CrossRef]
13. Ivanov, D.; Kapustyan, S.; Kalyaev, I. Method of Spheres for Solving 3D Formation Task in a Group of Quadrotors. In *Interactive Collaborative Robotics*; Lecture Notes in Computer Science; Springer: Cham, Switzerland, 2016; Volume 9812, pp. 124–132. ISBN 9783319439549.
14. Remmersmann, T.; Tiderko, A.; Schade, U. Interacting with Multi-Robot Systems using BML. In Proceedings of the 18th International Command and Control Research and Technology Symposium (ICCRTS), Alexandria, VA, USA, 19–21 June 2013.
15. Kryuchkov, B.; Karpov, A.; Usov, V. Promising Approaches for the Use of Service Robots in the Domain of Manned Space Exploration. *SPIIRAS Proc.* **2014**, *1*, 125–151. [CrossRef]

16. Merkel, L.; Atug, J.; Merhar, L.; Schultz, C.; Braunreuther, S.; Reinhart, G. Teaching Smart Production: An Insight into the Learning Factory for Cyber-Physical Production Systems (LVP). *Procedia Manuf.* **2017**, *9*, 269–274. [CrossRef]

17. Fiorini, S.R.; Carbonera, J.L.; Gonçalves, P.; Jorge, V.A.M.; Rey, V.F.; Haidegger, T.; Abel, M.; Redfield, S.A.; Balakirsky, S.; Ragavan, V.; et al. Extensions to the core ontology for robotics and automation. *Robot. Comput. Integr. Manuf.* **2015**, *33*, 3–11. [CrossRef]

18. Haidegger, T.; Barreto, M.; Gonçalves, P.; Habib, M.K.; Ragavan, S.K.V.; Li, H.; Vaccarella, A.; Perrone, R. Applied ontologies and standards for service robots. *Robot. Auton. Syst.* **2013**, *61*, 1215–1223. [CrossRef]

19. Jorge, V.A.M.; Rey, V.F.; Maffei, R.; Fiorini, S.R.; Carbonera, J.L.; Branchi, F.; Meireles, J.P.; Franco, G.S.; Farina, F.; da Silva, T.S.; et al. Exploring the IEEE ontology for robotics and automation for heterogeneous agent interaction. *Robot. Comput. Integr. Manuf.* **2015**, *33*, 12–20. [CrossRef]

20. Zhang, Y.; Wen, J. The IoT electric business model: Using blockchain technology for the internet of things. *Peer-to-Peer Netw. Appl.* **2017**, *10*, 983–994. [CrossRef]

21. Dorri, A.; Kanhere, S.S.; Jurdak, R. Towards an Optimized BlockChain for IoT. In Proceedings of the 2017 Second International Conference on Internet-of-Things Design and Implementation (IoTDI), Pittsburgh, PA, USA, 18–21 April 2017; pp. 173–178. [CrossRef]

22. Ferrer, E.C. The blockchain: A new framework for robotic swarm systems. *arXiv*, **2016**. [CrossRef]

23. Cachin, C.; Vukolić, M. Blockchain Consensus Protocols in the Wild. *arXiv*, **2017**. [CrossRef]

24. Dey, A.; Abowd, G.; Salber, D. A Conceptual Framework and a Toolkit for Supporting the Rapid Prototyping of Context-Aware Applications. *Hum.-Comput. Interact.* **2001**, *16*, 97–166. [CrossRef]

25. Smirnov, A.; Kashevnik, A.; Petrov, M.; Parfenov, V. Context-Based Coalition Creation in Human-Robot Systems: Approach and Case Study. In *Interactive Collaborative Robotics, Proceedings of the International Conference on Interactive Collaborative Robotics (ICR 2017), Hatfield, UK, 12–16 September 2017*; Lecture Notes in Computer Science; Springer: Cham, Switzerland, 2017; Volume 10459, pp. 229–238. ISBN 978-3-319-66470-5.

26. Smirnov, A.; Kashevnik, A.; Shilov, N.; Balandin, S.; Oliver, I.; Boldyrev, S. On-the-fly ontology matching in smart spaces: A multi-model approach. In *Smart Spaces and Next Generation Wired/Wireless Networking*; Lecture Notes in Computer Science; Springer: Berlin/Heidelberg, Germany, 2010; Volume 6294, pp. 72–83.

27. IEEE Robotics and Automation Society. *IEEE Standard Ontologies for Robotics and Automation*; IEEE: Piscataway, NJ, USA, 2015; ISBN 9780738196503.

28. Korzun, D.; Kashevnik, A.; Balandin, S. *Novel Design and the Applications of Smart-M3 Platform in the Internet of Things*; Advances in Web Technologies and Engineering; IGI Global: Hershey, PA, USA, 2017; ISBN 9781522526537.

29. Lamport, L.; Shostak, R.; Pease, M. The Byzantine Generals Problem. *ACM Trans. Program. Lang. Syst.* **1982**, *4*, 382–401. [CrossRef]

30. Androulaki, E.; Barger, A.; Bortnikov, V.; Cachin, C.; Christidis, K.; De Caro, A.; Enyeart, D.; Ferris, C.; Laventman, G.; Manevich, Y.; et al. Hyperledger Fabric: A Distributed Operating System for Permissioned Blockchains. *arXiv*, **2018**. [CrossRef]

31. Bessani, A.; Sousa, J.; Alchieri, E.E.P. State Machine Replication for the Masses with BFT-SMART. In Proceedings of the 2014 44th Annual IEEE/IFIP International Conference on Dependable Systems and Networks, Atlanta, GA, USA, 23–26 June 2014; pp. 355–362.

electronics

MDPI

Article

Creation and Detection of Hardware Trojans Using Non-Invasive Off-The-Shelf Technologies

Catherine Rooney [1], Amar Seeam [2] and Xavier Bellekens [1,*]

[1]   Division of Cyber Security, Abertay University, Dundee DD1 1HG, UK; c.rooney@abertay.ac.uk
[2]   Department of Computer Science, Middlesex University, Mauritius Campus, Flic en Flac, Mauritius;
      a.seeam@mdx.ac.mu
*   Correspondence: x.bellekens@abertay.ac.uk; Tel.: +44-(0)-1382-30-8482

Received: 6 June 2018; Accepted: 20 July 2018; Published: 22 July 2018

Abstract: As a result of the globalisation of the semiconductor design and fabrication processes, integrated circuits are becoming increasingly vulnerable to malicious attacks. The most concerning threats are hardware trojans. A hardware trojan is a malicious inclusion or alteration to the existing design of an integrated circuit, with the possible effects ranging from leakage of sensitive information to the complete destruction of the integrated circuit itself. While the majority of existing detection schemes focus on test-time, they all require expensive methodologies to detect hardware trojans. Off-the-shelf approaches have often been overlooked due to limited hardware resources and detection accuracy. With the advances in technologies and the democratisation of open-source hardware, however, these tools enable the detection of hardware trojans at reduced costs during or after production. In this manuscript, a hardware trojan is created and emulated on a consumer FPGA board. The experiments to detect the trojan in a dormant and active state are made using off-the-shelf technologies taking advantage of different techniques such as Power Analysis Reports, Side Channel Analysis and Thermal Measurements. Furthermore, multiple attempts to detect the trojan are demonstrated and benchmarked. Our simulations result in a state-of-the-art methodology to accurately detect the trojan in both dormant and active states using off-the-shelf hardware.

Keywords: hardware trojan taxonomy; thermal imaging; side channel analysis; infrared; FPGA

1. Introduction

A hardware trojan can be described as a malicious alteration or inclusion to an integrated circuit (IC) that will either alter its intended function or cause it to perform an additional malicious function. These malicious inclusions or alterations are generally programmed to activate only under a specific set of circumstances created by an attacker and are extremely hard to detect when in their dormant state [1]. As technology advances, so does the demand for IC boards leaving many technology companies without the resources to produce secure enough ICs to meet current demands. This has pushed companies into the 'fabless' trend prevalent in today's semi-conductor industry, where companies are no longer attempting to produce the goods in their own factories, but instead are outsourcing the process to cheaper factories abroad [2,3].

This growth brings with it a significant rise in the level of threat posed by hardware trojans, a threat that directly affects all companies concerned with products that utilise ICs. This encompasses many different industries, including the military and telecommunications companies, and can potentially affect billions of devices from mobile phones and computers to military grade aviation and detection devices, particularly at a time when wireless devices are being introduced as links in critical infrastructure, compounding trust and security issues even further [4,5]. It is also a direct threat to the already vulnerable Internet of Things, meaning that wireless-enabled household devices also become potential targets [6].

The problem is such that even previously 'reputable' factories are vulnerable to attacks, since all that is required is one employee to alter the existing code to include a trojan [7,8]. As most IC designs are extremely large and contain a huge amount of hardware description, these inclusions are difficult to detect and the sheer size of the code can require many people having access to the code at production level.

Regarding military grade products utilising ICs, the problem of hardware trojans is critical with the threat level of the trojan being such that it could potentially be catastrophic. Malicious inclusions of code could cause life saving equipment to fail, missiles to lose control, and cryptography keys to be leaked. While incidents of hardware trojans, are not openly discussed there have been a few noted. In 2007, it was assumed that a backdoor built into a Syrian radar system was responsible for the system's failure [9]. There are also reports of trojans being used by the USSR to intercept American communications during the cold war [3].

The problem is aggravated further still when considered in relation to the growth in production of counterfeit goods. Such goods may be produced in less than reputable factories, so the inclusion of malicious code in the production process is far from unrealistic [10]. As counterfeit goods are not generally sold through trustworthy channels, it is impossible to recall products found to be unsafe or indeed to produce updated firmware to deal with emerging threats. This can expose consumers to a plethora of malicious attacks by hackers. For example, a trojan leaking cryptography keys in counterfeit IoT devices could potentially give hackers access to a network of devices that can be utilised in 'Mirai' like attacks and cannot be recalled or patched [11].

In this paper, a hardware trojan is created and emulated on a consumer FPGA board. The experiments to detect the trojan in a dormant and active state are made using off-the-shelf technologies which rely on thermal imaging, power monitoring, and side-channel analysis. Furthermore, three attempts to detect the trojan are demonstrated and benchmarked. Finally, a state-of-the-art methodology is presented which allows accurate detection of the trojan in both dormant and active states.

Other researchers have attempted to detect trojans using similar techniques to the ones presented in this paper, using thermal imaging [12–14] and side channel attack and power analysis [15,16] However, to the best knowledge of the authors this is the first manuscript successfully proposing a practical approach using off-the-shelf hardware for the detection of hardware trojans.

The rest of this paper is organised as follows, Section 2 provides an overview of the different hardware threats, the industries affected and a detailed hardware taxonomy, while Section 3 provides an overview on the testing methodology, the design of the trojan, its functions and details of its implementation. In Section 4 an overview of the different techniques used to detect the hardware trojan are provided. Section 5 describes the results obtained using the different off-the-shelf devices. The results and techniques are further discussed in Section 6 and the paper finishes in Section 7 with the conclusion.

2. Background

Whilst the ongoing war between hackers and software developers has been raging since the 1980s, the underlying hardware being utilised was always considered to be secure [17]. Over the last decade, however, the technological progress has been such that the demand for IC boards has grown to unprecedented levels. To cope with such demand, many technology companies have evolved into the so-called 'fabless' companies, meaning that their designs are outsourced to cheaper, usually foreign, foundries for production [18]. As the number of companies involved in the production chain has grown, so have concerns, and the problem of hardware trojans has escalated.

As the full potential of the threat posed by the hardware trojan has been realised and acknowledged by the electronics industries many different conspiracy theories have emerged. Unfortunately, those theories are not in nature groundless or out-with the realms of the possible. In fact several of them have already been instantiated. One such rumour of 'kill switches' being hidden in commercial processors was confirmed by an anonymous U.S. defence contractor who indicated the culprit to be a 'European Chip Maker' [17]. The potential consequence of the existence of such a switch could be catastrophic. Indeed, as previously highlighted, this particular hardware trojan was blamed for the failure of a Syrian radar to detect an incoming air strike [19].

2.1. The Threat of the Hardware Trojan

2.1.1. At Design Level

The complexity and cost of the design of ICs has grown exponentially over the last decade as the semiconductor industry has scaled to sub-micron levels. A typical IC board will go through a rigorous process consisting of several stages.

Firstly, the specifications must be translated into a behavioural description, usually in a hardware description language such as Verilog or VHDL. Once this has been completed, the next phase is to perform synthesis to transform the behavioural description into a design implementation using logic gates such as a netlist. Once the synthesis has been completed, the netlist is implemented as a layout design and the digital files are passed to the foundry for fabrication [17].

As well as outsourcing the production of ICs, many companies are also purchasing third party intellectual property (IP) cores, and utilising third party Electronic Design Automation (EDA) tools. Each use of a third-party software presents a new opportunity for attacks such as hardware trojan insertion, IP piracy, IC tampering, and IC cloning. Although these attacks are all of importance, the hardware trojan is by far the most dangerous attack, and, as such, has garnered much attention [17].

2.1.2. At Foundry Level

As semiconductor technology has advanced, the cost of owning foundry has increased dramatically. In 2015, the cost was estimated to be in the region of 5 billion USD [20]. As a direct result of this, many companies can no longer afford to fund the production process from start to finish, and are outsourcing their production to cheaper foreign foundries [17].

Whilst undesirable modifications to ICs should ideally be detectable by pre-silicon verification and simulation, this would require a specific model of the entire IC design and this is not always readily available particularly where third party IP cores or EDA tools have been used. In addition, large multi module designs are rarely compliant with exhaustive verification [21].

Post silicone approaches to design verification include destructive de-packaging and reverse engineering of the IC. However, current techniques do not allow destructive verification of ICs to be scalable [22]. It is also possible for an attacker to infect only a portion of the produced ICs, making these tests futile [23].

Most post silicone logical testing techniques are also unsuitable for detecting hardware trojans. This is attributed to the stealthy nature of the hardware trojan and to the large numbers of differing taxonomys that can be employed by the attackers. Most hardware trojans are programmed to activate under a specific set of conditions, and a skilled attacker would ensure that these conditions were undetectable by the testing routine. This is particularly true of trojans targeting sequential finite state machines [24].

2.2. Industries Affected

2.2.1. Military

Hardware trojans are a huge threat to many industries. However, security conscious industries, such as the military, are in a particularly high risk bracket and defence departments are very aware of this. The U.S. Department of Defense (DoD) has created a "Trusted Foundry Program" to ensure its military equipment remains free of hardware trojans by using only accredited foundries. This means that only American foundries which are located on the Americal soil and which underwent the strictest vetting process are allowed to work on the chips for the U.S. DoD. In addition to vetting the foundries, close attention is being paid to the other links in the design and supply chain [9].

While this approach may seem effective, it has its limitations. The majority of western foundries are woefully behind their foreign counterparts when it comes to the level of technology they can provide. This seriously limits access to more advanced chips which are required for modern avionics and weapons systems [9].

2.2.2. Financial Infrastructures

In 2008, an experiment was carried out by the University of Illinois in which researchers designed and inserted a small backdoor circuit that gave access to privileged areas of the chip's memory [9]. This trojan could then be used to change process identifiers allowing attackers to access all data contained on the chip's memory. It is easy to see why this could be catastrophic in settings such as critical infrastructures. Trojans such as the one described are usually small and are nigh impossible to detect.

2.2.3. Consumer Industries

Security industries are not the only potential targets of a hardware trojan attack. There exists the possibility of utilising a hardware trojan by rival firms in industrial sabotage. The potential damage that could be caused by such an attack could be enough to disable even global corporations, particularly in industries such as telecommunications.

Aside from industrial sabotage, the potential threat to consumer privacy is also of major importance. Devices such as smartphones and tablets could be targeted by trojans designed to leak private encryption keys or private information [9].

2.3. *Hardware Trojan Taxonomy*

Although the terms and definitions used to classify different hardware trojans can vary between different authors, the general taxonomy is universally agreed to consist of the physical representation, the behavioural phase or trigger, and the action phase in which the trojan will execute its payload.

2.3.1. Physical Representation

When designing malicious circuitry, there are several characteristics that must be considered. The first of these characteristics is the 'type' of the hardware trojan, which can be defined as functional or parametric.

A trojan is categorised as functional when the attacker adds components such as logic gates to the original design. Accordingly, the deletion of components to cause a malicious function would also place it in this category [25].

If the attacker creates the trojan through the modification of the existing code, then it will be classified as a parametric. Typically, this can be achieved by thinning wires or weakening transistors and flip flops. This type of trojan is notoriously hard to detect as the alteration can be minuscule.

The next physical characteristic the attacker would have to consider would be the size of the hardware trojan. In this context, the size refers to the physical extension of the hardware trojan or the number of components it consists of. In case of a large trojan consisting of many components, an attacker can distribute these across the IC, placing components where they are necessary to execute their payload in accordance with the functions of the hardware trojan. This is known as loose distribution [25].

In contrast, a smaller hardware trojan consisting of only a few components allows for the components to be placed together as they will occupy only a small part of the layout of the IC. This is known as tight distribution.

On rare occasions, a determined attacker could regenerate the layout to encompass the hardware trojan, moving the components of the IC to accommodate the components of the hardware trojan. This is referred to as a structural alteration [25].

2.3.2. Activation Characteristics

Typically, a hardware trojan will be condition-based, meaning that its activation will be dependent on a trigger defined by the attacker. The trigger itself will generally consist of either a predefined input pattern, or specific internal logic state, or counter value, and can be triggered both internally and externally.

An externally triggered hardware trojan will usually consist of malicious logic within the IC that utilises an external sensor such as a radio antenna. The attacker will then communicate via the compromised component enabling them to trigger the antenna. It is easy to see why this can be extremely dangerous when it comes to security conscious industries such as the military. It is not out-with the realms of the believable to postulate that an attacker could feasibly re-route or switch off a missile via a radio signal as suggested in [22]. Conversely, an internally triggered hardware trojan will look within the circuitry for the set of conditions that will cause it to activate. A typical example of this would be a countdown logic.

In contrast to the condition-based trojan that will only activate when its trigger conditions are met, the "always-on" trojan is active from the moment of insertion, and relies on internal signals. This type of hardware trojan is generally split into two categories; combinational and sequential. A combinational trojan will activate upon detection of a specific set of circumstances within the internal signals of the IC. Sequential trojans will also monitor the internal signals of the IC. However, instead of looking for a specific condition, they activate when a specific sequence of events occurs [26].

2.3.3. Action Characteristics

The action characteristics of a hardware trojan refer to the effect the trojan will have on the execution of its payload. Hardware trojans will typically fall into one of two categories: implicit or explicit [27]. Implicit trojans will not change the board's circuitry of the IC; instead, they will perform their malicious function in tandem with the intended function of the board. This makes these trojans easier to detect as they tend to cause small path delays on activation and consume more power whilst active.

In contrast, an explicit trojan will change the function of the board's circuitry on activation. This can come in the form of a signal alteration or even leaking of information via predefined board pins. These trojans tend to cause distinct path delays as well as large changes in circuit's capacity [27].

2.4. Hardware Trojan Detection

Detecting a hardware trojan requires overcoming numerous challenges. Namely:

1. Handling large architectures.
2. Being non-destructive to the IC.
3. Being cost effective.
4. Ability to Detect trojans of all sizes.
5. Authenticating chips in as small a time frame as possible.
6. Dealing with variations in manufacturing processes.
7. Detecting all trojan classifications.
8. Detecting trojans in a reasonable time frame.

To the best of the authors' knowledge, there is no single method capable of detecting all types of hardware trojans, nor overcoming all the challenges described here-above. Over the years, several methods have been developed to detect different types of trojans. These methods are described here-after.

2.4.1. Physical Inspection

One of the most obvious method of detection is physical inspection of the board itself. This method is sometimes classified as a failure analysis based technique. Those techniques usually comprise two steps: (1) cutting and lifting the molding coat to expose the circuitry; and (2) performing various scans [25,28].

2.4.2. Functional Testing

Often referred to as Automatic Test Pattern Generation (ATPG) this technique is more commonly used to locate manufacturing faults; it has been shown to be effective in detecting hardware trojans. ATPG involves inputs of ports are stimulated and then the output ports are monitored for variations that may indicate a hardware trojan has been activated. Functional testing techniques can also be useful when attempting to determine the trigger patterns of conditional trojans [25,27].

2.4.3. Built-In-Self-Test Techniques

Built-In-Self-Test (BIST) techniques are commonly used to detect manufacturing faults and are present in many chips. If unknown or malicious logic is detected during these tests a bad checksum result is given, although designed to detect manufacturing faults on some occasions these tests can detect hardware trojans [25].

2.4.4. Side Channel Analysis

Side channel analysis techniques are some of the most commonly used procedures in hardware trojan detection. These techniques generally measure signals such as power and path delay, looking for fluctuations potentially caused by trojans. Side channel analysis can have a high success rate as even in a dormant state the trojans trigger signal will cause some current leakage [25].

3. Methodology

In order to carry out the investigation our trojan was designed and loaded onto an Basys 3 Artix 7 FPGA board. Three different detection techniques are demonstrated, the first utilises power analysis techniques as well as side channel analysis, allowing security investigators to measure both the power variance, traces and current leakage, followed by a concentrated heat measurements using an infrared thermometer, and finally a thermal camera test is carried out. The three experiments are carried out using off-the-shelf hardware and are applied to both the trojan-free and trojan-inserted designs. Attempts are then made to detect the trojan in its dormant form. While in in their dormant form trojans do not perform any malicious actions, however, wait to be activated, through an activation signal, this can be done through the push of a button, or through a specific set of instructions. It is however important to be able to detect trojans in their dormant form, before they activate and perform malicious actions. In the last experiment, the thermal camera is also used to measure the impact of the trojan in its active form. Table 1 provides information on the type of devices used to carry out the different experiments.

Table 1. Description of Off-The-Shelf Hardware for Hardware Trojan Detection.

Type	Device	Price
Oscilloscope	Infinium series 1 GHz	£4000
Oscilloscope	Digilent OpenScope	£60.00
Infrared Thermometer Gun	Powefix	£17.99
Heat Camera	Flir C2 Camera	£550.00

3.1. Design

Whilst there is much literature available on the topic of hardware trojans and many papers aimed at the various taxonomies of hardware trojans, there are few practical examples to be found. This paper aims at providing thorough details on the creation of a tailored hardware trojan. The factors considered in the experiments and the resultant decisions are shown in the list below;

Main Function The algorithm as a main function running continuously. This function serves by providing information on the board, and executing tasks in a round robin fashion, these tasks run in a loop, this ensures steady power readings.

Physical Representation The trojan designed for the experiments is in the form of a functional trojan as opposed to a parametric trojan. Functional trojans are categorised as a malicious inclusion of components and code as opposed to an alteration to existing code.

Physical Size The trojan designed for the experiments is required to be small and consists of no more than three components, this allows a tight distribution on the board. The creation of a complex design is deemed to be out of the scope of this paper.

Characteristics The trojan is activated by an external trigger to preserve the integrity of the testing results. This provides assurances for the different test results, and allows the monitoring of the free, dormant, and active states. This test falls within the parameters of a conditional trojan.

Malicious Function The trojan is required to have a malicious impact on the board, while more advanced functions such as the leaking of cryptography keys are within the scope of this investigation, simpler functions are privileged to demonstrate the process. Two distinct functions are chosen, the first one causes the board to overheat, while the second one performs an unwanted action on the board.

The design was carried out using the Vivado WebPACK Design Suite, this allowed for high level design, synthesis and implementation.

Currently the use of third party Intellectual Property (IP) Cores to construct large and complex designs is standard practice industry wide. An IP core is a block of logic or code that can be utilised in the design of an ASIC or FPGA, most common components can be found the form of an IP Core and this method of design drastically reduces the amount of time the designer spends writing code. It is equally likely that an attacker inserting malicious code would have had the skill to create and insert their own hand crafted code. A block diagram is shown in Figure 1, the block diagram represent the counter without the trojan inserted.

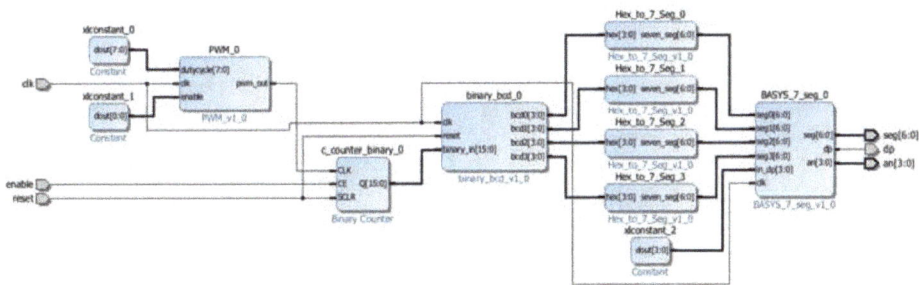

Figure 1. Configured IP Cores Block Diagram.

3.2. Trojan-Free Design

The main function of the trojan-free design was decided to be a counter program, counting from 1 to 9999 on a loop and output to the four seven segment displays present on the Basys 3 board.

The components used in this design operate at a steady rate, this ensures steady power tests taken whilst performing side channel analysis on the clean board, thereby guaranteeing that any fluctuations detected on the infected board are indeed caused by the presence of the trojan. The trojan-free design requirements for the counter can be found below;

- The refresh rate for the 7 segment displays is set to 5000 Hz
- The design utilises all four 7 segment displays
- The 7 segment displays output a count that beginning at 0 and incrementing up to 9999
- Upon reaching maxcount of 9999 the displays resets to 0
- The counter is enabled via a switch
- The counter can be reset via a pushbutton
- The counter will count at a rate of 5 Hz

After the final configuration, it was a matter of creating a top level VHDL wrapper to contain the inputs and outputs of the block design. The wrapper file maps all the I/O ports of the physical board with the I/O described in the constrain file. Finally, the constraints for the project needed to be added to successfully synthesise and implement the design. Constraints files are used to map the input and output ports to specific pins on the FPGA board and can be easily configured by uncommenting a master file and editing the port names to reflect those of your design. Figure 2 shows the trojan-free elaborated design schematic. The schematic is based the RTL file. The schematic identifies pieces of code representing hardware structure. The code further converted (Synthesis) into "technology cells" abstracting elements such as multiplexer, adder, etc....

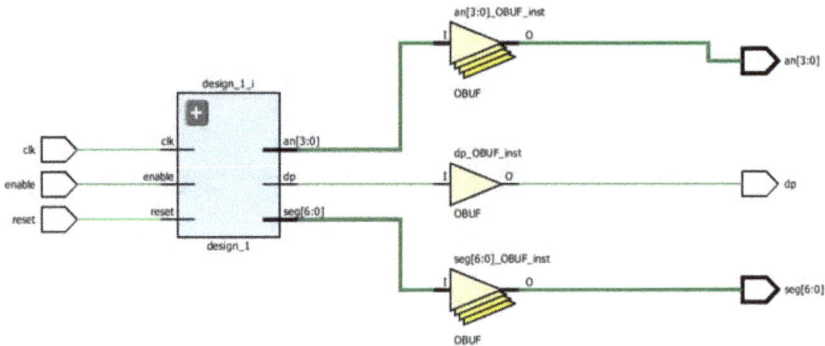

Figure 2. Trojan-free Elaborated Design Schematic Provided by Vivado.

3.3. Unsigned Multiplier

As previously discussed the trojan code used in this project was handcrafted to mimic the techniques that would be used by an attacker, although it is possible that malicious code could be contained within a third-party IP core.

The design to be used for the trojan is in the form of an unsigned multiplier that operates on a combination of buttons and switches and output onto the LED. The required inputs of both switches and a button can be said to represent the usually hard to guess trigger. These inputs are often used in conditional trojans.

The trojan was designed with these specifications in mind.

- There shall be 15 input switches and 1 input button
- There shall be a start button and a reset button
- The multiplier will output in binary via LEDs
- The multiplier shall utilise two finite state machines
- The multiplier should have a malicious effect on the board

FPGA boards do not tend to contain many dedicated multipliers with the Basys 3 containing only four. This means that when creating a design that requires a multiplier most designers will compensate for this by creating their own. As such, the presence of this code would not be likely cause suspicion. The junction temperature of the board is the highest temperature that a board can operate safely at, it is calculated using Equation (1).

$$T_j = T_a + (\theta J_a \times P_d) \tag{1}$$

Let T_a be the ambient temperature expressed in °C, let θJ_a be the Junction-to-ambient thermal resistance, let P_d be the core power, finally let T_j be the junction temperature. The relationship between the thermal parameters of the board can be expressed as show in Equation (2), let θJ_a be the junction-to-ambient thermal it defines the difference between junctions temperatures and the ambient temperature when the device dissipates 1Watt of power, hence is expressed in °C/W resistance, let θJ_c be the junction-to-case thermal resistance and is expressed in °C/W and θC_a be the case-to-ambient thermal resistance and is expressed in °C/W.

$$\theta J_a = \theta J_c + \theta C_a \tag{2}$$

The design for the trojan also utilises two finite state machines, This is necessary for the multiplier as it requires several inputs. This can be seen in Figure 3 (Highlighted in Red). For the purposes of this trojan a Mealy machine was deemed to be most appropriate. Generally Mealy machines have fewer states that Moore machines, the output changes at the clock edges in a Mealy machine and react faster to inputs. This fits well with the requirements of the trojan as it must have several inputs.

Once the trojan designed the next step is to create a working constraints file as this is necessary to test the functionality of the trojan.

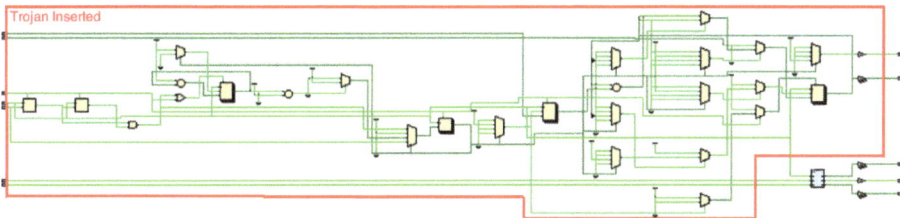

Figure 3. Trojan-inserted Elaborated Design Schematic.

3.4. The Trojan

Having designed and built both the counter and multiplier circuits the next step is to integrate the code from the multiplier into the top level VHDL wrapper created for the counter circuit. The counter signal is asynchronous. The counter activates on the first flip-flop. The subsequent flip-flops are clocked by using the output of the previous flip-flop.

The additional ports required by the multiplier was to be added to the original design for the counter. Whilst this is straight forward for most of the ports, the clk port in the multiplier was deleted as there was already a clock in place and they had been set to the same value, meaning that the "trojan" was now utilising the clk port of the counter.

The second amendment that had to be made was the reset port for the multiplier, as again the counter already had a reset port. While it may have been possible to allow both to use the same button, the ability to reset the trojan alone is important to the integrity of the testing and results. To circumvent this the reset port for the multiplier was renamed as $reset_2$ and would be mapped to a different push button. As the multiplier is unsigned it was also necessary to include the *IEEE.numeric_std.alllibrary* and is working synchronously.

As the code to be inserted was still referencing the reset port it was necessary to go through all the processes and change the reset port to the newly created $reset_2$ port. The next step was to insert the code for the processes that would be utilised by the trojan, this code was again inserted into the architecture body of the program.

Finally, the constraints file was updated to include the ports utilised by the 'trojan'. At this point it became evident that there were not enough input switches to accommodate both designs, the multiplier requires all sixteen switches as it takes inputs in binary form and the counter also requires an input switch.

As the switch was essential to the counters enable signal, a push button would only increment on each button press, the multipliers $input_1$ was mapped to a push button. This made the input pattern required by the trojan a combination of buttons and switches followed by a specific button press.

3.5. Elaborating the Designs

Although the code has now been created for both the trojan-free and trojan-inserted boards there are several more phases to the process of creating the final bitstream file that will be used to program the board. The first of these stages is that of elaborating the design. The elaboration process essentially takes the code provided and changes it from RTL into a variety of visual representations.

3.6. Synthesising the Designs

Synthesis refers to the process of transforming the design from an RTL design into a gate level representation. Synthesis is run after the relevant design and constraint files have been added to the project and will only be successful if the files added contain no syntax or constraints errors. Following a successful synthesis, it is possible to access the synthesised design and view the schematics of the design along with various other representations.

3.7. Implementing the Designs

Having successfully synthesised the design the next step was to run it through the implementation process. Implementation is essentially equivalent to routing and placing the design on the board. The implementation process places the netlist onto the board in a virtual environment and connects them in accordance with the constraints file. If it is possible to safely place the design on the board and it will then be possible to generate a bitstream file which can then be placed on the FPGA board.

3.8. Generating Bitstream File and Programming the Board

Once the design has passed the required phases it is possible to generate a bitstream file, this is the file that used to program the FPGA board.

Before programming the board, it was necessary to ensure that the board was correctly configured. The board can be programmed to accept a file or to perform a Built In Self Test (BIST), by default it is set to BIST to enable programming mode on the board.

4. Malware Detection Methodologies

The scenario presented in this manuscript for the trojan detection assumes that the board has been infected during production, hence the full design of the board is known. It is also assumed that we have a free trojan board at our disposal. This could be a prototype.

4.1. Side Channel Analysis

Side channel analysis techniques are employed in order to detect fluctuations in variables such as power and path delay, for this investigation it was decided to test power levels using an oscilloscope.

Two oscilloscopes were used for the investigation, the first one was an Infinium series 1 GHz 4 GSa/s oscilloscope from Agilent. While the second one was an OpenScope 2 scope channel with 12 bits at 2 MHz as shown in Table 1.

By analysing the schematics of the Artix 7 board, we identified the power in/out pins to the FPGA chip. As these pins are likely to be utilised by any design they were most likely to exhibit power fluctuations caused by the trojan. In a real world scenario, the authors of a circuit to be able to locate and test multiple pins and capacitors against a golden design in order to detect a trojan.

In order to detect power fluctuations, we set a trigger on the oscilloscope. The trigger allows the user to set a limit on readings. If the voltage measured goes above the set trigger level then the oscilloscope will stop the measurement. The trigger can be dragged and to fit measurements for pins on the trojan-free board, if the trojan boards readings exceed those of the trojan-free board, we could assume that the board was infected. This experiment was realised with the two oscilloscope, over fifteen times, to ensure accurate readings.

4.2. Temperature Readings

In order to obtain temperature readings during the testing phase of the investigation an infrared thermometer and a heat camera were used on the boards Artix 7 FPGA chip. With the trojan active, it is assumed that during normal operation, the trojan-free board will consume less power, hence the chip will be cooler. Moreover, even in its dormant form fluctuations caused by the presence of the trojan may be detectable.

The Infrared Thermometer Gun used is described in Table 1. The thermometer is an infrared thermometer which utilises sensors to acquire and measure infrared radiation from the surface it is aimed at, and from this radiation it determines temperatures. The device can detect and record temperatures ranging from $-50\,^{\circ}$C to $+380\,^{\circ}$C and as such it fits within the expected temperature range of the FPGA chip. The FLIR C2 heat camera used is described in Table 1. The Flir C2 Camera is a standard off-the-shelf heat camera, with a sensitivity of $0.10\,^{\circ}$C and works within the $-10\,^{\circ}$C to $+150\,^{\circ}$C (14 $^{\circ}$F to 302 $^{\circ}$F) The IR sensors provides 80×60 (4800 measurement pixels) and works within as spectral range of 7.5–14 µm.

4.3. Testing Process

There were several factors considered for the testing process, firstly after each test the board is left to cool down for fifteen minutes before being re-tested. This ensure that the results are not skewed by remnant heat. Misleading results could potentially affect the camera and heat gun readings, hence the cool down period. Secondly, ten readings where taken on both boards with 30 s intervals. All results are then compared. The infrared thermometer is being held at 15 cm from the board for all readings, while the heat camera is held at 20 cm from the board.

5. Results

5.1. Elaborated Design Results

Once the design has been elaborated there are several different graphical representations that can be accessed; the default is the schematic of the design. The schematic representation of the design is a pre-optimized design and is comprised of generic symbols such as AND gates, OR gates, adders and multipliers, it can be useful in helping uncover errors early in the design process. Figures 2 and 3 show the trojan-free design schematic and the trojan-inserted schematic. Another interesting representation that can be accessed is the design hierarchy, this is a graphical representation of the hierarchy of the design. The top level VHDL should always be at the top of the hierarchy. The final representation that is of interest at elaboration level is that of the I/O pin mapping. This is a graphical representation of the constraints files with each of the mapped pins being marked, as shown in Figures 4 and 5. Although the differences are subtle close inspection reveals the extra pins being utilised by the trojan in the bottom left and right sections of Figure 5 (Highlighted in Red). The differences in the figures lie in the I/O pins (highlighted in red). These pins represent the extra ports being used by the trojan. The I/O Planning view layout represents graphically the various I/O, clock and logic objects present in the design. Through these representation, one can make design decisions and optimisations. In these Figures, we aim at showing the little difference between the trojan-free and trojan-inserted design. Further highlighting the requirements of both an expert eye to differentiate between both I/O mappings and the need for an off-the-shelf approach to detect trojans.

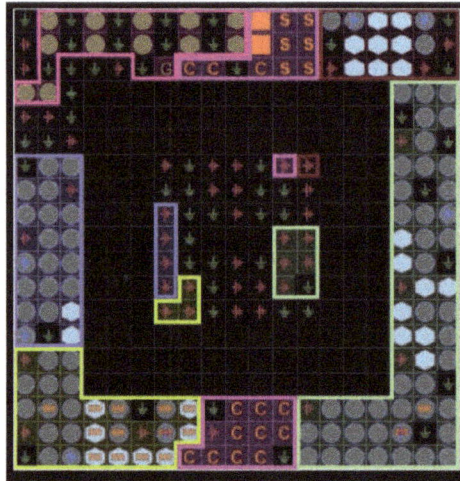

Figure 4. I/O pin mapping representation of the trojan-free design.

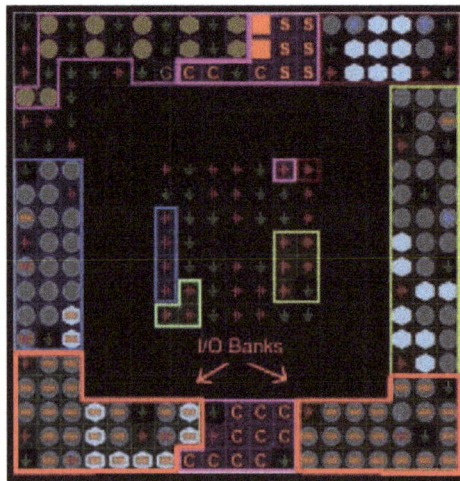

Figure 5. I/O pin mapping representation of trojan-inserted design.

5.2. Synthesised Design Results

Synthesis transforms the RTL design into a gate level netlist consisting of primitives such as Look-Up-Tables (LUTS) and as such offers new and updated representations. The default representation at this stage is a floor plan of the FPGA that can be zoomed into and examined in detail. The floorplans of the trojan-free and trojan-inserted circuits are shown in Figures 6 and 7. The presence of the trojan can be seen most clearly in Figure 7 sections X0Y0, X1Y0 and X1Y1. The differences between Figures 6 and 7 are the coloured sections, these indicate different layout with more or less cells being used. While it is unlikely to detect the trojan through the visualisation of the floor plan, the Figures aim at providing an overview of potential differences in the design.

The hierarchy of the design can also be accessed and examined; when examining the design it is found that both the trojan-free and trojan-inserted designs now contained significantly higher numbers of leaf cells. The hierarchal representation is shown in Figures 8 and 9. The Hierarchical Design enable to partition a design into small manageable modules that can further be processed independently from each other. This view also allows out-of-context (OOC) optimisation, when building a larger piece of software, or when building a more sophisticated trojan.

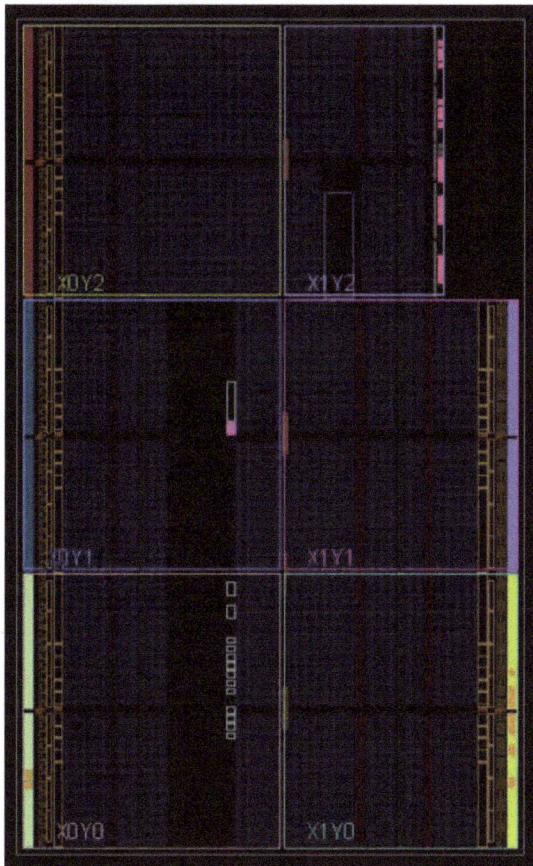

Figure 6. Floorplan-trojan-free.

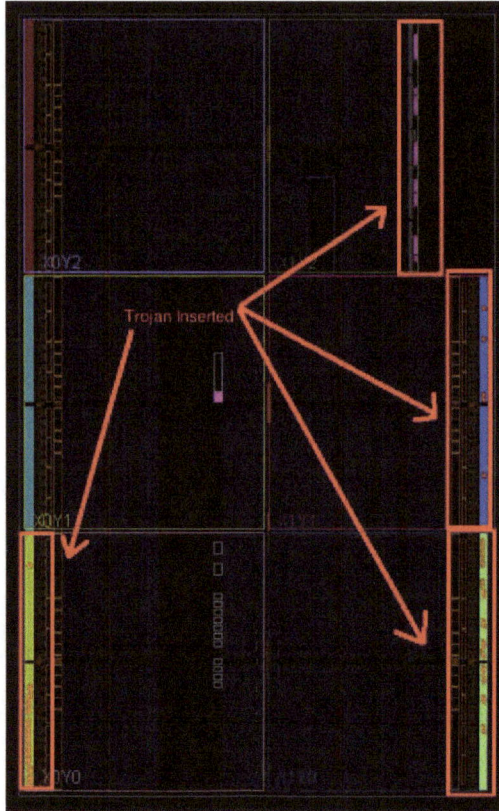

Figure 7. Floorplan-trojan-inserted.

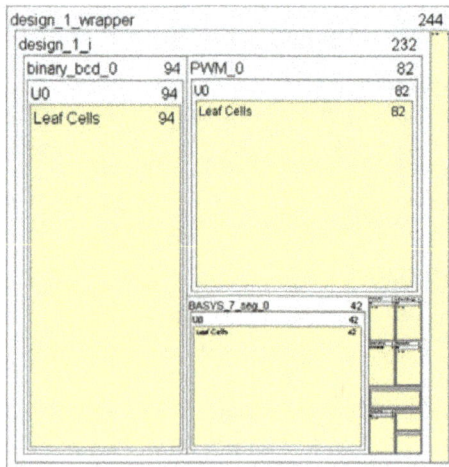

Figure 8. Design hierarchy-trojan-free.

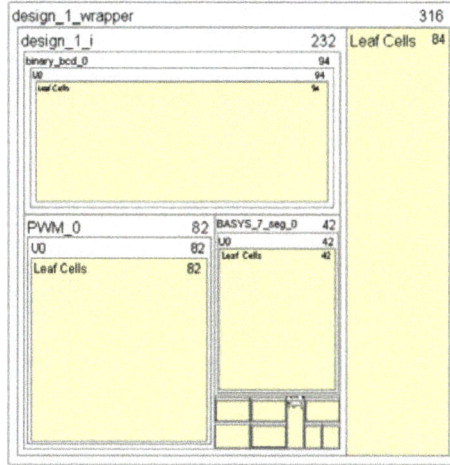

Figure 9. Design hierarchy-trojan-inserted.

5.3. Implemented Design Results

The successful completion of implementation allows access to further high level representations and the default presentation is a floorplan of the FPGA. The floorplans produced after implementation are more detailed that those produced during synthesis as the design has now been placed and routed in accordance with the constraints file.

5.4. Power Reports

The post implementation power reports summarise various factors such as on chip power and junction temperature. These reports were accessed to determine that the trojan-inserted board would indeed have the intended malicious purpose; to cause a junction overheat within the board. It is worth noting that the design is fully placed and routed during implementation so the report is indicative of the expected effect of the functioning design. The summary results for both the trojan-free and trojan-inserted boards are shown in Figures 10 and 11. These results can be used by the attacker to have an idea of the power required when designing the trojan. A high power consumption when active will render the stealthiness of the trojan useless, however, in this work we aim at detecting the trojan in a dormant state, hence, the power consumption during is active state does not affect the detection of the trojan, it is merely an indication of the stealthiness in an active state.

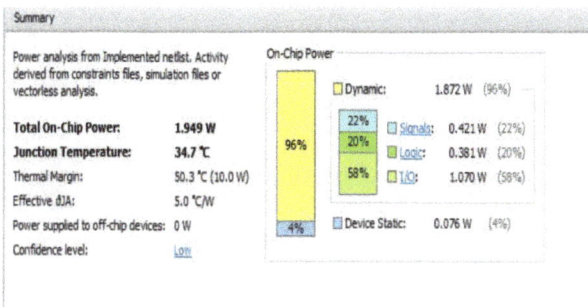

Figure 10. Trojan-free summary results.

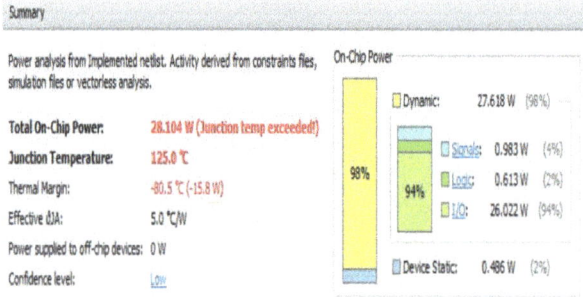

Figure 11. Trojan-inserted summary results.

5.5. Side Channel Analysis Results

During the side channel analysis a number of pin out were tested, the main difference was shown on the C101 capacitor. The C101 capacitor located next to the FPGA chip on the board [29]. The results are shown in Table 2. The fluctuations can also be seen in Figures 12 and 13 both figures shows oscilloscope readings. the trojan-free board requires 3.49 v while the trojan insert reading demonstrates that the board requires 3.6 v. The trojan was dormant at all times, during the readings, when active the trojan required over 3.9 v. This significant fluctuation was repeated over ten times, and provided us with the same reading during each iteration, essentially demonstrating that minimalist trojans do require more power and can be identified using side channel analysis with off-the-shelf oscilloscope.

Table 2. Results of C101 Capacitor.

C101 FPGA Power in	Trojan-Inserted	Trojan-Free
V amptd	3.665 V	3.493 V
V avg	−87.77 mV	−29.631 mV
V p-p	3.665 V	3.493 V

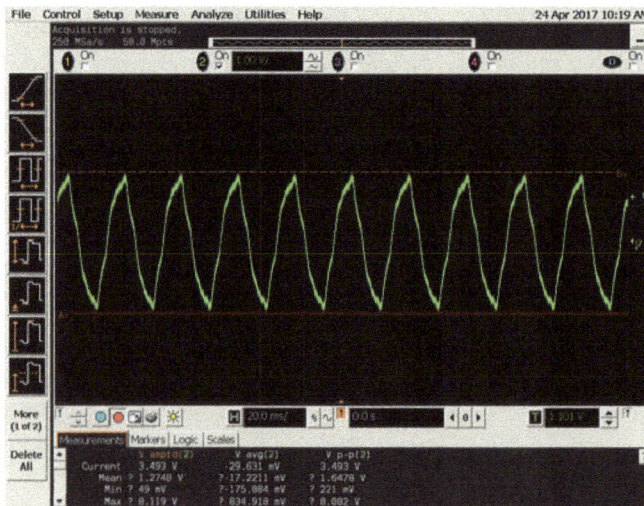

Figure 12. Trojan-free readings of the C101 Capacitor Charging and Discharging.

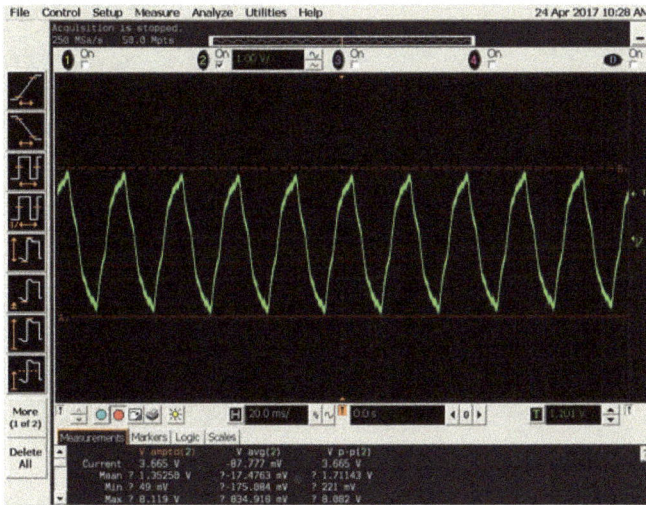

Figure 13. Trojan-inserted readings of the C101 Capacitor Charging and Discharging.

5.6. Infrared Thermometer Results

The results of the infrared thermometer showed a gradual increase in temperature over time for both boards, however the trojan infected board had a higher temperature, the results are shown in Table 3.

Table 3. Infrared thermometer results.

Run No.	Trojan in °C	Trojan-Free in °C
1	17.9	17.4
2	18.1	17.4
3	18.2	18.1
4	18.4	18.2
5	18.6	18.4
6	19.0	18.6
7	19.2	18.8
8	19.2	19.0
9	19.5	19.2
10	19.6	19.4

As shown in Table 3 the temperatures taken on the trojan-infected board are on average 0.2 °C above those of the trojan-free board. While the increases in temperature over time can be attributed to the counter program as it runs continuously, the fact that the trojan-inserted board was tested first negates this as a reason for the different temperatures. Prior to testing the trojan-inserted board, there had been no power to the board for over an hour so residual temperature from another test can also be ruled out. These factors combined with the recorded drop in temperature of the trojan-free board can be considered to be proof of the presence of the trojan.

5.7. Heat Camera

The board was also subject to a heat camera test in order to detect the trojan in three different states (Free, Sleeping and Active). The results of the heat camera confirmed the results obtained through the infrared thermometer. As shown in Figure 14 the temperature increases over time. Figure 14a from left to right presents the readings for the trojan-free, sleeping and active. Figure 14b provide a close up reading of the trojan in sleeping mode over a period of 10 min. Finally, Figure 14c provides close up reading of the active trojan board over a period of 10 min As previously observed in Section 5.6 it is possible to notice an obvious difference in temperature between the trojan-free and trojan sleeping boards.

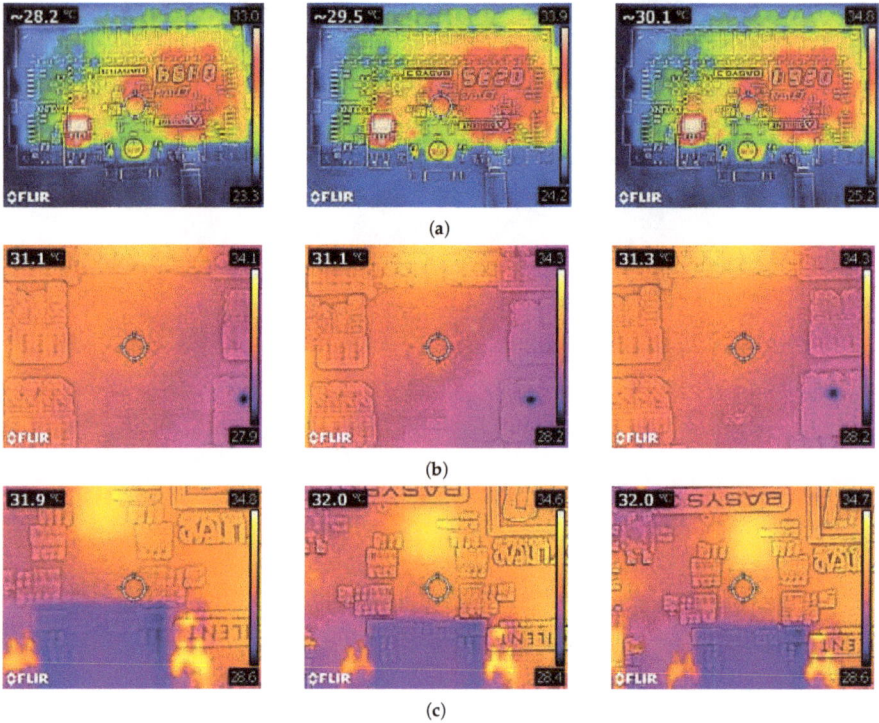

(a)

(b)

(c)

Figure 14. The set of Figure (**a**) from right to left represents the readings for the trojan free, sleeping and active respectively. The set of Figure (**b**) demonstrate a close up representation of the trojan free, sleeping and active on the board, while the set of Figure (**c**) provide close up temperature readings of the trojan active on the board over a period of 10 min.

Note that there is a difference between the readings of the heat gun and the heat camera of approximately 10 °C; we expect this to be due to the calibration of both the camera and the heat gun. While the camera had calibration options only a laboratory multi-point calibration can provide a high accuracy, moreover, the detector can experiences different temperatures due to ambient temperature changes and to the heat dissipation of the material. However, this does not make the results less relevant. The temperature difference remains constant across the devices and provides higher temperatures for trojan in sleep mode and in active mode.

6. Discussion

In this section, the results are discussed in detail. Our test provided the corner-stone for hardware trojan creation and detection using off-the-shelf technologies against a golden model. The practical work presented in this manuscript demonstrated the feasibility of hardware trojan detection using side-channel attacks, a heat camera, and a heat gun. While the settings of these tests where pre-defined, it is not un-common for this type of work as suggested in [30].

The three tests required to provide readings between the trojan-inserted and trojan-free boards w to determine the existence of the trojan. In an industial setting, it is unlikely to suffice, however, the results presented in this paper provide evidence that trojans can be detected using off-the-shelf hardware due to the increase in accuracy of the devices and the access to technologies previously reserved to specialized industries, and which are now becoming commodities.

6.1. Thermal Camera

The thermal camera provided extensive results of the three stages of the trojan. This technique provides a good visual representation, demonstrating the presence of the trojan on the board. However, as aforementioned, in order for the camera to provide accurate results a trojan-free board is needed. While this technique is accurate, when used as a stand-alone test it requires the camera to be well calibrated in order to provide the best results. In the setting presented in this manuscript, both the trojan taxonomy and its purpose where known, hence, during this white-box testing, the full calibration of the camera could be overlooked.

6.2. Heat Gun

The heat gun was able to demonstrate the presence of a a hardware trojan in different settings, while the heat gun was accurate enough to detect the temperature variation, as for the heat camera, this test requires to have free trojan board for comparison. As aforementioned, this is not uncommon for this type of work [30]. While the results provided by the heat gun are notable, they differ greatly from the thermal camera. It was expected from the start for the thermal camera to outperform the heat gun, due to the low accuracy of the heat gun.

6.3. Side Channel Analysis

The side channel analysis techniques was employed in attempts to locate the trojan whilst in its dormant form. Measurements were taken from several pins and capcitors during this process, whilst most of them showed little difference in reading only the C101 capacitor demonstrated a significant difference.

7. Conclusions

This work provides sharper bounds for the case of detection of hardware trojans using off-the-shelf devices, allowing to reduce the costs associated with trojan detection. The increasing number of devices being produced by untrusted foundries puts critical infrastructure at the center of attention. In this manuscript, we highlighted the dilemma of finding a one fits all solution to the problem finding hardware trojans fitting different taxonomies. To this end, we presented the corner stone for the detection of hardware trojans using off-the-shelf devices. We successfully demonstrated the ability of off-the-shelf devices to detect trojans in different settings, namely: sleeping and active. We believe that our practical work has the enormous potential in the successful detection of hardware trojans. In the future we will aim at developing techniques to use thermal imaging for the detection of large scale hardware trojan infection and explore other trojan taxonomies in more intricate designs and with advanced malicious purposes. While we believe this technique is fully applicable to FPGAs, the technique might not be well suited for the denser ASICs and slight modifications might be required both in the methodology and in the off-the-shelf tool used. Moreover, we believe that this method could be used widely with the democratisation of specialised off-the-shelf hardware, following Moore's law with a higher detection accuracy and better thermal imaging capabilities. Future work will compare the technique proposed against smaller known trojans and the process variation and manufacturing variation will be taken into account. Furthermore, the number of test vectors for Vivado power estimator will be increased in order to increase its accuracy.

Author Contributions: Conceptualization, C.R. and X.B.; Methodology, C.R. and X.B.; Software, C.R.; Validation, C.R.; Formal Analysis, C.R.; Investigation, C.R., X.B. and A.S.; Resources, X.B.; Data Curation, C.R.; Writing—Original Draft Preparation, C.R.; Writing—Review & Editing, C.R., X.B. and A.S.; Visualization, C.R.; Supervision, X.B. and A.S.; Project Administration, X.B.

Funding: This research received no external funding.

Acknowledgments: The authors would like to thank Keysight Technologies for their valuable input and the oscilloscope used in this study.

Conflicts of Interest: The authors declare no conflict of interest.

References

1. Tehranipoor, M.; Koushanfar, F. A survey of hardware trojan taxonomy and detection. *IEEE Des. Test Comput.* **2010**, *27*. [CrossRef]
2. Huffmire, T.; Brotherton, B.; Sherwood, T.; Kastner, R.; Levin, T.; Nguyen, T.D.; Irvine, C. Managing security in FPGA-based embedded systems. *IEEE Des. Test Comput.* **2008**, *25*. [CrossRef]
3. Huffmire, T.; Irvine, C.; Nguyen, T.D.; Levin, T.; Kastner, R.; Sherwood, T. *Handbook of FPGA Design Security*; Springer Science & Business Media: Berlin/Heidelberg, Germany, 2010.
4. Akram, R.N.; Markantonakis, K.; Holloway, R.; Kariyawasam, S.; Ayub, S.; Seeam, A.; Atkinson, R. Challenges of security and trust in avionics wireless networks. In Proceedings of the 2015 IEEE/AIAA 34th Digital Avionics Systems Conference (DASC), Prague, Czech Republic, 13–17 September 2015; p. 4B1-1.
5. Omoogun, M.; Ramsurrun, V.; Guness, S.; Seeam, P.; Bellekens, X.; Seeam, A. Critical patient eHealth monitoring system using wearable sensors. In Proceedings of the 2017 1st International Conference on Next Generation Computing Applications (NextComp), Mauritius, 19–21 July 2017; pp. 169–174. [CrossRef]
6. Bellekens, X.; Seeam, A.; Nieradzinska, K.; Tachtatzis, C.; Cleary, A.; Atkinson, R.; Andonovic, I. Cyber-physical-security model for safety-critical iot infrastructures. In Proceedings of the Wireless World Research Forum Meeting, Santa Clara, CA, USA, 21–23 April 2015.
7. Robinson, W.H.; Reece, T.; Mahatme, N.N. Addressing the challenges of hardware assurance in reconfigurable systems. In Proceedings of the International Conference on Engineering of Reconfigurable Systems and Algorithms (ERSA). The Steering Committee of The World Congress in Computer Science, Computer Engineering and Applied Computing (WorldComp), Las Vegas, NV, USA, 22–25 July 2013; p. 1.
8. Department of Justice Press Release. VisionTech Administrator Sentenced to Prison for Role in Sales of Counterfeit Circuits Destined to US Military. Available online: https://www.ice.gov/news/releases/visiontech-administrator-sentenced-prison-role-sales-counterfeit-circuits-destined-us (accessed on 17 September 2017).
9. Mitra, S.; Wong, H.S.P.; Wong, S. Stopping hardware Trojans in their tracks. *IEEE Spectr.* **2015**. Available online: https://spectrum.ieee.org/semiconductors/design/stopping-hardware-trojans-in-their-tracks (accessed on 22 July 2018).
10. Narasimhan, S.; Du, D.; Chakraborty, R.S.; Paul, S.; Wolff, F.; Papachristou, C.; Roy, K.; Bhunia, S. Multiple-parameter side-channel analysis: A non-invasive hardware Trojan detection approach. In Proceedings of the 2010 IEEE International Symposium on Hardware-Oriented Security and Trust (HOST), Anaheim, CA, USA, 13–14 June 2010; pp. 13–18.
11. Cao, C.; Guan, L.; Liu, P.; Gao, N.; Lin, J.; Xiang, J. Hey, you, keep away from my device: Remotely implanting a virus expeller to defeat Mirai on IoT devices. *arXiv* **2017**, arXiv:1706.05779.
12. Pyrgas, L.; Pirpilidis, F.; Panayiotarou, A.; Kitsos, P. Thermal Sensor Based Hardware Trojan Detection in FPGAs. In Proceedings of the 2017 Euromicro Conference on Digital System Design (DSD), Vienna, Austria, 30 August–1 September 2017; pp. 268–273.
13. Nowroz, A.N.; Hu, K.; Koushanfar, F.; Reda, S. Novel Techniques for High-Sensitivity Hardware Trojan Detection Using Thermal and Power Maps. *IEEE Trans. Comput.-Aided Des. Integr. Circuits Syst.* **2014**, *33*, 1792–1805. [CrossRef]
14. Forte, D.; Bao, C.; Srivastava, A. Temperature tracking: An innovative run-time approach for hardware Trojan detection. In Proceedings of the 2013 IEEE/ACM International Conference on Computer-Aided Design (ICCAD), San Jose, CA, USA, 18–21 November 2013; pp. 532–539. [CrossRef]

15. He, J.; Zhao, Y.; Guo, X.; Jin, Y. Hardware Trojan Detection Through Chip-Free Electromagnetic Side-Channel Statistical Analysis. *IEEE Trans. Very Large Scale Integr. (VLSI) Syst.* **2017**, *25*, 2939–2948. [CrossRef]

16. Shende, R.; Ambawade, D.D. A side channel based power analysis technique for hardware trojan detection using statistical learning approach. In Proceedings of the 2016 Thirteenth International Conference on Wireless and Optical Communications Networks (WOCN), Hyderabad, India, 21–23 July 2016; pp. 1–4. [CrossRef]

17. Xiao, K.; Forte, D.; Jin, Y.; Karri, R.; Bhunia, S.; Tehranipoor, M. Hardware Trojans: Lessons learned after one decade of research. *ACM Trans. Des. Autom. Electron. Syst. (TODAES)* **2016**, *22*, 6. [CrossRef]

18. Kumar, R. *Fabless Semiconductor Implementation*; McGraw-Hill, Inc.: New York, NY, USA, 2008.

19. Adee, S. The hunt for the kill switch. *IEEE Spectr.* **2008**, *45*, 34–39. [CrossRef]

20. Yeh, A. Trends in the global IC design service market. *DIGITIMES Res.* **2012**. Available online: https://www.digitimes.com/news/a20120313RS400.html?chid=2 (accessed on 22 July 2018).

21. Abramovici, M.; Bradley, P. Integrated circuit security: New threats and solutions. In Proceedings of the 5th Annual Workshop on Cyber Security and Information Intelligence Research: Cyber Security and Information Intelligence Challenges and Strategies, Knoxville, TN, USA, 13–15 April 2009; p. 55.

22. Chakraborty, R.S.; Narasimhan, S.; Bhunia, S. Hardware Trojan: Threats and emerging solutions. In Proceedings of the High Level Design Validation and Test Workshop (HLDVT), San Francisco, CA, USA, 4–6 November 2009; pp. 166–171.

23. Collins, D.R. *Trust in Integrated Circuits*; Technical Report; Defense Advanced Research Projects Agency Arlington Va Microsystems Technology Office: Arlington, VA, USA, 2008.

24. Wolff, F.; Papachristou, C.; Bhunia, S.; Chakraborty, R.S. Towards Trojan-free trusted ICs: Problem analysis and detection scheme. In Proceedings of the Conference on Design, Automation and Test in Europe, Munich, Germany, 10–14 March 2008; pp. 1362–1365.

25. Sanno, B. *Detecting Hardware Trojans*; Ruhr-University: Bochum, Germany, 2009.

26. Banga, M.; Hsiao, M.S. A region based approach for the identification of hardware Trojans. In Proceedings of the 2008 IEEE International Workshop on Hardware-Oriented Security and Trust (HOST), Anaheim, CA, USA, 9 June 2008; pp. 40–47.

27. Jin, Y.; Makris, Y. Hardware Trojan detection using path delay fingerprint. In Proceedings of the 2008 IEEE International Workshop on Hardware-Oriented Security and Trust (HOST), Anaheim, CA, USA, 9 June 2008; pp. 51–57.

28. Wang, L.W.; Luo, H.W. A power analysis based approach to detect Trojan circuits. In Proceedings of the 2011 International Conference on Quality, Reliability, Risk, Maintenance, and Safety Engineering, Xi'an, China, 17–19 June 2011; pp. 380–384. [CrossRef]

29. Digilent Basys 3 Artix7 FPGA Board. Available online: https://reference.digilentinc.com/_media/basys3/basys3_sch.pdf (accessed on 7 July 2018).

30. Bao, C.; Forte, D.; Srivastava, A. Temperature tracking: Toward robust run-time detection of hardware Trojans. *IEEE Trans. Comput.-Aided Des. Integr. Circuits Syst.* **2015**, *34*, 1577–1585. [CrossRef]

electronics

MDPI

Article

Impact of Quality of Service on Cloud Based Industrial IoT Applications with OPC UA

Paolo Ferrari [1,*] , **Alessandra Flammini** [1], **Stefano Rinaldi** [1] , **Emiliano Sisinni** [1] , **Davide Maffei** [2] **and Matteo Malara** [2]

[1]   Department of Information Engineering, University of Brescia, Brescia 25123, Italy;
     alessandra.flammini@unibs.it (A.F.); stefano.rinaldi@unibs.it (S.R.); emiliano.sisinni@unibs.it (E.S.)
[2]   Siemens Spa, 20128 Milano, Italy; davide.maffei@siemens.com (D.M.); matteo.malara@siemens.com (M.M.)
*   Correspondence: paolo.ferrari@unibs.it; Tel.: +39-030-371-5445

Received: 9 June 2018; Accepted: 5 July 2018; Published: 9 July 2018

Abstract: The Industrial Internet of Things (IIoT) is becoming a reality thanks to Industry 4.0, which requires the Internet connection of as many industrial devices as possible. The sharing and storing of a huge amount of data in the Cloud allows the implementation of new analysis algorithms and the delivery of new "services" with added value. From an economical point of view, several factors can decide the success of Industry 4.0 new services but, among others, the "short latency" can be one of the most interesting, especially in the industrial market that is used to the "real-time" concept. For these reasons, this work proposes an experimental methodology to investigate the impact of quality of service parameters on the communication delay from the production line to the Cloud and vice versa, when gateways with OPC UA (Open Platform Communications Unified Architecture) are used for accessing data directly in the production line. In this work, the feasibility of the proposed test methodology has been demonstrated by means of a use case with a Siemens S7 1500 controller exchanging data with the IBM Bluemix platform. The experimental results show that, thanks to the proposed method, the solutions based on OPC UA for the implementation of industrial IoT gateways can be easily evaluated, compared and optimized. For instance, during the 14-day observation period of the considered use case, the great impact on performance of the Quality of Service parameters emerged. Indeed, the average communication delay from the production line to the Cloud may vary from less than 90 ms to about 300 ms.

Keywords: industry 4.0; distributed measurement systems; automation networks; node-RED; cloud computing; OPC UA

1. Introduction

Nowadays, the industrial world is using many technologies created for the consumer market and Internet: low-cost sensors, advanced computing and analytics [1]. The surprising level of connectivity supports the so called fourth industrial revolution, promising greater speed and increased efficiency. New embedded sensors/instruments with connectivity features move data from the production site (i.e., the machines) to the Cloud. The collection of data takes place from every production site in the world and, then, the giant quantity of gathered data is analysed. In this way, new services, derived from the analysis of the data, are offered with the intention of improving both the general system and the single machine performance [2]. Internet of Things (IoT) and Industrial Internet of Things (IIoT) are the keywords that label this ongoing evolution process in industrial automation [3]. The concept of Industry 4.0 (which has been addressed as the "fourth industrial (revolution) is involved as well but the terms are often misused and additional remarks are reported in the next section.

Even if there is not a clear distinction between IoT and IIoT (because, formally IIoT is a subset of the IoT), what is usually named as IoT can be considered the "consumer IoT", as opposed to "Industrial

IoT". Consumer IoT is mainly centred around the human beings; indeed the "things" typically are smart appliances interconnected with one another in order to provide improved user awareness of the surrounding environment. On the other side, it is usually said that the aim of IIoT is to integrate Operational Technology (OT) with Information Technology (IT) domains. Differently from consumer IoT, IIoT communications are mainly machine oriented and can involve very different market sectors and activities. As a consequence, despite the most general communication requirements of IoT and IIoT are aimed at large scale connectivity, the specific needs can be very different. For instance, industrial scenarios pay attention to the Quality of Service (QoS, e.g., in terms of determinism and communication delays), the availability and the reliability. Very generally and roughly speaking, it is possible to resume that the IIoT originates when the IoT approach crosses the manufacturing stage of the production cycle.

Unfortunately, the ubiquity of Internet is only one of the aspects of the new era, not even the main one. The most researched subject is the utopic "single protocol" (i.e., accepted by any application market, industry and consumer) that could describe methods and data in a smart and flexible way. There are several examples of shared and widely used protocols in specific application markets and, probably, in industry, the most accepted protocol which harmonizes the machine to machine (M2M) interaction is OPC UA (Open Process Communications Unified Architecture). The OPC Foundation in the past had a great success with the "OPC Classic" and today it is proposing the OPC UA protocol as more powerful successor for its platform independent architecture. OPC UA makes use of most recent concepts to include key features like security, structured information model and auto discovery functions. Thanks to these important foundations, M2M can become "smart" enabling highly flexible scenario where machines could be self-organized. In contrast, the major Cloud platforms for data analytics and sales of services (e.g., Amazon S3, IBM Bluemix, Microsoft Azure, etc.) are still strongly related to the consumer market; they do not natively interface with OPC UA. Usually, gateways/proxies are required because their information model is different. Moreover, OPC UA is for industrial applications and it timestamps any data modification, while, on the contrary, the timely delivering of information among very different applications in Internet with the needed level of flexibility, scalability and geographic coverage is still not possible [4,5]. Considering all these situations, the delay/latency of services may be one of the most important advantages of new services in the near future, especially in the industrial market [6] that is used to real-time. However, despite the fact that it is typically affirmed that an IIoT communication framework is "generally" designed to fulfil timing requirements and packet losses minimization (e.g., for supporting functional safety and self-healing), very few, if any, research exists on experimental verification and tests in real-world scenarios. In particular, we focus on time-related performance, which are considered the most demanding for industrial applications [7].

Accordingly, the target of this paper is to overcome this gap, proposing an experimental methodology to investigate the impact on the delays of the various Quality of Service (QoS) parameters offered by the current Cloud platform, when they are used as sink/source points of data coming from machines with native OPC UA. The paper is voluntary focused only on data transfer passing through usual gateway toward Cloud platform; time for data elaboration in the Cloud is out of the scope of this work. The structure of the paper is the following: Section 2 introduces IIoT and Industry 4.0 scenario and the considered application; Section 3 details the proposed methodology, while Section 4 shows the experimental use case, the experimental results and the discussion about more general considerations. Finally, the conclusions are presented.

2. The Industrial IoT

The "Industry 4.0" tries to increase efficiency in the industry through the exchange/collection of information in the course of the entire product lifecycle. This concept requires the creation of the so called "digital twin" of the product (or manufacturing process), a virtual repository where all the information about each product instance are stored. The union of physical and cyber components is called Cyber-Physical System (CPS) [8]. On the other hand, the consumer market uses IoT devices (smart devices) for providing better efficiency, comfort and safety by means of data and services exchange on a common infrastructure (Internet) with standard interfaces. The IIoT defines the overlapping area where IoT approach is used inside the Industry 4.0 architecture: this circumstance occurs especially when the product is physically created (manufactured), that is in the so called "operation phase".

Industrial automation sector was always a pioneer of innovation and most of industrial systems (machines and plants) have a lot of installed smart field devices. Consequently, there are many connectivity options that can be used today but often, only data related to control are used, while the great amount of other available data is simply dropped. Industry 4.0 proposes: to collect data with consistent data models across the whole plant; to analyse that data extracting the meaningful information; and, lastly, to use services based on the revealed information. In the Industry 4.0 world the field devices and the Cloud can directly communicate. Thus, the new industrial communication hierarchy is flat: services are available for any participants of the automation system [3,9]. This approach changes the situation of applications that now use only local parameters (i.e., related to just one plant or machine e.g., [10,11]), allowing for the scope widening of the input data, since now they can come from other machines/plants worldwide. In addition, the high computational power in the Cloud can be exploited to aggregate/analyse/optimize data before their use in the field.

In the near future, Industry 4.0 applications will activate an endless loop with four main blocks: (1) Device in the field send measurements to the Cloud; (2) measurements are analysed with distributed algorithms in the Cloud; (3) system in the field receives back from the Cloud the optimized parameters; (4) those parameters are used to adapt/improve performance and efficiency of the production system.

2.1. The OPC UA in Industry

Up to very few years ago, the communication systems for industrial automation applications were oriented only to real-time performance suitable for industry and maintainability based on "international standards". Among others, the most diffused industrial protocols today are, for instance; the wired based EtherNET/IP, PROFINET, Powerlink and EtherCAT; and the wireless based IEEE802.11, ISA100.11a and Wireless HART. However, it is known that interoperability between systems of different vendors with different protocols is always difficult because of the incompatible information models for data and services. Industry 4.0 manufacturing systems cannot rely only on such legacy approaches in order to reach the required flexibility level.

The most promising solution for this challenge is the OPC UA of the OPC Foundation [12,13]: OPC UA defines the mode for exchanging information between industrial engineering systems. OPC UA enhances the old OPC Classic with extended features in terms of data modelling, address space architecture, discovery functionalities and security. In OPC UA, servers contain the structured information model that represents the data and the communication model is client-server. Until today, OPC UA has been included in large number of machines and systems and it is a "de facto" reference method for process to process communication. Being more specific, the usual way to proceed is to offer OPC UA as the interface to export data from systems while, internally, the automation at field level is still implemented with traditional fieldbus networks. With OPC UA the data in the address space (described in the following) can be modelled without constrain to a specific communication protocol, obtaining information flows between heterogeneous systems with different data models. In conclusion, OPC UA is the major aspirant to be the backbone protocol for the harmonization of different industrial automation networks and systems [14].

2.2. OPC UA Outline

OPC UA has been designed to facilitate the exchange of information across the hierarchy of systems that commonly coexist in industry: enterprise resource planning (ERP); manufacturing execution systems (MES); control systems; and, last but not least, field devices. OPC UA has a message based communication and a service oriented architecture (SOA) with clients and servers connected to any types of networks.

A client application may use the OPC UA client API (application program interface) in order to send/receive OPC UA service requests/responses to/from the OPC UA server. From the programmer point of view, the OPC UA client API is like an interface that decouples the client application code from the client OPC UA communication stack. In the OPC UA API, there is a discovery service that can be used to find available OPC UA servers and to explore their address space. Clearly, the OPC UA communication stack converts the calls to the OPC UA API to proper messages for the underlying network layers.

In the servers, the OPC UA server API and the OPC UA communication stack are very like the client ones. As additional feature, the server has the so called "address space" in which it can expose the object to be exchanged. In OPC UA, a multiplicity of data structures (called "nodes") can exist, representing, for instance: variables, complex objects, methods (i.e., remotely called functions) and definitions of new types for creating new OPC UA metadata. A hierarchical structure of arbitrary complexity can be created with OPC UA since an object node may contain other variables, objects, methods and so on. In other words, the OPC UA address space is the information model for the communication: real hardware devices or real software "objects" (sensors, actuators, software applications, etc.) are available for OPC UA communication only if they are modelled, added to the address space and finally discovered by the OPC UA clients.

2.3. Quality of Service

The definition of "Quality of Service—QoS" is usually depending on the application field because the "service" may vary from case to case [15]. Generally speaking, in industry, the QoS is often related to timeliness of services or to the guaranteed availability grade of a given service but it should be remembered that QoS timeliness does not necessarily imply a guaranteed delivery. Some examples are:

- the IEC 61850, where the QoS is related to the achievable class of latency in transferring data from the different parts of an electrical Substation Automation System;
- the Ethernet Time Sensitive Networking (TSN) group of standards, where the QoS is referred to the maximum allowed delay for a stream of transferred data;
- the MQTT (Message Queuing Telemetry Transport), where the QoS is tied to the confirmation that a message is delivered to any subscribers (which in turn impact again on latency).

As a consequence, the definition of "quality of service parameters" is even wider, since it refers to any parameters that can affect the QoS of the desired service.

The task of measure the effect of the variation of some parameters on the QoS of a given service could be cumbersome, requiring specific experimental setup for each different application. On the contrary, in this paper a general approach, focused on measuring latency and jitter of services, is proposed for general cases and not bound to any specific industrial applications.

3. The Proposed Approach

In any communication systems, the delay of a data transfer is related to the path of the data. Specifically, the user cannot modify most of the parts within a Cloud based architecture, which appears as "black box". There are two methods for the estimation of the overall delay and its sub-components: simulations can be done as described in [16,17], or experiments can be setup as in [18,19]. However, it has to be highlighted that simulations offer the possibility to finely control the impact of varying the quantity of interest but they can suffer from over simplified models, that do not accurately represent real-world scenarios. For this reason, in this work, a general-purpose test methodology based on experimental approach is designed and discussed that can fit several situations that are actually found in real plants. In particular, as previously stated, the focus is on the evaluation of time-related metrics.

Description of the Experimental Method

In the typical IIoT service scenario there is a machine that sends its data to the Cloud; there, data are elaborated together with other information of other machines; then, the outcome is "sold" to be used in some other field application/machine and, for this reason, it has to come back to the production field. It should be noted that, compared to IoT applications for mobile devices [20], the considered IIoT scenario is simpler.

The proposed method is tailored for the described situation and it is intended to measure the delay of the communication between: a data originator in the field (machine) and a data destination in the Cloud; and, on the reverse route, a data originator in the Cloud and a data destination in the field. The focus of this paper is to evaluate only the delay due to communication without taking into account elaboration of data; for this reason, the originator and the destination are the same machine/cloud application. Moreover, the sample data are sent with a loopback as soon as they reach the Cloud.

The block diagram of the proposed approach is shown in Figure 1. There are three players: the Machine; the IIoT Gateway; and the Cloud Application. Two OPC UA nodes A and B are created such that in their properties they include an array of timestamp values (*T1* to *T8*). Nodes are sent from the Machine, the IIoT Gateway with OPC UA protocol and from the IIoT Gateway to the Cloud Application using one of the most diffused IoT messaging protocol (e.g., MQTT-Message Queuing Telemetry Transport). As soon as the nodes A and B pass in one of the timestamping points (labelled in the picture with the name of the corresponding timestamp), the matching timestamp value is loaded with the time of the operating system local clock. The Machine publishes data A at time *T1*; the IIoT Gateway gets it at time *T2*; closely after, at time *T3 > T2*, the IIoT Gateway sends A to the Cloud Application that finally receives it at time *T4*. Immediately, the cloud application copies back the object A to object B and it starts the reverse path sending back B at *T5* to the IIoT Gateway. When B is received at *T6* in the IIOT Gateway, it is routed to the Machine as a OPC UA node with departure time *T7* and arrival time *T8*. After a complete roundtrip, the data structure contains all the timestamps (which are useful for the estimation of the delays).

The first important metric is the OPC UA end-to-end delay (OD) that can be calculated in the two directions using the timestamps *T1* and *T2* from Machine to Gateway and using timestamps *T7* and *T8* from Gateway to Machine:

$$OD_{MG} = T2 - T1 \tag{1}$$

$$OD_{GM} = T8 - T7 \tag{2}$$

The second important metric is the Cloud messaging protocol end-to-end delay (MD) that is obtained for the two directions using the timestamps *T3* and *T4* from Gateway to Cloud and using timestamps *T5* and *T6* from Cloud to Gateway:

$$MD_{GC} = T4 - T3 \tag{3}$$

$$MD_{CG} = T6 - T5 \tag{4}$$

The last metric is the total end-to-end communication delay ED from Machine to the Cloud and from Cloud to Machine, which is just the sum of the previously calculated partial paths:

$$ED_{MC} = OD_{MG} + MD_{GC} \tag{5}$$

$$ED_{CM} = MD_{CG} + OD_{GM} \tag{6}$$

Summarizing, the proposed experimental approach allows for the characterization of the communication delay of a generic industrial IoT application that uses OPC UA to access generic Cloud services. The three metrics presented above are simultaneously available in the experiments and can be used as performance indexes for the comparison of different setups.

In particular, the effect of different QoS parameter settings on the performance indexes of the communication delay will be studied in this paper, by means of a suitable use case.

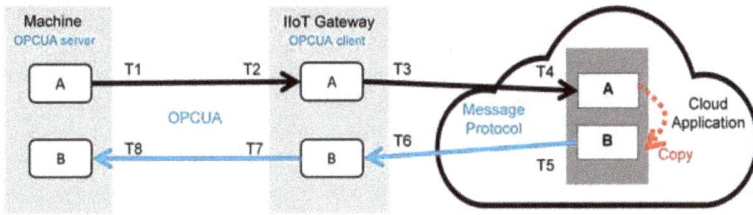

Figure 1. The block diagram of the setup for the experiment about the impact of quality of service parameters on the communication delay between a Machine with OPC UA interface and a Cloud platform. Black arrows show the path from the Machine to the Cloud, while blue arrows represent the reverse path. The red arrow is the software loop in the cloud that enable automatic bidirectional experiments.

4. An OPCUA Use Case

In order to show the applicability of the proposed measurement methodology, a sample use case is considered. It is clear that the obtained results depend on all the blocks, components and software composing this specific use case.

The experimental setup is realistically based on commercially available components and software for industrial automation. The Machine in the field uses a Siemens S7-1516 controller with embedded OPC UA communication stack. The IIoT Gateway has been built with a Siemens IOT2040 (embedded device with Yocto Linux-kernel 2.2.1, Intel Quark X1020, 1 GB RAM, 8GB SD disk, 2 × Ethernet ports, 2 × RS232/485 interfaces, battery-backup RTC). The considered Cloud platform is the IBM Bluemix; it runs the "Internet of Things Platfom-m8" service, while it uses "Node-RED" framework for programming the data transfer/elaboration. IBM Bluemix has several access/use profiles: in this paper, the free access version is used, resulting in limited feature in terms of Cloud computational resources (which are out of the scope of this paper). However, no limitation from the communication features are mentioned in the contract.

The IOT2040 Gateway is attached to the same network of the Machine (in Milan at the Siemens Spa headquarter), hence the local area network introduces a negligible delay. Last, the Siemens network connection to the Internet has a very high quality with extended availability. As an additional remark, it has to be considered that network paths are not always guaranteed to be the same, due to the well know internet asymmetry [21]; however, the proposed methodology can identify such an asymmetry as well.

4.1. Node-RED Flows for the Experimental Setup

The experiments have been carried out using specific Node-RED flows. Node-RED is the graphic tool established by IBM for "wiring together hardware devices, APIs and online services in new and interesting ways" (from Node-RED website, https://nodered.org/). Node-RED uses the famous JavaScript runtime Node.js and by means of a browser based editor, allows for drag and drop, connection and configuration of "nodes" (i.e., open-source functions and interfaces). Thus, a program in Node-RED describes the data flow from source to destination passing through elaboration steps; for this reason, it is simply called "flow". Node-RED is an efficient option for applications that are aimed to prototype some IoT connectivity. In the specific application of the paper, Node.js has been considered as reference platform for the experimental implementation of the use case due to a trade-off between effectiveness and cost of human resources (programmers). The impact of Node.js can be estimated around 10% of the processing power of the gateway used in the demonstration use case.

The number of devices connected to the gateway linearly increases the CPU and memory usage. As a consequence, Node.js with embedded devices is only recommended for small projects, with not demanding requirements, or for experimental test environment, as in the case of the current research.

In this paper, Node-RED is used both in the IIoT Gateway and in the Cloud application. The flows that implement the formulas/algorithms described in the previous section are shown in Figure 2 for IIoT Gateway along the path from Machine to Cloud, Figure 3 for IIoT Gateway along the path from Cloud to Machine and Figure 4 for the Cloud application.

It should be noted that the main limit of the OPC UA solution is currently the high computational power required by OPC UA stacks. Commonly available PLCs support a limited number of concurrent connections. In addition, the OPC UA implementation for Node RED in the use case is resource consuming also in the Gateway. Moreover, the Internet connection bandwidth could also affect the delays. Since this paper is focused on the measurement methodology and on the demonstration of the feasibility of such methodology by means of a sample use case, further investigations about different use cases are out of scope.

The flows are voluntarily kept as simple as possible to reduce computational overhead. The Gateway is forwarding the data object (containing the timestamp values as properties) to the Cloud using nodes of the OPC UA Library and IBM Bluemix Library. The properties of the data object are updated at the corresponding timestamping points. The Cloud application sends back any incoming data object after a fixed delay of 60 s. The data object is regularly sampled in the flows and saved to files for backup purpose; in case the complete path is not available, partial metrics can still be computed.

In this paper, for the configuration of the connection to the IBM Cloud, several QoS settings have been used and compared:

- "quickstart" mode (unregistered user)
- QoS = 0 with the "registered user" mode
- QoS = 1 with the "registered user" mode.

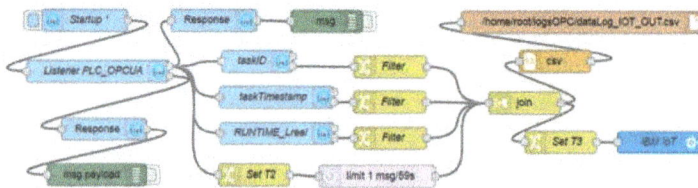

Figure 2. Node-RED flow for the IIoT Gateway related to the path from Machine to Cloud. The timestamps *T2* and *T3* are taken in this flow.

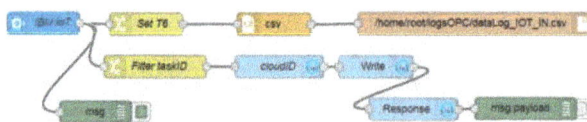

Figure 3. Node-RED flow for the IIoT Gateway related to the path from Cloud to Machine. The timestamps *T6* and *T7* (*T7* = *T6*) are taken in this flow.

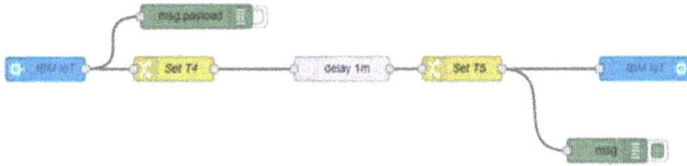

Figure 4. Node-RED flow for the Cloud application. Data is received from the IIoT Gateway and sent back after a fixed delay (one minute). The timestamp *T4* and *T5* are taken in this flow.

4.2. Synchronization Uncertainty in the Experimental Setup

In the proposed experimental methodology, the desired metrics are obtained combining timestamps taken by different actors working in a distributed system. Hence, the measurement uncertainty depends on several factors; among those, the major contributions are: the synchronization uncertainty among actors; the frequency uncertainty of the local oscillator in the physical device that takes the timestamp; and the uncertainty of the software delay of routines that correlate the event and the snapshot of the local clock to obtain the "timestamp".

Fortunately, the entire transaction (from Machine to Cloud and back) is expected to last few seconds; as a result, the contribution of the local oscillator uncertainty can be neglected (usually crystal deviation on short period is in the order of few parts per million).

In this paper, the Machine, the IIoT Gateway and the Cloud platform use the Coordinated Universal Time (UTC) as time reference. The NTP (Network Transfer Protocol) has been used to synchronize each system with a local time server locked to GPS clock. The synchronization uncertainty using NTP may vary with the quality of the network connection in terms of latency [22,23] but all the considered systems are connected via a local area network with time server, reducing the variability to the minimum. Nevertheless, the synchronization uncertainty of the devices involved in this use case is estimated in different ways. For the IIoT Gateway, the time synchronization is experimentally measured considering the residual time offset after compensation, as listed in the NTP statistics [23]. For the Machine, the equivalent synchronization uncertainty is derived from Siemens documentation that sets the maximum error to 0.2 ms and by supposing the error distribution is uniform. Only the synchronization uncertainty of the IBM Cloud platform is difficult to be estimated, since no documents or literature is available on this topic. Anyway, it is hard to think that IBM cloud servers are worse than the other actors of the experiment case. For these reasons, the synchronization uncertainty of the IBM Cloud platform has been considered equal to the contribution of the IIoT Gateway.

The synchronization standard uncertainty results are shown in Table 1, where the experimental standard uncertainty is evaluated as the worst case $u_{sn} = \sqrt{\mu_{sn}^2 + \sigma_{sn}^2}$ because the systematic error μ_{sn} introduced by the operating system is not compensated. Anyway, the resulting synchronization uncertainty is always less than 0.1 ms with respect to UTC.

The timestamping uncertainty has been experimentally estimated at application level in the three actors. In details, a suitable software routine triggers a software delay at time *T9* and takes a timestamp *T10* when such a delay expires. Since timestamps and delay are obtained with the local system clock, the quantity $\Delta = T10 - T9$ should be theoretically identical to the imposed delay value. In truth, the timestamp uncertainty u_{tn} affects both T9 and T10. Including the systematic error μ_Δ, the timestamping uncertainty is $u_{tn} = \sqrt{\frac{\mu_\Delta^2 + \sigma_\Delta^2}{2}}$. Table 2 shows the timestamp standard uncertainty u_{tn} of the considered system.

In particular, if the proposed approach is used (Equation 1 to Equation 6 are considered), the standard uncertainty u_{mn} of any delay calculated between any two points (*n* and *m*), in the flow in Figure 1, is modelled as $u_{mn}^2 = u_{sm}^2 + u_{tm}^2 + u_{sn}^2 + u_{tn}^2$. It is clear that it is always dominated by the timestamp uncertainty.

Table 1. Synchronization uncertainty over an observation time of 14 days (ms).

Source	u_{sn}	μ_{sn}	σ_{sn}
Machine (theor.)	0.06	0	0.06
Gateway (exper.)	0.07	0.001	0.07
Cloud (supposed)	0.07	-	-

Table 2. Timestamp uncertainty over an observation time of 14 days (ms).

Source	u_{tn}	μ_Δ	σ_Δ
Machine (exper.)	0.01	0.016	0.01
Gateway (exper.)	4.2	4.589	3.8
Cloud (exper.)	5.9	4.214	7.3

4.3. Use Case Results

The experimental setup was aimed at the estimation of all the metrics used as performance indicators by means of a single experiment. The experiment has been repeated changing the QoS settings as described in Section 4.1. The measurement campaigns took 14 days. A new measure loop is started every minute; when the roundtrip ends, all the delay values regarding the run are stored. Each experimental campaign has more than 8000 valid samples and the resulting performance indicators are reported in Table 3. The results of Table 3 are discussed in the following Sections 4.4 and 4.5, where the probability density functions of the measurements are also shown.

Table 3. Delay for the considered use case over an observation time of 14 days (ms).

Path		"Quickstart"		QoS = 0 "Registered User"		QoS = 1 "Registered User"	
		Mean	St. Dev.	Mean	St. Dev.	Mean	St. Dev.
Machine to Cloud	OD_{MG}	135	53	173	61	176	64
	MD_{GC}	155	39	47	20	89	33
	ED_{MC}	289	67	219	62	265	69
Cloud to Machine	MD_{CG}	152	17	64	97	99	24
	OD_{GM}	16	6	18	12	17	10
	ED_{CM}	168	18	82	97	116	26

4.4. Discussion of the Use Case Result about Overall Delays

The detailed discussion of the overall delays is carried out only under the QoS setting called "quickstart", that is the basic setting offered by the Cloud platform of this use case.

The probability density function estimates of the OPC UA delay in the two directions (Machine to Gateway, OD_{MG} e Gateway to Machine, OD_{GM}) are shown in Figure 5. In the Gateway to Machine direction there is a single, well defined, peak centred in the mean value; probably, the reception timestamp is assigned directly in the reception interrupt inside the Machine PLC, because the mean time is low. On the other hand, in the Machine to Gateway, direction the distribution shows three peaks with equal inter-distance of almost exactly 100 ms (please note that the peak centred at 280 ms is almost not visible in Figure 5). Given the analogy with similar situations [24], a multimodal shape of the OD_{MG} distribution signifies that at least one of the tasks that manage the OPC UA data in the Machine to Gateway operates on the basis of discrete cycle times. Under such a hypothesis, it may be inferred that the OPC UA task inside the Machine PLC is executed cyclically every 100 ms and timestamps are assigned inside that task.

The probability density function estimates of Cloud messaging protocol end-to-end delay (MD) in the two directions (Machine to Gateway e Gateway to Machine) are shown in Figure 6. The behaviour is similar and the two distributions have only two main peaks. Finally, the probability density function estimates of the total end-to-end communication delay ED from Machine to the Cloud and from Cloud to Machine are shown in Figure 7. As expected, the distributions show several peaks, because they are the convolution of the distributions in Figures 5 and 6.

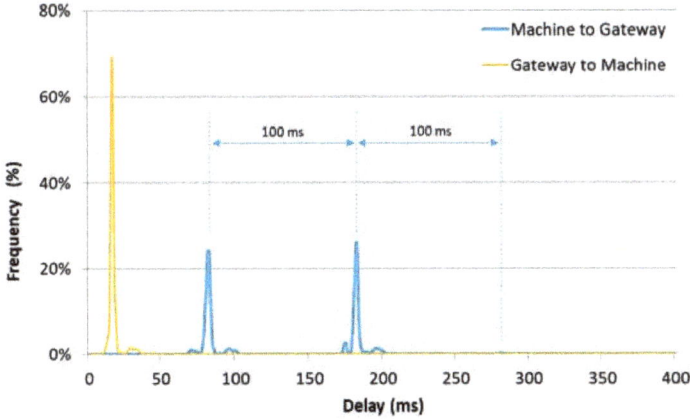

Figure 5. Probability density function estimate of the OPC UA end-to-end delay (OD) in the two directions (Machine to Gateway e Gateway to Machine).

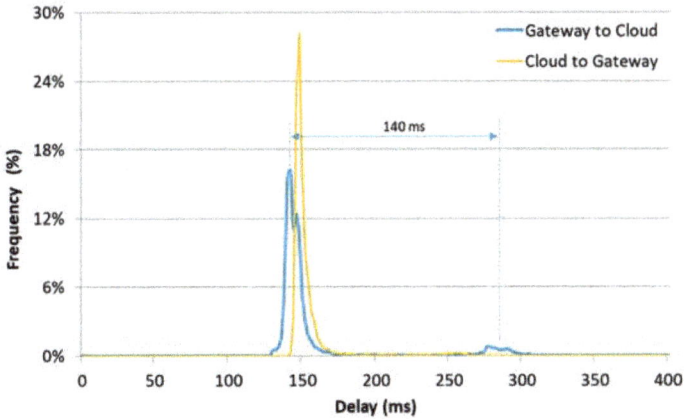

Figure 6. Probability density function estimate of the Cloud messaging protocol end-to-end delay (MD) in the two directions (Machine to Gateway e Gateway to Machine).

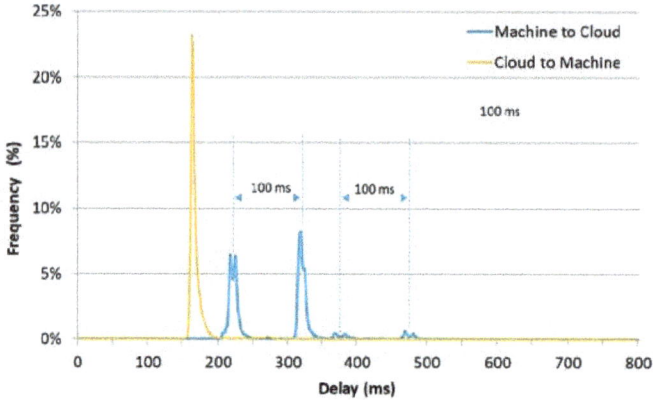

Figure 7. Probability density function estimate of the total end-to-end communication delay ED from Machine to the Cloud and from Cloud to Machine.

4.5. Discussion of the Use Case Result Varying QoS Settings

The three QoS settings described in Section 4.1 can be compared.

The delay OD_{MG} and OD_{GM} vary of few milliseconds in the three situations. The QoS settings (that only influence the Gateway to Cloud and Cloud to Gateway paths) do not affect the OPC UA drivers of the Machine and of the Gateway. For sake of completeness, it should be said that the OD_{MG} distribution maintain the equidistant peaks with any QoS setting.

The probability density function estimates of the MD_{GC} are shown in Figure 8, while the probability density function estimates of MD_{CG} are shown in Figure 9. Here, the three distributions are clearly different. The behaviour of QoS 1 and QoS 0 in registered mode are typical of MQTT data transfers across Internet (as shown in [25–27]): the QoS 0 is faster but it has a larger standard deviation, while QoS 1 distribution is narrow with an higher mean value. The probability density function of the "quickstart" mode has the same shape of the QoS 1 in registered mode but has the highest mean value.

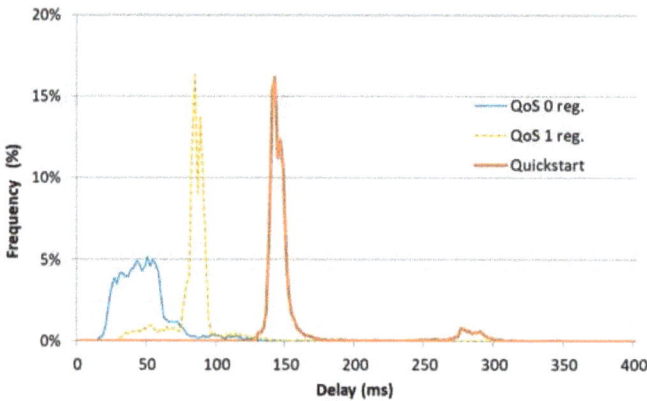

Figure 8. Probability density function estimate of the Gateway to Cloud delay (MD_{GC}) using different QoS parameters in IBM Bluemix.

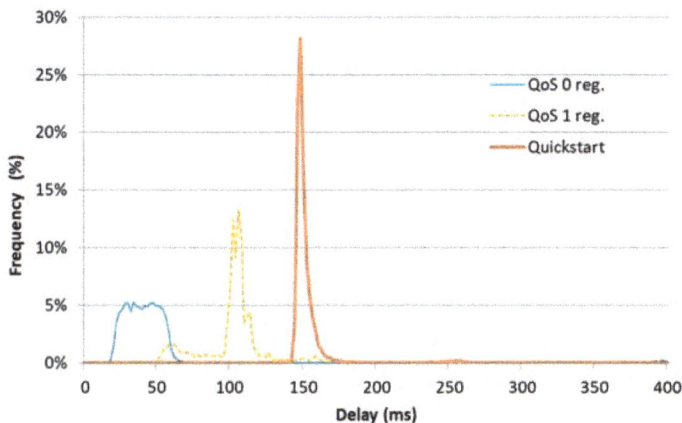

Figure 9. Probability density function estimate of the Cloud to Gateway delay (MD_{CG}) using different QoS parameters in IBM Bluemix.

Moreover, during the experiments, there were some anomalous delay values for MD_{GC} and MD_{CG}; few samples (i.e., <0.2%) were well above three times the standard deviation and were marked as outliers. A careful analysis revealed that the anomalies appear only at the same hour during the night on different days and the number of outliers were greater in the "quickstart" mode than in the QoS 1 registered. In conclusion, the sporadic anomalous delays may be due to the "free access" version of IBM Bluemix platform that has no guaranteed quality of service but it is clear that, among others, the "quickstart" mode nodes have the lower priority.

4.6. Generalization of the Use Case Results

The sample use case demonstrates the feasibility of the proposed measurement methodology, giving also some directions for more general applications.

First of all, the used PLC (Siemens S7 1516) is a medium performance PLC whose characteristics can be found also in PLCs of other producers, including the OPC UA support; for this reason, the proposed approach can be implemented also with other control systems.

The gateway architecture that has been implemented with a Node.js platform can be easily ported to different hardware (e.g., Raspberry PI devices) and improved, from the performance point of view, by optimizing the OPC UA library. However, the logic flow of the measurement methodology remains the same.

Last, the cloud platform can be changed (e.g., Microsoft Azure, or Amazon S3) provided that a suitable messaging protocol can be supported. MQTT is generally available but AMQP (Advanced Message Queuing Protocol) can be also considered. Anyway, the proposed methodology is not bonded to a specific messaging protocol, guaranteeing the consistency of the results.

Moreover, three important general observations arise from analysing the results of the use case.

The QoS of the OPC UA stack is clearly related to the implementation inside the PLC and, thus, their up/down scale is expected depending on the ratio between PLC computational performance and computational load.

The concern about performance of the Cloud is more about the availability (lack of timely response in some cases) than latency. The QoS at this level is closely related to the messaging protocol performance through the Internet, while Cloud computational power practically does not affect the results (since the proposed methodology correctly decouples it).

The performance of the Gateway is not stressed in the considered use case but a reasonable dependency of the results from the throughput (in terms of message per second) is expected. In large applications, the gateway must scale his performance accordingly with the desired number of data exchanges with the cloud.

5. Conclusions

The success of Industrial Internet of Things (IIoT), from the economical point of view, depends on several factors. Among others, the "short latency" can be one of the most interesting, especially in the industrial market that is used to the "real-time" concept. This paper deals with a methodology to measure time delay metrics in OPC UA systems in order to study the impact that quality of service parameters have on the communication delay from the production line to the Cloud and vice versa. By means of a sample use case, the proposed method was applied, its feasibility was demonstrated and the results are generalized. In the use case, a Gateway exploiting the widely accepted OPC UA was used for data access directly in the devices inside the production line. The experimental results show that the overall delay is always bound to less than 300 ms, while the impact of the QoS parameters on the communication delay is clearly visible. The major experimental evidence in the medium term (14 days) is that the average delay from production line to Cloud is tightly related to the QoS settings of the IBM Bluemix platform. The "quickstart" mode has the worst performance with an average delay of 290 ms from Machine to Cloud and 170 ms in the opposite direction. If QoS 0 "registered user" mode is used, the average delays decrease, respectively, down to 220 ms and 80 ms. The QoS 1 "registered user" mode has higher delays but a lower standard deviation. Finally, it should be highlighted that the "free access" version of IBM Bluemix platform has no guaranteed quality of service: some samples (>0.2%) may be delayed by several seconds.

Author Contributions: Conceptualization, P.F.; Data curation, S.R. and E.S.; Formal analysis, E.S.; Funding acquisition, A.F.; Investigation, M.M.; Methodology, P.F.; Project administration, P.F. and A.F.; Software, S.R., D.M. and M.M.; Supervision, A.F.; Validation, P.F. and D.M.; Writing–original draft, P.F., S.R. and E.S.

Funding: The research has been partially funded by research grant MIUR SCN00416, "Brescia Smart Living: Integrated energy and services for the enhancement of the welfare" and by University of Brescia H&W grant "AQMaSC"

Acknowledgments: The authors would like to thank Siemens Italy Spa for hosting the experimental setup.

Conflicts of Interest: The authors declare no conflict of interest.

References

1. Xu, L.D.; He, W.; Li, S. Internet of things in industries: A survey. *IEEE Trans. Ind. Inf.* **2014**, *10*, 2233–2243. [CrossRef]
2. Tao, F.; Zuo, Y.; Xu, L.D.; Zhang, L. IoT-Based intelligent perception and access of manufacturing resource toward Cloud manufacturing. *IEEE Trans. Ind. Inf.* **2014**, *10*, 1547–1557.
3. Liu, Y.; Xu, X. Industry 4.0 and Cloud manufacturing: A comparative analysis. *J. Manuf. Sci. Eng.* **2017**, *139*, 034701. [CrossRef]
4. Yang, C.; Shen, W.; Wang, X. Applications of Internet of Things in manufacturing. In Proceedings of the 2016 20th IEEE International Conference on Computer Supported Cooperative Work in Design (CSCWD), Nanchang, China, 4–6 May 2016.
5. Bellagente, P.; Ferrari, P.; Flammini, A.; Rinaldi, S.; Sisinni, E. Enabling PROFINET devices to work in IoT: Characterization and requirements. In Proceedings of the 2016 IEEE Instrumentation and Measurement Technology Conference Proceedings (I2MTC), Taipei, Taiwan, 23–26 May 2016.
6. Szymanski, T.H. Supporting consumer services in a deterministic industrial internet core network. *IEEE Commun. Mag.* **2016**, *54*, 110–117. [CrossRef]
7. Wollschlaeger, M.; Sauter, T.; Jasperneite, J. The Future of Industrial Communication: Automation Networks in the Era of the Internet of Things and Industry 4.0. *IEEE Ind. Electron. Mag.* **2017**, *11*, 17–27. [CrossRef]
8. Stojmenovic, I. Machine-to-machine communications with in-network data aggregation, processing, and actuation for large-scale cyber-physical systems. *IEEE Internet Things J.* **2014**, *1*, 122–128. [CrossRef]

9. Battaglia, F.; Iannizzotto, G.; Lo Bello, L. JxActinium: A runtime manager for secure REST-ful CoAP applications working over JXTA. In Proceedings of the 31st Annual ACM Symposium on Applied Computing, Pisa, Italy, 4–8 April 2016; pp. 1611–1618.
10. Rinaldi, S.; Ferrari, P.; Brandao, D.; Sulis, S. Software defined networking applied to the heterogeneous infrastructure of Smart Grid. In Proceedings of the 2015 IEEE World Conference on Factory Communication Systems (WFCS), Palma de Mallorca, Spain, 27–29 May 2015.
11. Fernandes, F.; Sestito, G.S.; Dias, A.L.; Brandao, D.; Ferrari, P. Influence of network parameters on the recovery time of a ring topology PROFINET network. *IFAC Pap. Online* **2016**, *49*, 278–283. [CrossRef]
12. *OPC Unified Architecture-Part 1: Overview and Concepts*; OPC Foundation: Scottsdale, AZ, USA, 2015.
13. *OPC Unified Architecture-Part 3: Address Space Model*; OPC Foundation: Scottsdale, AZ, USA, 2015.
14. Lee, B.; Kim, D.K.; Yang, H.; Oh, S. Model transformation between OPC UA and UML. *Comput. Standard. Interfaces* **2017**, *50*, 236–250. [CrossRef]
15. Liao, Y.; Leeson, M.S.; Higgins, M.D.; Bai, C. Analysis of In-to-Out Wireless Body Area Network Systems: Towards QoS-Aware Health Internet of Things Applications. *Electronics* **2016**, *5*, 38. [CrossRef]
16. Collina, M.; Bartolucci, M.; Vanelli-Coralli, A.; Corazza, G.E. Internet of Things application layer protocol analysis over error and delay prone links. In Proceedings of the 2014 7th Advanced Satellite Multimedia Systems Conference and the 13th Signal Processing for Space Communications Workshop (ASMS/SPSC), Livorno, Italy, 8–10 September 2014; pp. 398–404.
17. Govindan, K.; Azad, A.P. End-to-end service assurance in IoT MQTT-SN. In Proceedings of the 2015 12th Annual IEEE Consumer Communications and Networking Conference (CCNC), Las Vegas, NV, USA, 9–12 January 2015; pp. 290–296.
18. Mijovic, S.; Shehu, E.; Buratti, C. Comparing application layer protocols for the Internet of Things via experimentation. In Proceedings of the 2016 IEEE 2nd International Forum on Research and Technologies for Society and Industry Leveraging a better tomorrow (RTSI), Bologna, Italy, 7–9 September 2016; pp. 1–5.
19. Forsstrom, S.; Jennehag, U. A performance and cost evaluation of combining OPC-UA and Microsoft Azure IoT Hub into an industrial Internet-of-Things system. In Proceedings of the 2017 Global Internet of Things Summit (GIoTS), Geneva, Switzerland, 6–9 June 2017.
20. Pereira, C.; Pinto, A.; Ferreira, D.; Aguiar, A. Experimental Characterization of Mobile IoT Application Latency. *IEEE Internet Things J.* **2017**, *4*, 1082–1094. [CrossRef]
21. He, Y.; Faloutsos, M.; Krishnamurthy, S.; Huffaker, B. On routing asymmetry in the Internet. In Proceedings of the GLOBECOM'05 IEEE Global Telecommunications Conference 2005, St. Louis, MO, USA, 28 November–2 December 2005.
22. Rinaldi, S.; Della Giustina, D.; Ferrari, P.; Flammini, A.; Sisinni, E. Time synchronization over heterogeneous network for smart grid application: Design and characterization of a real case. *Ad Hoc Netw.* **2016**, *50*, 41–45. [CrossRef]
23. Sherman, J.A.; Levine, J. Usage Analysis of the NIST Internet Time Service. *J. Res. Nat. Inst. Stand. Technol.* **2016**, *121*, 33–46. [CrossRef]
24. Ferrari, P.; Flammini, A.; Marioli, D.; Taroni, A.; Venturini, F. Experimental analysis to estimate jitter in PROFINET IO Class 1 networks. In Proceedings of the 2006 IEEE Conference on Emerging Technologies and Factory Automation (ETFA), Prague, Czech Republic, 20–22 September 2006; pp. 429–432.
25. Ferrari, P.; Sisinni, E.; Brandao, D.; Rocha, M. Evaluation of communication latency in industrial IoT applications. In Proceedings of the 2017 IEEE International Workshop on Measurements and Networking (M&N), Naples, Italy, 27–29 September 2017; pp. 17–22.
26. Persico, V.; Botta, A.; Marchetta, P.; Montieri, A.; Pescapé, A. On the performance of the wide-area networks interconnecting public-cloud datacenters around the globe. *Comput. Networks* **2017**, *112*, 67–83. [CrossRef]
27. Ferrari, P.; Flammini, A.; Sisinni, E.; Rinaldi, S.; Brandão, D.; Rocha, M.S. Delay Estimation of Industrial IoT Applications Based on Messaging Protocols. *IEEE Trans. Instrum. Meas.* **2018**, *PP*, 1–12. [CrossRef]

electronics

MDPI

Article

Open-Source Hardware Platforms for Smart Converters with Cloud Connectivity

Massimo Merenda [1,2,*], **Demetrio Iero** [1,2], **Giovanni Pangallo** [1], **Paolo Falduto** [1], **Giovanna Adinolfi** [3], **Angelo Merola** [3], **Giorgio Graditi** [3] and **Francesco G. Della Corte** [1,2]

[1] DIIES Department, University Mediterranea of Reggio Calabria, 89126 Reggio Calabria, Italy; demetrio.iero@unirc.it (D.I.); giovanni.pangallo@unirc.it (G.P.); paolo.falduto.572@studenti.unirc.it (P.F.); francesco.dellacorte@unirc.it (F.G.D.C.)

[2] HWA srl, Spin-off University Mediterranea of Reggio Calabria, Via R. Campi II tr. 135, 89126 Reggio Calabria, Italy

[3] ENEA C.R. Portici, Piazzale Enrico Fermi 1, Portici, 80055 Naples, Italy; giovanna.adinolfi@enea.it (G.A.); angelo.merola@enea.it (A.M.); giorgio.graditi@enea.it (G.G.)

* Correspondence: massimo.merenda@unirc.it; Tel.: +39-0965-1693-441

Received: 24 February 2019; Accepted: 22 March 2019; Published: 26 March 2019

Abstract: This paper presents the design and hardware implementation of open-source hardware dedicated to smart converter systems development. Smart converters are simple or interleaved converters. They are equipped with controllers that are able to online impedance match for the maximum power transfer. These conversion systems are particularly feasible for photovoltaic and all renewable energies systems working in continuous changing operating conditions. Smart converters represent promising solutions in recent energetic scenarios, in fact their application is deepening and widening. In this context, the availability of a hardware platform could represent a useful tool. The platform was conceived and released as an open hardware instrument for academy and industry to benefit from the improvements brought by the researchers' community. The usage of a novel, open-source platform would allow many developers to design smart converters, focusing on algorithms instead of electronics, which could result in a better overall development ecosystem and rapid growth in the number of smart converter applications. The platform itself is proposed as a benchmark in the development and testing of different maximum power point tracking algorithms. The designed system is capable of accurate code implementations, allowing the testing of different current and voltage-controlled algorithms for different renewable energies systems. The circuit features a bi-directional radio frequency communication channel that enables real-time reading of measurements and parameters, and remote modification of both algorithm types and settings. The proposed system was developed and successfully tested in laboratory with a solar module simulator and with real photovoltaic generators. Experimental results indicate state-of-art performances as a converter, while enhanced smart features pave the way to system-level management, real-time diagnostics, and on-the-flight parameters change. Furthermore, the deployment feasibility allows different combinations and arrangements of several energy sources, converters (both single and multi-converters), and modulation strategies. To our knowledge, this project remains the only open-source hardware smart converter platform used for educational, research, and industrial purposes so far.

Keywords: smart converter; maximum power point tracking (MPPT); photovoltaic (PV) system; Field Programmable Gate Array (FPGA); Digital Signal Processor (DSP); interleaved; DC/DC converter; distributed energy resource

1. Introduction

The inexorable reduction of fossil fuels and non-renewable energy sources feeds the wind of change that leads to substantial use of renewable energy to such a point that many countries have now defined a roadmap for replacing polluting traditional energy sources with clean and renewable plants [1]. Among renewable sources, photovoltaic energy is a primary choice because of the energy density that affects every part of the Earth, the maturity of conversion technologies, and the low cost/watt.

Despite its advantages, photovoltaics suffers from the discontinuity of the resource and the inability of large accumulations, if not at very high costs. Furthermore, existing solar technologies have very low returns compared to other production facilities. Therefore, the need to maximize the yield of individual generators is of primary importance, regardless of the grid connection, in order to reduce the need to use large and extensive surfaces of ground or roofing, as in the case of solar systems, or expensive and, in general, over-dimensioned plants.

With this in mind, smart converter devices provide a suitable response to the need to maximize the produced energy. They work increasing the instantaneous production also responding to variations in weather and working operating conditions [2]. There are many open points still to be solved or improved to ensure an optimal system response, even in variable conditions such as partial shading, non-uniformity of PV panel temperatures, dust effects for solar panels [3], and the discontinuity of the wind [4].

The efficiency of various MPPT algorithms has been compared in several papers [5], showing that perturbation and observation (P&O) and incremental conductance (IC) methods are simple and often applied, despite their slow tracking and low utilization efficiency. Minor modifications of the P&O and IC have been proposed by several authors [6] to reduce the known drawbacks of the aforementioned algorithms. Fuzzy and neural network techniques are used nowadays and show an increased efficiency [7].

In this scenario, algorithms are mainly computer-simulated, e.g., using specific engineering software tools (Matlab [8], Spice [9]) and environments that provide easily deployable algorithms; discussions on efficiency and results comparison are always pursued on different hardware platforms, leading to barely comparable conclusions and results due to the heterogeneity of the hardware used and the analysis conditions.

A possible solution to overcome these problems consists in the development of a flexible platform exploitable for the implementation, debugging, and the testing of smart converters algorithms. This topic has acquired prominence since these conversion systems are widely applied in current energetic contexts. In the literature, different papers propose open-access platforms. Currently, none of them have been dedicated to smart DC/DC converters.

In [10], a control platform for soft-switching multi-level inverters is proposed. The paper presents an FPGA-based prototyping tool for modular DC/AC converters. A verification of concept case is reported; the implemented hardware is not open or accessible.

In this work, we propose a novel open-source electronics platform for research, industrial, and educational purposes. The platform has been conceived and released as an open hardware and software instrument for the academic and industrial communities. It can be used by students for a deep comprehension of DC/DC converters operating and control modes. Researchers can take advantage of the proposed architecture for modulation and control strategies validation. It can be also used by designers for a rapid project check. The platform itself is proposed as a benchmark in the development and testing of different smart converters algorithms.

This hardware configuration and embedded software functions are freely available to benefit from users' community improvements.

The novelty of this approach also consists in the cloud connectivity feature. In fact, the platform is equipped with a Wi-Fi communication module to enable data gathering and dispatching. Users can access a suitable application programming interface to validate algorithms online and control techniques changing the smart converter functional parameters (duty cycle, switching frequency, dead time, algorithms, etc). In addition, the Wi-Fi communication module permits one to implement diagnostic functions to verify smart converters behavior during the design process, but also to monitor and acquire operating data in functioning mode. This feature is crucial, especially when the system is being tested in the field.

So, the proposed platform represents an innovative hardware and a valid pre-prototyping tool for DC/DC converters.

The board is populated with a 5000 LookUp Table (LUT) Complementary Metal-Oxide Semiconductor (CMOS) FPGA integrated circuit (iCE40UP5K UltraPlus™, Lattice Semiconductor, [11]), an embedded DSPic (dsPIC33EPXXGS50X, Microchip Technology INC., [12]), and a Silicon Labs (WGM110 [13]) Wi-Fi module.

The VHDL (Very High Speed Integrated Circuits Hardware Description), the C-code for the DSPic, and the Wi-Fi module are provided under the terms and conditions of the MIT License [14].

The schematic, layout, and bill of materials are also provided on the same public repository on the popular platform Github [15], at the web address [16], released under Creative Commons Share-alike 4.0 International License [17].

The FPGA generates high-frequency PWM signals for controlling Metal-Oxide-Semiconductor Field-Effect Transistors (MOSFETs) with appropriate frequency, dead-time, and duty cycle, and implements bootstrap and interleaved control logic.

The microcontroller acquires currents and voltages from the power circuit and has a bidirectional communication with FPGA and Wi-Fi module.

Both the two logical blocks could implement power conversion algorithms: the FPGA by using high-speed circuits that implement logic at a hardware level and the microcontroller by executing a firmware code within its own CPU. This novel architecture allows the flexibility of algorithm implementation, deployable in hardware, software, or in combined mode.

The separation between the power board and the logic board allows the comparison of different architectures, different device technologies, and the usage of interleaving options.

The sum of the already developed software parts constitute a new software framework, based on FPGA, uC, and Wi-Fi module source codes and their interoperability.

The basic motivation behind the conception and release as an open hardware and software instrument for the academic and industrial communities is to benefit from the researchers' community improvements. An open platform, according to the opinion of the authors, would allow more developers to design smart converters and MPPTs, focusing on algorithms instead of electronics.

The authors focused, as a proof of concept, on the deployment of a solution tailored for photovoltaic applications up to 450W. The designed system is capable of implementing MPPs accurately, allowing for the testing of different current and voltage-controlled MPPT algorithms.

The versatility of the solution, despite this implementation, does not limit the possibility to be adapted to several renewable energies, with the only limits being the maximum power of 450 W and wide voltage input and output ranges of 15–70 V.

The maximum pulse width modulation (PWM) frequency used for feeding the gates of the H-Bridge topology is 250 kHz, and the minimum dead time is 83 ns. These values are limited by MOSFETs used in the power circuit (IPB038N12N3G by Infineon [18,19]). However, the platform can work up to a maximum frequency of about 500 kHz with a minimum dead time of 21 ns.

The paper is organized as follows: Section 2 describes the architecture of the provided open hardware platform, the circuital blocks, and the most important electronic components used. Section 3 and 4 discuss the design phase and the layout outcome of the main logic board and the power shield board, respectively. In Section 5, experimental results are shown for the P&O algorithm. Section 6 provides a better interpretation and guideline of the platform use, while conclusions are drawn in Section 7.

2. Architecture

Figure 1 depicts the circuital blocks of the system. The system is conceived as a main logic board (MLB) with a power shield (PS). In the version presented hereafter, the PS is a buck-boost converter populated with Si MOSFET devices. Proper connectors allow for the implementation of an interleaved configuration [20], adding one more PS on the top.

The MLB consists of the following three main components:

1. A microcontroller that acquires currents and voltages from the power circuit and has a bidirectional communication with FPGA and Wi-Fi module;
2. An FPGA whose main functions are the generation of PWM signals for controlling MOSFETs with appropriate frequency, dead-time, and duty cycle; implementation of the bootstrap and interleaved control logic; and bidirectional communication with microcontroller;
3. A Wi-Fi module that transmits and receives the working parameters of the MPPT and has bidirectional communication with microcontroller.

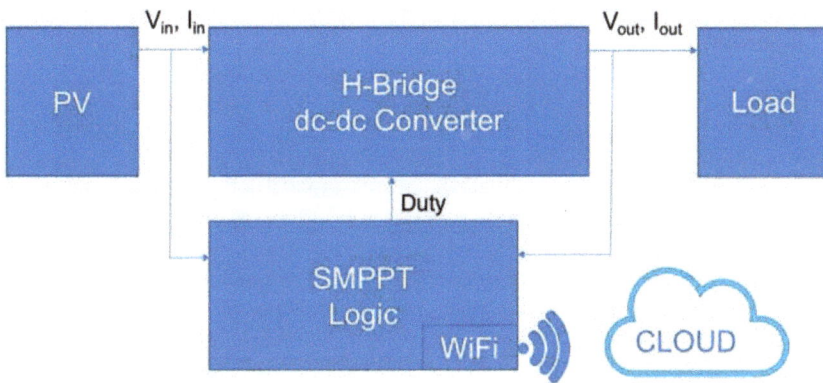

Figure 1. Functional blocks of the open-source hardware smart converter platform.

2.1. Logical Blocks

2.1.1. DC/DC Converter

Figure 2 shows a simplified diagram of a full-bridge converter topology realized with four MOSFETs: M_1, M_2, M_3, and M_4. It can work in three different modes: buck, boost, and buck/boost. For example, when V_{IN} is significantly higher than V_{OUT}, the circuit operates in buck mode where M_3 and M_4 are always off and on, respectively. Moreover, the MOSFETs M_1 and M_2 alternate the on and off state with a variable duty cycle, behaving like a typical synchronous buck regulator.

When V_{OUT} is significantly higher than V_{IN}; the circuit operates in the boost region where M_1 is always on and M_2 is always off, while the MOSFETs M_3 and M_4 alternate the on and off condition with a variable duty cycle, behaving like a typical synchronous boost regulator.

When V_{in} is close to V_{out}, the circuits operates in the buck/boost region where all four MOSFETs (M_1, M_2, M_3, and M_4) turned off and on with variable duty cycle.

Moreover, the circuit has two additional MOSFETs (M_5, M_6) between input and output, used to bypass the DC/DC converter when an output voltage almost equal to the input voltage ($V_{out} = V_{in} \pm 5\% \, V_{in}$) is required.

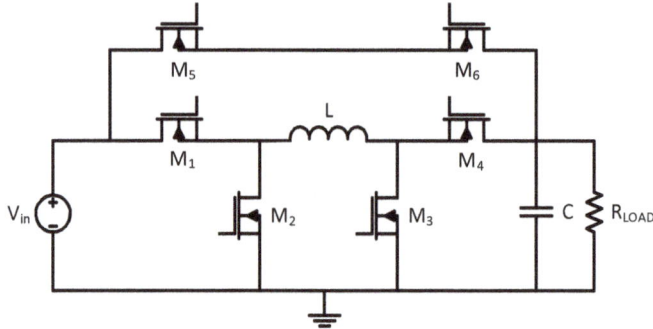

Figure 2. Full-bridge converter simplified diagram.

2.1.2. MPPT

As an example of realization, the FPGA is used to implement the perturbation and observation (P&O) algorithm for maximum power point tracking (MPPT). The P&O algorithm [21], shown in Figure 3, is based on the application of continuous variations to the output voltage (by changing the converter duty cycle) of a photovoltaic system until the MPP is reached. After the acquisition of the signal $V_{in}(t)$ and $I_{in}(t)$, the algorithm calculates the power $P_{in}(t)$. If the calculated power is higher than the power calculated at the previous time $P_{in}(t-1)$, the voltage value $V_{in}(t)$ is compared with $V_{in}(t-1)$, representing the actual input voltage and the input voltage received at the previous time, respectively. If $V_{in}(t)$ is higher than $V_{in}(t-1)$, the value of the duty cycle (D) is decreased by an amount equal to ΔD; otherwise it is increased by the same amount. However, if $P_{in}(t)$ is smaller than $P_{in}(t-1)$ and if $V_{in}(t)$ is greater than $V_{in}(t-1)$, the duty cycle value is increased; otherwise it is decreased.

The above algorithm, also shown in Figure 3, has been translated in VHDL code and synthesized with Lattice Diamond software [22].

However, the platform is adaptable to different MPPT algorithms that can be implemented in the FPGA or in the microcontroller.

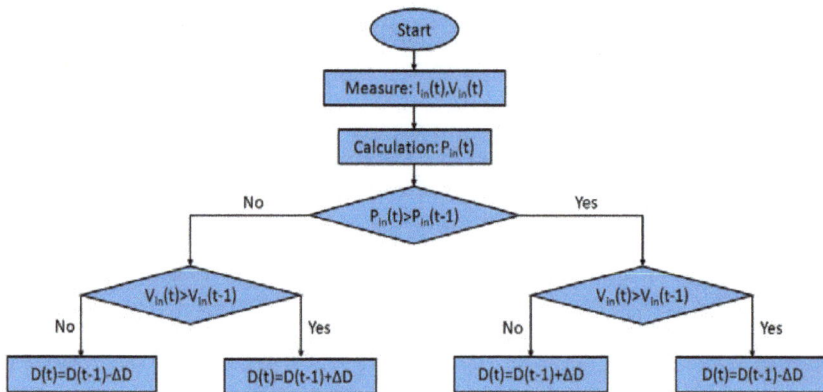

Figure 3. P&O algorithm diagram.

2.1.3. RF Transmission

Considering an application scenario of multiple DER units, it becomes of fundamental importance the feature that enables real-time reading of measurements and parameters, and remote modification of both algorithms type and settings. This result can be achieved thanks to the Wi-Fi module (Silicon

Labs WGM110 [13]) that could act both as an access point or Wi-Fi client, featuring an 802.11b/g/n radio and an embedded microcontroller.

By means of a serial communication with the master microcontroller, it sends and receives working parameters of the whole system, and the current and voltage measurements.

3. Main Logic Board Design and Layout

In this section, design and realization of the main logic board (MLB) are described. The board, shown in Figure 4, is used to control the power shield (PS) and the Wi-Fi communication. It consists of a power section where there are

- A step-down converter (LM5161 [23]) that allows to generate 12 V from the input voltage (output voltage of the photovoltaic source);
- A step-down converter (TPS561208 [24]) that converts the 12 V into 3.3 V;
- A CMOS low dropout (LDO) voltage regulator (MCP1700-1202 [25]) that generates 1.2V from 3.3V for the FPGA core.

Another section consists of the components for the programming of the FPGA. It is realized with an USB connector and an FTDI chip (FT2232 [26]) that allows one to program an external FLASH memory (SST25VF080B [27]), used for storing the bitmap file of the FPGA, while, as previously mentioned, the main components of this board are a Lattice Semiconductor FPGA (iCE40UP5K [11]), a Microchip Technology INC. dsPIC (dsPIC33EP64GS504 [12]), and a Silicon Labs (WGM110 [13]) Wi-Fi module.

In the following, these three main components and their firmware are described more in detail.

Figure 4. Main logic board top view.

3.1. FPGA

iCE40UP5K is an ultra-low power FPGA with a sleep current of about 35 µA and 1–10 mA active current. The characteristics of the device are reported in [11]. In the FPGA, commonly used to make high-frequency circuits, it is possible to instantiate a control unit that works as a coprocessor of the microcontroller. Its use allows one to obtain a high data processing speed and the addition of logical blocks without hardware changes. The FPGA firmware starts with the acquisition of the digital values of input and output current and voltage of the MPPT, received from the microcontroller through a serial communication. Direction and number of connections are detailed in Table 1. Moreover, it acquires some PWM parameters, i.e., frequency, duty cycle, and deadtime. Subsequently, it calculates the MOSFETs duty cycles through the implementation of the MPPT's algorithm and it generates the PWM signals of (M_1, M_2, M_3, and M_4). In the meanwhile, proper values of frequency, duty cycle, and dead time, and some debug data, are sent to the microcontroller.

Table 1. Connections between microcontroller (µC), FPGA, and WiFi.

Connection Directions	from/to µC to/from FPGA	from/to µC to/from WiFi	µC Analog Input	µC Debug	FPGA LED	FPGAI/O	WiFi Debug
# Connections	15	5	8	2	3	8	2

The minimum resolution of the signals generated by the FPGA is 21 ns, related to the maximum frequency of the used hardware (50 MHz). Using a 7-bit, mod-100 counter, the generation of PWM could lead to a 500 kHz signal for feeding the MOSFET's gate. In practice, using a Si power board, the maximum PWM frequency has been limited to 250kHz, using an 8-bit, mod-200 counter. The switching frequency can be calculated as $f_{PWM} = \frac{f_{CLK}}{2^n}$.

At the same time, the FPGA controls the two MOSFET M_5 and M_6 used to bypass the DC/DC converter. In particular, the two MOSFETs are on for 1 min when the voltage output of the converter remains almost equal to the voltage input ($0.95\ V_{in} < V_{out} < 1.05\ V_{in}$) for almost 15 s. Moreover, the microcontroller can force the bypass mode when the circuit has a malfunction. Finally, the FPGA allows the control of one or more power shield (interleaved condition) in order to improve the performance of the circuit, e.g., reducing the voltage output ripple and improving the circuit reliability.

3.2. Microcontroller

The microcontroller is the control center of the logic board and interfaces with the FPGA chip, which generates the fast control signals of the MOSFETs, and with the Wi-Fi module for communicating parameters and measurements to other devices.

The microcontroller acquires the analog signals from the sensors and sends them, through a custom communication protocol, to the FPGA, managing moreover its reset and enable signals.

The microcontroller chosen for the realization is the dsPIC33EP64GS504 [12]. It is a 16-bit microcontroller from Microchip and is part of the digital signal processing family, and it has some specific functions for the control and management of switching circuits and power conversion, including high speed 12-bit Analog to Digital Converter (ADC) with 5 ADC SAR cores, flexible and independent ADC trigger sources, and comparators with dedicated 12-bit digital-to-analogue converter (DAC). The microcontroller is set to use its internal RC oscillator and PLL to achieve a working frequency of 42.38 MHz.

In the microcontroller choice, particular importance has been given to the speed and accuracy characteristics of the analog to digital controller module. In fact, the specific application requires the simultaneous operation of several ADC modules, to measure accordingly the current and voltage and then calculate the relative power.

The microcontroller has four independent modules plus a shared one; the dedicated modules can sample the signal at the same time, using a common trigger signal to trigger them simultaneously; this allows us to perform a synchronous and instantaneous sampling of the various power card measurements. The readings of the ADC modules are appropriately averaged with a high number of samples to reduce the measurement noise. The dedicated cores of the ADC have been used for the most important signals for the operation of the MPPT, and in particular currents and input voltage. The analog inputs were set as follows:

AN0: VOUT (core 0)
AN1: VIN (core 1)
AN2: I_IN_1 (core 2)
AN3: I_IN_2 (core 3)
AN4: TEMP_1 (shared core)
AN5: TEMP_2 (shared core)
AN12: I_OUT_1 (shared core)
AN14: I_OUT_2 (shared core)

The analog comparator is used to detect when input and/or output voltages and/or currents exceed the design limits. When the limits are exceeded, an interrupt is generated that allows one to intervene in the duty cycle of the converter, limiting current/voltage and bringing the system back to the expected operating range. Each comparator is internally connected to the same inputs as the ADC converter, and is always active, even during normal operation of the ADC module.

To transfer to the FPGA the information obtained from the ADC, a communication protocol is implemented that transfers five words (VIN, VOUT, I_IN, IOUT_1, and IOUT_2) of 16 bits each on five separate pins, and two control signals (START and CLOCK). Transmission of the operating parameters is made serially with a 32-bit word that contains duty cycle MOSFET M1, duty cycle MOSFET M2, 4-bit dead time selection, three frequency selection bits, and one bit for forcing the duty from the outside.

The microcontroller receives the operating status of the FPGA via the SPI protocol. Three 16-bits words are transferred, the first four bits of which indicate the type of data sent, while the remaining 12 bits contains the actual data (MOSFETs duty cycle, dead time, frequency, and operating mode).

The microcontroller also sends three additional control signals to the FPGA to (1) reset the FPGA chip, (2) enable the FPGA, and (3) force the bypass condition.

The microcontroller and the Wi-Fi module communicate via the UART serial interface. Periodically, the microcontroller sends to the Wi-Fi module all the current, voltage, temperature, operating mode, and system parameters measurements, which are stored in the Wi-Fi module and displayed in its internal web server interface, or communicated via appropriate APIs to other devices.

In addition, the Wi-Fi module can send, through the same interface, operating parameters to the microcontroller (duty cycle, frequency, dead time, and duty cycle) to customize the operation of the device remotely.

To debug the system, a UART transmission and reception protocol has been implemented, through which, with a connection to the PC, it is possible to set the operating parameters of the card (duty cycle, frequency, dead time, duty cycle, and bootstrap) and observe the trend of the current, voltage, power, and temperature measurements of the power cards.

For programming the microcontroller, the MPLAB X software, an integrated development environment from Microchip, was used. The source code was written in C and compiled using the XC16 compiler.

3.3. API Wi-Fi

A web server was implemented, to enable querying of the Wi-Fi module for data gathering and dispatching coming from/to another device on the same network. Thanks to custom developed application programming interface (API), using GET/POST command over hypertext transfer protocol (HTTP), whose result is a JSON object, it is possible to establish different policy and strategies of data collection, supervision, and network optimization in a smart and effectively way. The APIs implementation and explanation are provided on the public repository [16].

The read status API has been implemented to retrieve the status of the system, thus responding with eight different parameters and measurements.

For the sake of completeness, we provide the following detailed CURL description:

curl -i -H "Accept: application/json" -H "Content-Type: application/json" -X GET http://192.168.1.xxx/api/smppt/status.

The query response, provided in JSON format, is the following:

```
{
"mac": "xxxxxxxxxxxx",
"mode": [int, bit0=BOOST, bit1=BUCK, bit2=BYPASS],
"temp1": [signed int, °C x 100],
"temp2": [signed int, °C x 100],
"meas": {
"Vin": [int, Volt x 100],
"Vout": [int, Volt x 100],
"Iin": [int, Ampere x 1000],
"Iout": [int, Ampere x 1000]
}
}
```

The *read settings* API has been implemented as a GET request too, providing information about the actual PWM frequency, dead time, and duty cycles.

The *write setting* API allows the users to modify the settings in use and it is recalled using the following:

curl - -data "value" http://192.168.1.xxx/api/smppt/settings/item

with the value to write that should be put in the body of the request.

"item" can be:

mppt	MPPT ON/OFF	[0–1] -> 0 = dc/dc, 1 = mppt
freq	switching frequency	[0–7] -> {100,80,120,150,175,200,225,250} kHz
dt	dead time	[0–15] -> {104,21,41,62,83,104,125,145,166,187,208,229,250,271,291,312} ns
m1	duty cycle MOSFET M1	[0–4095]
m3	duty cycle MOSFET M3	[0–4095]

In the Figure 5, a possible graphical representation of the information obtained from the *read status* and *read setting* APIs is provided.

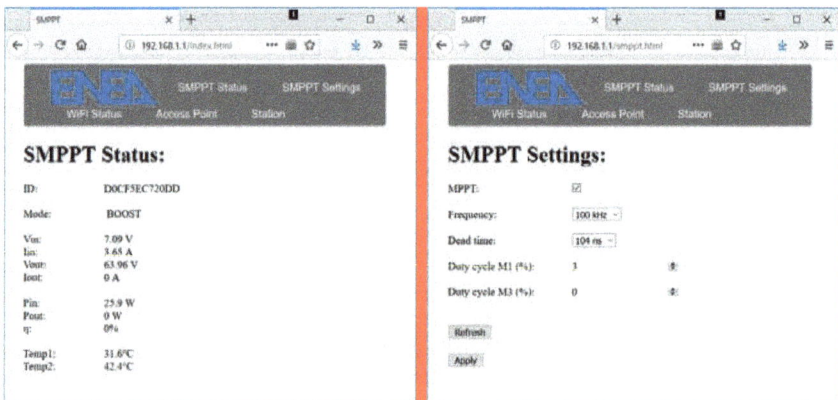

Figure 5. Dashboard examples.

4. Power Shield Design and Layout

The schematic diagram of the PS has been described in Section 2.1.1, providing also the use of six power MOSFETs. The selected devices used for the realization of the board are IPB038N12N3G MOSFETs by Infineon [18]. These N-channel transistors have a very low drain-source on-state resistance (R_{DS}), which allows one to reduce dissipated static power. Moreover, they have a small total gate charge (Q_g) that contributes to reduce the values of dissipated dynamic power.

Two UCC27211A-Q1 [28] high-side and low-side drivers with bootstrap circuits have been used for the control of the two half-bridge, while an isolated FET driver (Si8752 of Silicon LAB [29]) has been used for the control of bypass MOSFETs (M_5, M_6). Moreover, two current-sense amplifiers (INA240A4 [30]) with enhanced PWM rejection have been used for the measurement of input and output currents. The top view of power shield is shown in Figure 6.

Figure 6. Power shield top view.

5. Experimental Results

Experimental results are shown for P&O and IC algorithms and compared to state-of-the-art outcomes from similar platforms.

In Figure 7 two images of the realized MPPT are shown. The bottom board is the main logic board, on which the power board is fitted by strip line connectors.

(a)   (b)

Figure 7. Maximum Power Point Tracking (MPPT) front (**a**) and back (**b**) view.

A series of tests were performed using the modular solar array simulator from Agilent Technologies in order to characterize the goodness of the MPP tracking, the speed of convergence, and stability over time.

The results were obtained with various MPP from 50 W to 450 W, with different combinations of V_{OC} and I_{SC}. In Figure 8, the tracking measurements at 125 and 250 W are shown. The MPP is achieved with good accuracy between 99.5% and 100%, showing a convergence speed of about ten seconds. Moreover, the response to the variation of the on-state condition is of the order of a few seconds.

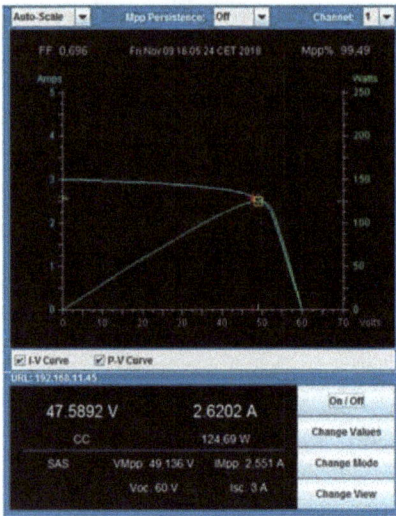 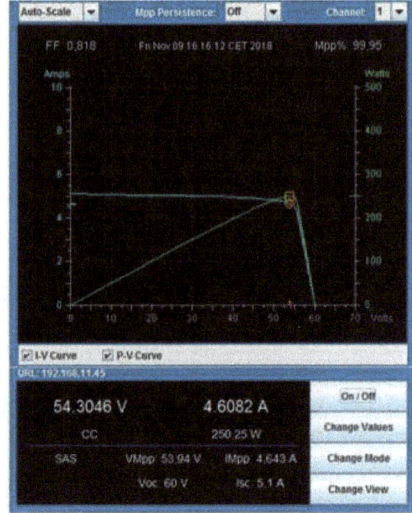

(a)   (b)

Figure 8. MPP tracking measurements for input power of 124 W (**a**) and 250 W (**b**).

The power shield, using silicon MOSFETs, shows good performances, reported in this document for input powers from 200 up to 450 W.

We considered the power conversion efficiency, defined as $\eta = \frac{I_{out} \cdot V_{out}}{I_{in} \cdot V_{in}}$.

As shown in Figure 9a, the efficiency (η) is higher than 97% in the range of considered input powers, higher than 98% for power above 300 W. In Figure 9b it can be seen how η presents its maximum value for a dead-time equal to 166 ns, using a switching frequency of 100 kHz, with a maximum value of 98.7% at 400 W of input power.

The efficiency decreases sharply as the switching frequency increase, as reported in Figure 9c. The measured values of efficiency, reported in Figure 9, are considered using the measured optimal load impedance, for every input power, detailed in Figure 9d. The operation of the DC/DC converter in bypass mode shows an efficiency higher than 99%.

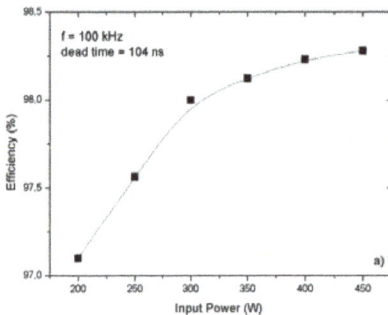 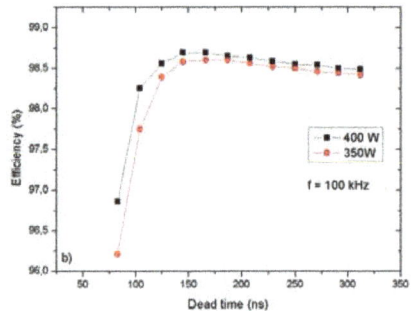

(a)   (b)

Figure 9. *Cont.*

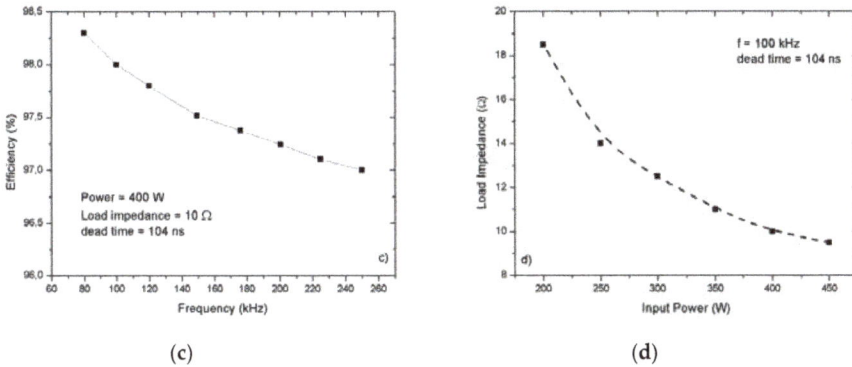

(c)   (d)

Figure 9. Efficiency (η) for different values of input power (**a**), of dead-time (**b**) and of switching frequency (**c**). In figure (**d**) is reported the optimal load impedance as a function of the input power.

The Figure 10 reports the curve fitting of the measured efficiency values in the 250–450 W range of input powers, for 50 W step. The optimal value, i.e., the maximum value of every single curve, moves to a lower value of load impedance as a response to the increment of the input power.

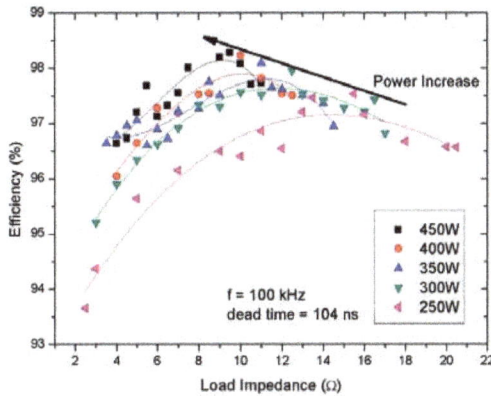

Figure 10. Smart converter efficiency vs. various load impedance and input power from 250 W to 450 W, in steps of 50 W.

6. Discussion

The designed solution is characterized by limits in terms of maximum current, voltage, and power that could be applied to the circuit.

Nevertheless, the overcome of the actual limits is possible but it has the huge drawback of the rapid increase of the bill of material (BOM) cost of the boards, thus reducing the adopting rate from contributors.

Being the repository enabled to host forks of the system, authors welcome contribution that, considering different technological limits and power ranges, could enable the use of a modified version of the system with other renewable energies as, for example, fuel cells.

The actual limits of the platform are a maximum power of 450 W, maximum input, and output voltage of 70 V, 10 A not simultaneously. The maximum PWM frequency used for feeding the gates of the H-Bridge topology is 250 kHz, the minimum dead time is 83 ns.

The undiscussed, provided value of the community improvements on the discussed open-source platform could rely on, but are not limited to, suggestion and test of different devices. For example, the used MOSFETs are in a quite common package, and could be modified with different part numbers thus exploring the overall effects of the parameter modifications. Both FPGA and µC have pinout compatible alternative parts that may differ in terms of LUTs, memory, and peripherals, and could be considered as possible alternatives to reduce the BOM cost or to explore different functionalities.

Nevertheless, different coding strategies could bring benefits on the FPGA and the microcontroller, increasing, e.g., the PWM frequency beyond the actual limit of 500 kHz, and developing further and efficient algorithm.

The user of the platform should be instructed and informed on the fact that he/she is managing high power, which is potentially dangerous, and he/she should use proper countermeasures to reduce every risk.

In case of working conditions that require the platform to perform close to its power limit of 450 W, a cooler device is recommended to help reduce the switching devices temperatures.

7. Conclusions

In this paper, the design and hardware implementation of a platform for the development of smart converter system have been described in detail. The system is conceived as an open-source hardware platform. Schematics, layout, bill of materials, and software programming codes are provided under the terms and conditions of the creative commons attribution (CC BY) license for hardware and MIT license for software.

The proposed system was simulated and constructed, and the functionalities of the suggested control algorithms P&O were proven and discussed. From the results acquired during the hardware experiments, the feasibility and functionality of the platform that allows for the implementation of MPPT algorithms and permits one to achieve an acceptable efficiency level have been confirmed.

A bi-directional radio frequency communication channel using Wi-Fi protocol enables the real-time reading of measurements and parameters, the remote modification of both algorithms type and settings, and the remote diagnostic of the board.

To our knowledge, this project remains the only open-source hardware platform for smart converter development used for educational, research, and industrial purposes so far.

In addition, it is conceived in a flexible and extensible manner so permitting further development for recent energetic scenarios.

Author Contributions: M.M., G.P., and D.I. conceived and designed the system and performed the experimental tests; P.F. contributed to the experimental tests; and F.G.D.C. conceived the experiment and supervised the design. G.A. and G.G. conceived the smart converter specifications, and A.M. and G.A. contributed to the prototype test validation. All authors contributed equally in writing the paper.

Funding: This work was supported by the Italian Ministry of Economic Development in the framework of the Operating Agreement with ENEA for Research on the Electric System.

Conflicts of Interest: The authors declare no conflict of interest.

References

1. Marques, A.C.; Fuinhas, J.A.; Pereira, D.A. Have fossil fuels been substituted by renewables? An empirical assessment for 10 European countries. *Energy Policy* **2018**, *116*, 257–265. [CrossRef]
2. Adinolfi, G.; Graditi, G.; Siano, P.; Piccolo, A. Multiobjective Optimal Design of Photovoltaic Synchronous Boost Converters Assessing Efficiency, Reliability, and Cost Savings. *IEEE Trans. Ind. Inform.* **2015**, *11*, 1038–1048. [CrossRef]
3. Pakkiraiah, B.; Sukumar, G.D. Research Survey on Various MPPT Performance Issues to Improve the Solar PV System Efficiency. *J. Sol. Energy* **2016**, *2016*, 1–20. [CrossRef]
4. Abdullah, M.A.; Yatim, A.H.M.; Tan, C.W.; Saidur, R. A review of maximum power point tracking algorithms for wind energy systems. *Renew. Sustain. Energy Rev.* **2012**, *16*, 3220–3227. [CrossRef]

5. Houssamo, I.; Locment, F.; Sechilariu, M. Maximum power tracking for photovoltaic power system: Development and experimental comparison of two algorithms. *Renew. Energy* **2010**, *35*, 2381–2387. [CrossRef]
6. Tey, K.S.; Mekhilef, S. Modified incremental conductance MPPT algorithm to mitigate inaccurate responses under fast-changing solar irradiation level. *Sol. Energy* **2014**, *101*, 333–342. [CrossRef]
7. Alajmi, B.N.; Ahmed, K.H.; Finney, S.J.; Williams, B.W. Fuzzy-Logic-Control Approach of a Modified Hill-Climbing Method for Maximum Power Point in Microgrid Standalone Photovoltaic System. *IEEE Trans. Power Electron.* **2011**, *26*, 1022–1030. [CrossRef]
8. *MATLAB and Statistics Toolbox*, The MathWorks, Inc.: Natick, MA, USA.
9. Iero, D.; Carbone, R.; Carotenuto, R.; Felini, C.; Merenda, M.; Pangallo, G.; Della Corte, F.G. SPICE modelling and experiments on a complete photovoltaic system including cells, storage elements, inverter and load. In Proceedings of the 2016 IEEE International Energy Conference (ENERGYCON), Leuven, Belgium, 4–8 April 2016; pp. 1–5.
10. Köllensperger, P.; Lenke, R.U.; Schröder, S.; De Doncker, R.W. Design of a flexible control platform for soft-switching multilevel inverters. *IEEE Trans. Power Electron.* **2007**, *22*, 1778–1785. [CrossRef]
11. Lattice Semiconductors iCE40 UltraPlus TM Family Datasheet. Available online: http://www.latticesemi.com/-/media/LatticeSemi/Documents/DataSheets/iCE/iCE40-UltraPlus-Family-Data-Sheet.ashx (accessed on 16 March 2019).
12. Microchip Technology Inc. dsPIC33EPXXGS50X FAMILY Datasheet. Available online: http://ww1.microchip.com/downloads/en/DeviceDoc/70005127d.pdf (accessed on 23 December 2018).
13. Silicon Labs WGM110 Datasheet. Available online: https://www.silabs.com/documents/login/data-sheets/wgm110-datasheet.pdf (accessed on 16 March 2019).
14. MIT License Open Source. Available online: https://opensource.org/licenses/MIT (accessed on 16 March 2019).
15. GitHub Inc. GitHub. Available online: https://github.com (accessed on 23 December 2018).
16. GitHub Code Repository. Available online: https://github.com/maxomous80/open-source-SMPPT (accessed on 16 March 2019).
17. Creative Common License Attribution-ShareAlike 4.0 International (CC BY-SA 4.0). Available online: https://creativecommons.org/licenses/by-sa/4.0/ (accessed on 16 March 2019).
18. Infineon IPB038N12N3G Datasheet. Available online: https://www.infineon.com/dgdl/Infineon-IPP_I_B041N12N3-DS-v02_03-en.pdf (accessed on 16 March 2019).
19. Iero, D.; Della Corte, F.G.; Merenda, M.; Felini, C. A PTAT-based heat-flux sensor for the measurement of power losses through a calorimetric apparatus. In Proceedings of the 30th Eurosensors Conference, EUROSENSORS 2016, Budapest, Hungary, 4–7 September 2016.
20. Chang, C.; Knights, M.A. Interleaving technique in distributed power conversion systems. *IEEE Trans. Circuits Syst. I Fundam. Theory Appl.* **1995**, *42*. [CrossRef]
21. Reza, A.; Hassan, M.; Jamasb, S. Classification and comparison of maximum power point tracking techniques for photovoltaic system: A review. *Renew. Sustain. Energy Rev.* **2013**, *19*, 433–443. [CrossRef]
22. Semiconductor, L. Lattice Diamond Software. Available online: http://www.latticesemi.com/latticediamond (accessed on 16 March 2019).
23. Texas Instrument LM5161 Datasheet. Available online: http://www.ti.com/lit/ds/symlink/lm5161.pdf (accessed on 16 March 2019).
24. Texas Instrument TPS56120x Datasheet. Available online: http://www.ti.com/lit/ds/symlink/tps561201.pdf (accessed on 16 March 2019).
25. Microchip Technology Inc. MCP1700 Datasheet. Available online: http://ww1.microchip.com/downloads/en/DeviceDoc/20001826D.pdf (accessed on 16 March 2019).
26. Future Technology Devices International Limited FT2232D Datasheet. Available online: https://www.ftdichip.com/Support/Documents/DataSheets/ICs/DS_FT2232D.pdf (accessed on 16 March 2019).
27. Microchip Technology Inc. SST25VF080B Datasheet. Available online: http://ww1.microchip.com/downloads/en/DeviceDoc/20005045C.pdf. (accessed on 16 March 2019).
28. Texas Instrument UCC27211A-Q1 Datasheet. Available online: http://www.ti.com/lit/ds/sluscg0a/sluscg0a.pdf (accessed on 16 March 2019).

29. Silicon Labs Si8752 Datasheet. Available online: https://www.silabs.com/documents/public/data-sheets/Si8751-2.pdf (accessed on 16 March 2019).
30. Texas Instrument INA240A4 Datasheet. Available online: http://www.ti.com/lit/ds/symlink/ina240.pdf (accessed on 16 March 2019).

electronics

Article

Development of EOG-Based Human Computer Interface (HCI) System Using Piecewise Linear Approximation (PLA) and Support Vector Regression (SVR)

Jung-Jin Yang [1], Gyeong Woo Gang [2] and Tae Seon Kim [2,*]

[1] School of Computer Science and Information Engineering, Catholic University of Korea, Bucheon 14662, Korea; jungjin@catholic.ac.kr
[2] School of Information, Communications and Electronics Engineering, Catholic University of Korea, Bucheon 14662, Korea; kkang719@nate.com
* Correspondence: tkim@catholic.ac.kr; Tel.: +82-2-2164-4367

Received: 1 February 2018; Accepted: 8 March 2018; Published: 9 March 2018

Abstract: Electrooculogram (EOG)-based human-computer interfaces (HCIs) are widely researched and considered to be a good HCI option for disabled people. However, conventional systems can only detect eye direction or blinking action. In this paper, we developed a bio-signal-based HCI that can quantitatively estimate the horizontal position of eyeball. A designed bio-signal acquisition system can measure EOG and temporalis electromyogram (EMG) signals simultaneously without additional electrodes. For real-time processing for practical application, modified sliding window algorithms are designed and applied for piecewise linear approximation (PLA). To find the eyeball position, support vector regression (SVR) is applied as a curve-fitting model. The average tracking error for target circle with a diameter of 3 cm showed only 1.4 cm difference on the screen with a width of 51 cm. A developed HCI system can perform operations similar to dragging and dropping used in a mouse interface in less than 5 s with only eyeball movement and bite action. Compare to conventional EOG-based HCI that detects the position of the eyeball only in 0 and 1 levels, a developed system can continuously track the eyeball position in less than 0.2 s. In addition, compared to conventional EOG-based HCI, the reduced number of electrodes can enhance the interface usability.

Keywords: human-computer interface (HCI); electrooculogram (EOG); electromyogram (EMG); modified sliding window algorithm; piecewise linear approximation (PLA); support vector regression; eye tracking

---

1. Introduction

The human-computer interface (HCI) is one of the key technologies used to improve the interaction between human and computer. In terms of usability of HCI, it is very important to study interfaces that can give commands to computers in a more human-friendly way. There are various interface approaches for HCI including voice, gesture, electroencephalogram (EEG) from brain, eye movement, etc., and the characteristics of each type are clear [1]. First, the voice-based HCI is a technology that receives the voice of a user by using a microphone and recognizes the user's intention through language processing. This technology has been steadily evolving since recognizing certain commands and has evolved into technologies such as voice typing and natural language processing [2,3]. This method is very attractive since it is very intuitive and can directly transmit the user's command. However, the performance of this system is very dependent on the type of users, ambient noise, and performance of natural language processing. Gesture-based HCI can recognize movements of the hands or fingers through a camera or electromyogram (EMG). Recently, the recognition rate and recognition speed have been greatly improved [4,5]. However, the use of this type of HCI method seems to be very limited since the conventional interfaces using hands including keyboard, mouse, remote control, and touch screen are much comfortable and reliable. The EEG from brain-based interface is a technique that recognizes the user's intention by receiving bio-signals through electrodes attached to the scalp near the brain [6,7]. Although it is ideal in the sense that it can transfer the user's intuition to computer more directly than voice, but the practical use of EEG-based HCI or brain computer interface (BCI) is expected to be very limited due to the lack of usability. It is very difficult to attach electrodes to various parts of the scalp and maintain the attachment position during measurement. The last type is eye movement-based HCI. Eye movement-based HCI technologies are mainly divided into two types, camera-based HCI and EOG-based HCI. In general, camera-based eye tracking systems require a small camera to capture an image of the eyeball and find the gazing position by mapping the eyeball position to the screen. Camera-based eye movement tracking systems have the advantage of directly determining the position that the user is currently gazing at. Camera-based systems have been continuously researched and their performance is continuously improving. Recently, commercialized camera-based eye-mouse products have been introduced [8]. However, despite several advantages, there is a problem in implementing click operation using blinking since the user may be blinking regardless of a click and may result in a malfunction. Additionally, the cost of system and portability are also considered to be disadvantages. EOG-based HCI systems have also been researched continuously. Wu et al. developed an EOG-based wireless HCI system for detecting eyes movements in eight directions with a simple system and short computational time [9]. Mulam et al. developed a HCI system to recognize four kinds of eye movement and blinking action using neural networks [10]. The EOG signals are decomposed by empirical mean curve decomposition (EMCD) process and principal component analysis (PCA) is used to reduce the signal dimension. Compared to similar approaches, these methods showed enhanced accuracy and precision. Most of EOG-based HCI systems including Wu and Mulam's system detect the up, down, left, and right movements or blinking of eyes using pattern recognition algorithms and uses them as commands for HCI. In other words, in conventional ways, only a series of discrete commands can be executed, and it cannot recognize the variance of gazing positions since it cannot determine how much the eyeball has moved.

Ang et al. developed an EOG-based HCI using a commercial wearable device (NeuroSky Mind Wave headset) [11]. They applied their system to a virtual on-screen keyboard to control the cursor by detecting double blinking action. They used L1-norm, Kurtosis, and Entropy as features and SVM classifier is used for activity detection. Similar to Ang's system, He et al. also used blinking action for HCI, but could produce commands by the user's eye blinks that are performed in synchrony with button flashes and could increase the number of possible commands. However, both systems can only make the selection command by blinking and they need to wait to select a specific command (or character on keyboard) until it flashes.

In this paper, we propose an EOG-based HCI system using single channel bio-signals that can measure EOG signal and temporalis EMG signal simultaneously without additional electrodes. The motivation of the proposed system is to make a novel interface method that can work in real-time for practical applications with higher usability. For this, we need to enhance the system performance in terms of usability, speed, and accuracy. To enhance usability, we designed a signal acquisition system that can measure EOG and EMG with only two electrodes. To enhance the system speed, we developed a modified sliding window algorithm-based PLA to reduce the complexity of signal. To enhance the accuracy, we designed our own signal acquisition system, optimized the PLA performance, and applied SVR after finding kernel function that shows better generalization performance than others in this application. The main difference between the proposed system and the conventional methods is that the proposed system can quantitatively estimate the horizontal position of eyeball without delaying time. For this purpose, we developed a single channel EOG and EMG acquisition system with only two electrodes. After measurement, the signals are separated to EOG and EMG signal based on frequency division. To reduce the signal complexity for real time eyeball tracking, modified sliding window algorithms are applied for piecewise linear approximation (PLA). The width and height of the line segment and slope and length of the line segment are selected as a feature set for eyeball tracking. To find the eyeball position, support vector regression (SVR) is applied as a curve-fitting model. A developed system is applied to an eyeball mouse interface to select horizontally listed buttons on the computer screen and this can be a practical alternative interface particularly for a disabled person with quadriplegia to access computer or mobile devices. Through the PLA and parameter optimization, proposed system minimized the consumption of hardware resources and it can be easily implemented on open-source platforms including Arduino and Raspberry-compatible modules.

2. Materials and Methods

2.1. EOG and EMG Signal Sensing System

2.1.1. Electrooculogram (EOG)

Electrooculogram is a signal that measures and records the resting potential of the retina. The human eye works as a dipole in which the front cornea is an anode and the rear part is a cathode. Therefore, when the eyeball moves and the cornea approaches one electrode, the potential of the corresponding electrode rises, and the potential of the opposite electrode falls. The most commonly used electrode attachment position for EOG is near the outer angle of both eyes as shown in Figure 1 and the position information of horizontal direction can be obtained by the pair of electrodes.

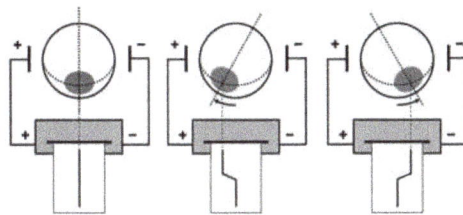

Figure 1. Potential changes according to the eyeball movement for EOG measurement.

2.1.2. Temporalis Electromyogram (EMG)

Electromyogram is a measure of activity potential produced by the activity of the skeletal muscles and it usually has an amplitude ranging from 50 µV to 30 mV. The skeletal muscles under the control of one motor neuron contract when the particular motor neuron is excited, and the muscles are excited or contracted almost simultaneously through connected motor neurons. Since the excitation of shock repeats at a frequency of 7 to 20 Hz, EMG is also measured in this frequency band. The temporal muscle or temporalis is the muscle located from the eyebrows to the cheekbone on both sides of the head. It is classified as a muscle of mastication and serves to move the lower jaw to chew food. Therefore, when the EMG is measured in the temporal muscle, a signal with large amplitude is generated in the case of bending on the muscle for bite, and a signal with small amplitude is generated when the bending force is released.

2.1.3. Design of Hardware System for EOG and EMG Signal Acquisition

The structure of the designed hardware for signal acquisition is shown in Figure 2. It amplifies the biological signals measured through the electrodes attached to both temples and transmits the signals to the PC. Measured signals are differential signals obtained by differential amplification of signals received from both electrodes. In this case, the common mode signal is treated as noise. In conventional bio-signal acquisition hardware, a ground electrode is added to effectively remove the interference of the common mode signal using signal feedback. However, in this case, the number of electrodes would need to be increased, which is not desirable for usability. In this paper, a pre-processing filter is applied as a method for measuring bio-signals without adding an additional ground electrode. The 1-pole active high-pass filter with a cut-off frequency of 0.008 Hz is used as a pre-processing filter. Here, the pre-processing filter is a kind of high-pass filter that removes low-frequency components before the signal is applied to the differential amplifier prior to passing the instrumentation amplifier. This pre-filter has two roles, the first is to ensure that the input signal is within the common mode input voltage range of the instrumentation amplifier of the measurement circuit. Since the designed system and the human body do not have a common ground, for the measurement circuit, it is not possible to predict the level of common mode potential of the human body signal. Therefore, in some cases, there is a risk of exceeding the common mode input voltage range of the instrumentation amplifier and it may lead to severe signal distortion and must be avoided. The second role of the pre-filter is to eliminate the DC offset. During the differential amplification of signals between the two electrodes, if a DC offset voltage exists between the two signals, they are not removed from differential amplifier. In the case of the bio-signal measuring circuit, since the amplification gain is usually several hundred times higher, the amplifier is easily saturated due to the DC component. At this time, the high-pass filter is equally applied to the signals transmitted from both electrodes, so that the DC voltages of the two signals are equalized and it can solve the two problems described above.

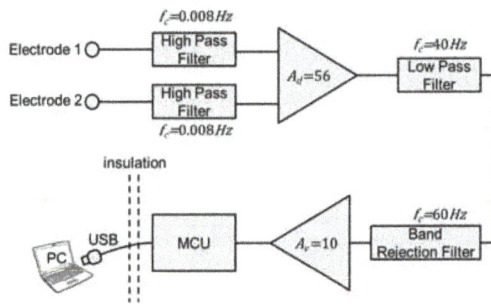

Figure 2. Structure diagram of hardware system for signal acquisition.

Although high-pass filters have essential roles, they must be handled with care in handling EOG signals. When gazing at one spot without eyeball movement, it resembles a nearly DC signal form. Therefore, the shape of the EOG signal is greatly affected by cut-off frequency and the selection of level of the high-pass filter is very critical. Figure 3 shows the filtered EOG signal according to the cut-off frequency of the high-pass filter.

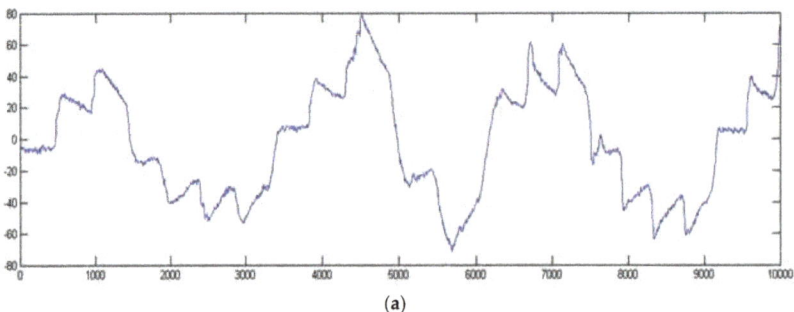

(a)

Figure 3. *Cont.*

(b)

(c)

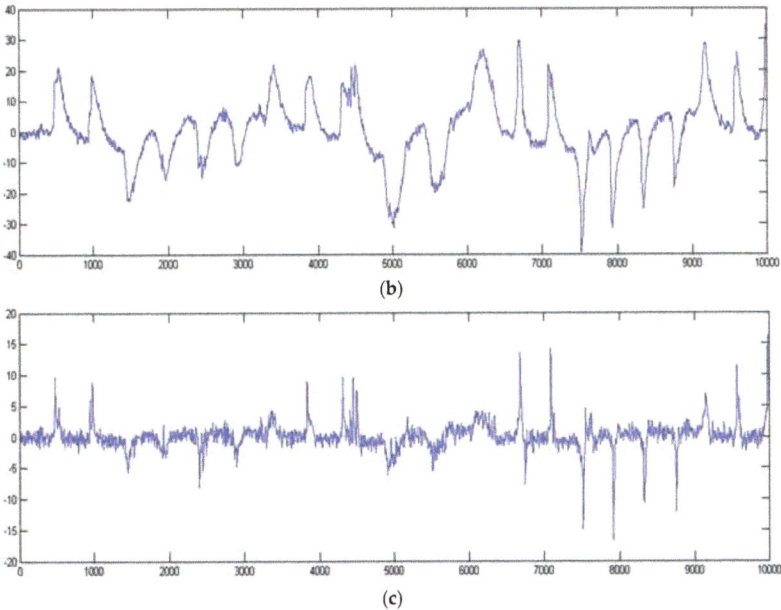

Figure 3. Effects of cut-off frequency of high pass filter on EOG signal. (**a**) filtered signal with 0.008 Hz cut-off frequency; (**b**) filtered signal with 0.1 Hz cut-off frequency; (**c**) filtered signal with 3.0 Hz cut-off frequency.

Based on the time point t_1, the left side shows the signal from slow eye movements and the right side shows the signal from fast eye movements. When using a high-pass filter with 3.0 Hz cut-off frequency, it is advantageous to detect the time when the eyeball moves since a waveform having large amplitude appears only in a section in which the eyeball moves, similar to the case of taking a derivative. However, it can be seen that the filtered signal varies greatly according to the velocity of the eyeball movement, and since the filter attenuates a large part of the frequency band of the signal, the size of the waveform itself is reduced and it is vulnerable to noise. On the other hand, when using a high-pass filter with 0.008 Hz cut-off frequency, when the eyeball does not move, it shows a typical EOG signal pattern showing a flat waveform, so that the change in the eyeball motion speed does not greatly affect the signal shape changes. Therefore, it is suitable to analyze the moving distance of the eyeball using the degree of the rise or fall of the signals. Also, since the signal strength of the desired portion of the signal is big enough, it is more robust against noise. However, since it requires a relatively long time to stabilize the signal close to '0', there is a disadvantage that it is difficult to recover the waveform if a large noise is added during the measurement such as when an impact is applied to an electrode or a cable. When a high-pass filter with 0.1 Hz cut-off frequency is used, the characteristics of the two cases mentioned above are shown together and can be selected as a compromise. Based on this, to consider the fact that the system requires precise analysis of eye movement distance, high pass filter with 0.008 Hz cut-off frequency was selected and applied.

The signal passed through the preprocessing filter is fed to the instrumentation amplifier for differential amplification. The instrumentation amplifier requires high common mode rejection ratio since the amplitude of the common mode signal to be removed is several ten times bigger than the EOG signal to be obtained from this system. Therefore, we applied 'INA128', which is a widely used IC chip for amplification of bio-signals and the designed circuit has 120 dB of common mode rejection ratio. In addition, the gain of the instrumentation amplifier is set to 56 V/V to obtain a magnitude of $\pm 50 \sim \pm 150$ mV which is sufficiently larger than the white noise that can be added in the subsequent signal processing. The differential amplified bio-signal is filtered through the post-processing filter to remove residual noise and amplified once more. The noise elimination of the post-processing filter starts with the low-pass filtering process to only pass the band less than 40 Hz which is enough to observe the EOG and temporalis EMG signals. In addition, since the low-pass filter limits the bandwidth of the signal, the sampling frequency in the analog-to-digital conversion (ADC) can be set low and it can reduce the computation complexity during digital signal processing. The second stage of noise cancellation is the elimination of 60 Hz noise using a band rejection filter. Often, the 60 Hz noise generated from various factors during the measurement environment cannot be removed solely by the low pass filter, so band elimination filter is added so that the system can be utilized even in a harsh measurement environment. In the final stage of the post-processing filter, signal amplification of 10 V/V level is performed so that the size of the bio-signal is increased to $\pm0.5\sim\pm1.50$ V to be suitable for analog-to-digital conversion.

After all analog signal processing steps have been performed, the ADC processed signals are sent to the PC. The ADC process is performed using a microcontroller with a 10-bit resolution and a sampling frequency of 300 Hz. Then, the converted digital data is transferred to USART to USB conversion IC chip through USART communication method and data can be transmitted through a USB cable connected to the computer. The USB cable also supports the 5 V power to the measurement system. To prevent electric shock from unexpected damage or leakage, the micro controller and the USART to USB conversion IC chip are electrically isolated. The ATMEGA8 microcontroller is used in this hardware and input voltage of this system is 5 V. This system consumes 1.8 mA at 5 V, so the power consumption of this system is around 9 mW.

2.2. Modified Sliding Window Algorithm for Piecewise Linear Approxamation (PLA)

The measured EOG signal using the proposed system showed potential difference only at the time of eyeball movement. Even if the speed of eyeball is very slow, still it showed very stiff potential difference like a step function. In this case, only the time of eyeball movement and the position of the eyeball are important data and we do not need to consider small signal change information during eyeball fixation. Therefore, PLA would be a good way to reduce signal complexity. There are various PLA algorithms including sliding window [12], scan along [13], top down [14], and bottom up [12]. Several hybrid algorithms are also developed to take advantage of both algorithms at the same time. In this paper, we designed a modified sliding window-based PLA algorithm that is best for real-time signal processing. Top-down- and bottom-up-based algorithms showed good approximation performance but were not appropriate for real-time signal processing applications since they require batch processing. Some research results showed modified top down or bottom up-based algorithms for real-time applications but still required some time delay to get the responses. Compared to these methods, sliding window and scan along-based algorithms showed quick processing responses. Another advantage of the sliding window-based method is that it can change the signals shapes to fit to our applications during calculation of error between the original signals and approximated signals.

Figure 4 shows the pseudo code of the standard sliding window algorithm. As shown in this code, when we increase the count number of P_2, we need calculate the error at every point between P_1 and P_2 which causes calculation burden. In addition, error, and threshold (*Th*)-based creation of vertices at every point sometimes showed poor approximation performances.

```
P₁= 0; P₂= 0; cnt= 0;
function SlidingWindow(Data)
        while P₂ is less than count of Data
              P₂ = P₂+ 1;
              err = ErrorFnc( Data, P₁, P₂);
              if err is greater than Th
                              P₁ = P₂;              cnt = cnt + 1;
                              Pₓ[cnt] = P₂;        P_y[cnt] = Data[P₂];
              end if
        end while
end function
```

Figure 4. Pseudo codes of standard sliding window method.

To enhance the calculation efficiency and approximation performance, we developed a modified sliding window algorithm as shown in Figure 5. First, as Koski suggested [15], we used k increment instead of an increment 1. By increasing the step point from 1 to k, we can reduce the calculation complexity but need to find the optimal k value for acceptable approximation performance. Also, we added a vertex position correction algorithm after adding a new vertex point. This correction algorithm performed if K more data are added right after new vertex point $P_x[cnt]$ was generated. At this point, if we generate temporal vertex point P_2, then we can make two line segments $\overline{P_x[cnt-1]P_x[cnt]}$ and $\overline{P_x[cnt]P_2}$, and we can find the new position of $P_x[cnt]$ that minimizes the sum of error. At this time, to consider computational time, we set the number of searching points to be less than $P_x[cnt] - K$. In this work, to set the maximum response delay to be less than 0.2 s, we set that K to 55.

```
P₁= 0; P₂= 0; cnt= 0;
function SlidingWindowImproved(Data)
                while P₂ is less than count of Data
                      P₂ = P₂+ k;
                      err = ErrorFnc(Data, P₁, P₂);
                      if P₂ is between P_y[cnt] +K - 1 and P_y[cnt] +K + k
                         Compensation(); end if
                      if err is greater than Th
                              if P₂ is less than P_y[cnt] +K
                                 Compensation(); end if
                              P₁ = P₂;                  cnt = cnt + 1;
                              Pₓ[cnt] = P₂; P_y[cnt] = Data[P₂];
                      end if
                end while
end function
function Compensation()
                min = MaxValue;
                i = Pₓ[cnt];
                while i is greater than Pₓ[cnt - 1] and Pₓ[cnt] - K
                        err = ErrorFnc(Data, Pₓ[cnt - 1], i) + ErrorFnc(Data, i, Pₓ[cnt]);
                        if err is less than min
                              i_min = i; end if
                end while
                Pₓ[cnt] = i_min;
end function
```

Figure 5. Pseudo codes of modified sliding window method for PLA.

Although we used the same approximation algorithm, the performance varies on the selection of function *ErrorFnc*. There are various error functions and we considered three kinds of error function in this paper, root mean squared error (RMSE), Euclidean error and Euclidian distance as shown in Figure 6.

Figure 7 shows the approximation performance dependency on error function type. For this, we defined the efficiency as the ratio of optimized number of line segment (M_{opt}) to PLA-based number of the line segment (M). In this case, M_{opt} is calculated based on the original number of eye movement trials. In other words, if efficiency is bigger than 1, then it infers that PLA missed the vertex point that needs to be created. Therefore, the smaller the efficiency, the more the part that can be represented by one line segment is often represented by two or more line segments.

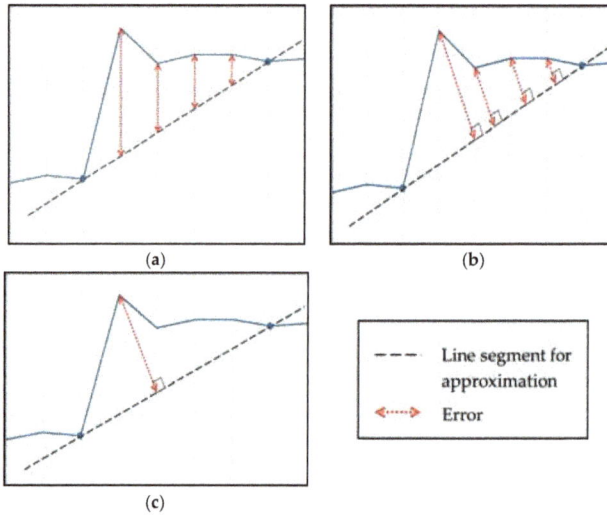

Figure 6. Three types of error function (*ErrorFnc*) for PLA. (**a**) RMSE; (**b**) Euclidean Error; (**c**) Euclidean Distance.

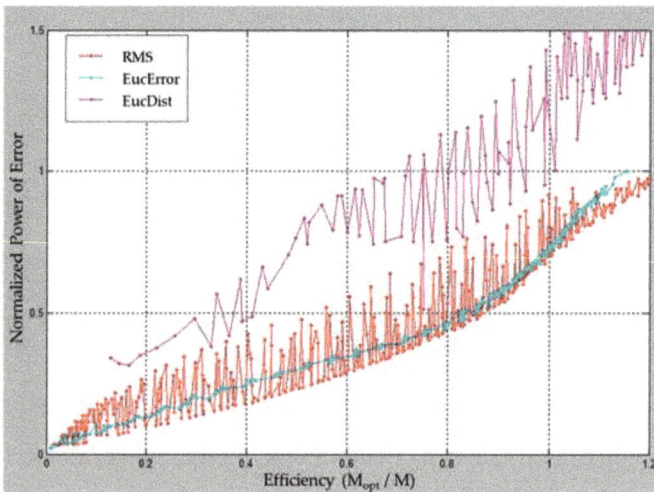

Figure 7. Performance variations according to types of error function.

In Figure 7, the graph of Euclidean Error shows gentle rises, while the graph of RMSE and Euclidean Distance shows irregular vibrations. The reason behind this difference can be found in Figure 8.

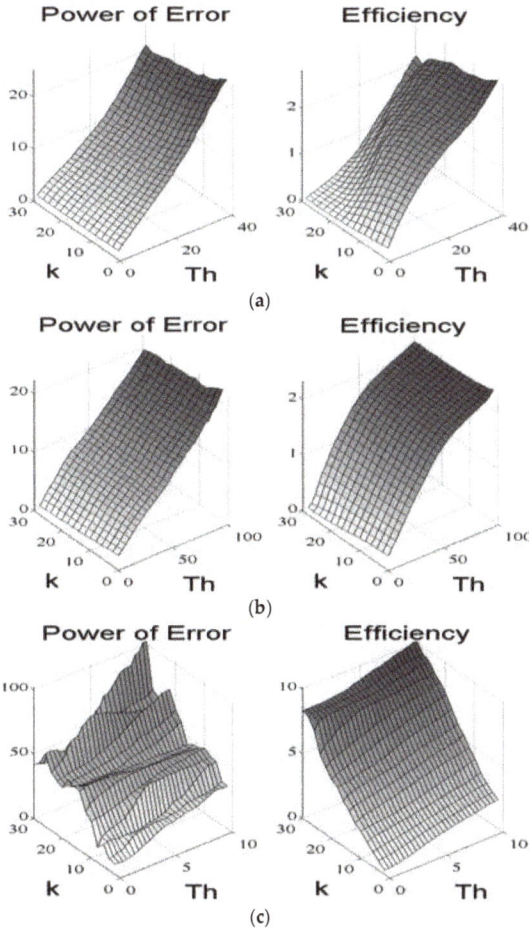

Figure 8. Effects of k, Th, *ErrorFnc* on approximation performances. (**a**) PLA using RMSE; (**b**) PLA using Euclidean Error; (**c**) PLA using Euclidean Distance.

In the case of Euclidean error, there is little change of power of error (P_{err}) and efficiency according to the change of k, so P_{err} and Efficiency gently rise as Th increases. In contrast to Euclidian Error, the RMSE shows an irregular efficiency change according to k and Euclidean Distance also shows irregular P_{err} changes according to k. In addition, in contrast to others, the efficiency is more sensitive to k than *Th* in Euclidean Distance. Therefore, we can increase the value of k without significant loss of approximation performance and we set k to 30 which is half of the buffer length K. Theoretically, it can be considered that *Th* is best optimized when the efficiency is 1, but in some practical cases, it is considered that there are cases where the extra vertices must be generate for better approximation. For this reason, 20% of the margin is considered despite lowered efficiency.

The Euclidean Error method also can overcome the overshooting phenomenon commonly found in EOG waveforms. The overshooting phenomenon occurs when the user attempts to move the eyeballs to the target position. If eyeballs move to the beyond the target position—in other words if eyeballs fail to move to the correct position at a time—then they return to the original target position quickly, creating an overshooting phenomenon. As shown in Figure 9, the Euclidean Error method can detect vertices effectively by ignoring overshoots in EOG waveforms.

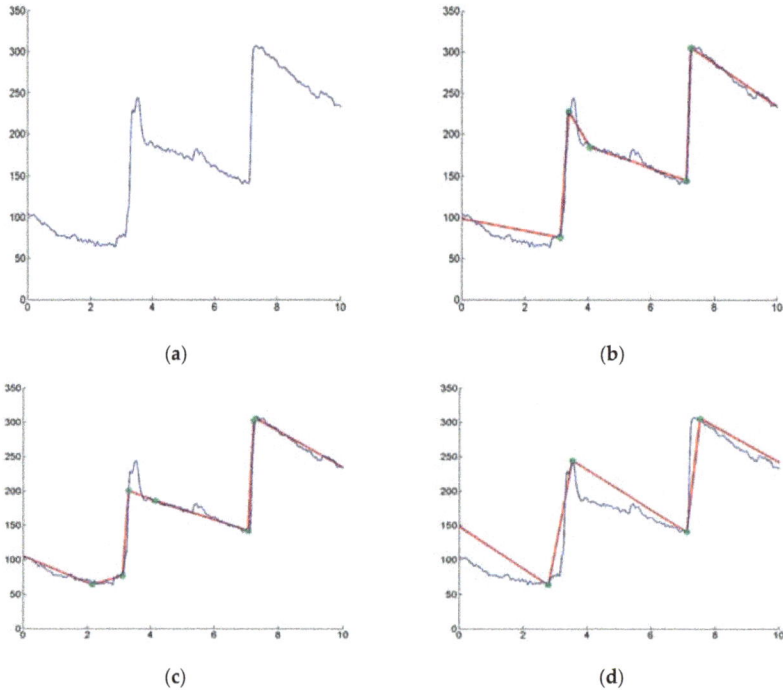

Figure 9. PLA results for EOG signal with overshooting phenomenon. (a) EOG signal with overshooting; (b) PLA using RMSE (c) PLA using Euclidean Error; (d) PLA using Euclidean Distance.

2.3. Curve-Fitting Model for Eye Tracking

Since the signal after PLA appears as a set of line segments, the width in the horizontal direction and the height, slope, or length in the vertical direction can be considered as feature parameters to represent the EOG signal. These features can be obtained by simple computation from the shape of the line segment and they can express the information of the line segment intuitively [16]. In this paper, we define the two-feature set F^{hv} that represents width and height of line segment and $F^{\theta l}$ that represents slope and length of the line segment.

Sometimes the information of the previous line segments may be useful to determine the characteristics of the current line segment. For example, if a steeply rising line segment appears after a flat signal has been sustained, it can be seen that the gaze is shifted when the EOG signal is rapidly rising from the flat region. In this case, it is desirable to use information of the previous line segment (flat line segment) to characterize the current line segment (rapidly rising line segment). The classification performance can depend on how many prior segments (N_p) are used to represent the current single segment. To find the optimal value of N_p, we tested for a range of 2 to 7 and we found that $N_p = 3$ is the optimal value in terms of performance and computation complexity. Another factor to consider for feature extraction is that the variation of the measured signal is different for each individual. To overcome these individual signal differences, the bio-signal data collected were analyzed and the average variation of EOG for each subject was set to a constant that is unique to each subject. This personalization constant is expressed as SF_u where u denotes the assigned unique number for each subject.

To improve the generalization ability of the system, we used the support vector regression (SVR) as a curve-fitting model. The curve-fitting model can represent a set of discrete data as a continuous function. Thus, by using the curve-fitting model to the proposed system, the proposed system is capable of coping with any arbitrary eye movements not included in the training data set. SVR is a kind of curve-fitting model used to solve the regression problem based on the existing Support Vector Machine (SVM) that has excellent modeling and prediction capabilities. In this paper, we used the Chang and Lin's model that can be found on "LibSVM' library [17]. Since the performance of SVR depends on the type of kernel functions, we tested three types of kernel functions including polynomial kernel function, sigmoid kernel function and Gaussian kernel function. For the performance test, we set 90% of data for training the SVR and 10% of data for test and repeated 10 times without duplication of the test data using a 10-fold cross validation method. Figure 10 shows the effects of the types of kernel functions on SVR performance. In these plots, the horizontal axis shows the actual moving distance of the gazing position and the vertical axis shows the prediction results of SVR. As shown in these plots, result plot of SVR with Gaussian kernel type showed the best results with 0.99317 of R value (multiple correlation coefficient value).

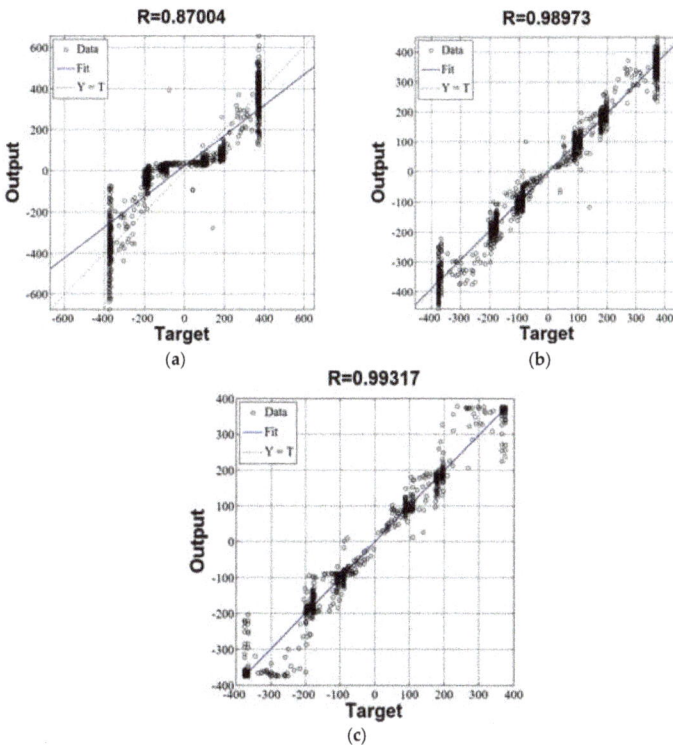

Figure 10. Performance plots of SVR for three types of kernel functions. (**a**) Plot of polynomial kernel function; (**b**) Plot of sigmoid kernel function (**c**) Plot of Gaussian kernel function.

2.4. Temporalis EMG Signal Processing for Recognition of Bite Action

EOG and EMG have different signal characteristics in terms of signal amplitude and frequency ranges. Therefore, the EMG of the temporalis muscle can be obtained by frequency division of the measured signal which is shown in Figure 11a. In this figure, we cannot recognize the EMG since the amplitude of EMG is much smaller than EOG. In general, EMG has a relatively bigger amplitude than EOG. However, in this case, the amplitude of EOG is bigger than the amplitude of EMG since the electrodes are attached to the temple on both sides which is relatively far from temporalis. The frequency division is performed using a Butterworth 4 pole 30 Hz high-pass filter designed by the IIR method, and the obtained temporal EMG signal by IIR filter is shown in Figure 11b. Since the frequency range of EMG signal does not overlap with EOG signals, we can use both signals independently at the same time. For this, the proposed system analyzes these temporal muscle EMG signals and determines if the line segment information obtained through the PLA process is generated in bite action or not. For this, when the PLA outputs new segment information, only the temporal muscle EMG of the same interval as the time segment of the corresponding line segment is separated and the signal amplitude within that interval is determined. When the temporal EMG according to the time is $EMG(t)$, the intra-segment power $Power(i)$ for the i-th segment can be expressed by the following equation.

$$Power(i) = \frac{1}{px_{i+1} - px_{i+1} + 1} \sum_{k=px_i}^{px_{i+1}} |EMG(k)|^2 \tag{1}$$

In the formula for calculating $Power(i)$, it can be seen that the temporal EMG signal is squared and then the time average is obtained between the two vertices px_i and px_{i+1} where the i-th line segment appears. As shown in Figure 11b, since the DC component is removed by the high pass filter and the temporal muscle EMG signal is concentrated around 0, the amplitude of the squared EMG signal becomes more prominent when the bite action occurs as shown in Figure 11c. Afterwards, the $Power(i)$ is compared with the threshold value Th, and if $Power(i)$ is greater than Th, we conclude the measurer to be in a state of bite and vice versa. Figure 11d conceptually shows that temporal muscle EMG is classified according to bite action by applying threshold value Th.

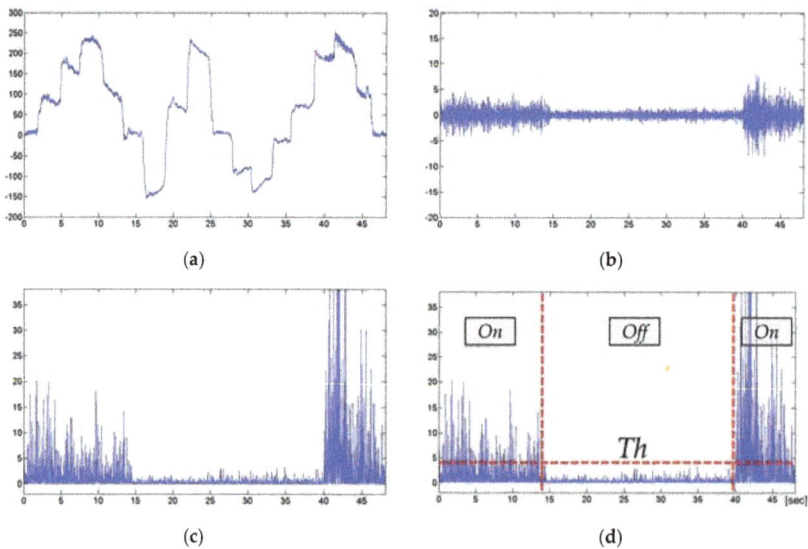

Figure 11. Temporalis EMG signal processing procedure for detection of temporal muscle movement. (a) Measured original signal; (b) EMG signal obtained by frequency division; (c) Squared EMG signal; (d) Bite action recognition by applying threshold value.

3. Results

In this paper, we designed two kinds of experiments for performance evaluation and real application to menu selection using the proposed system. For performance evaluation, the target circle located at the center of the screen randomly changes its horizontal position and we measured the EOG during that procedure. From this measurement, we calculated the difference between real horizontal position of target circle and expected gazing position from the developed system. In addition, this test also recognized the bite action that replaces the click action of the mouse interface. The second experiment is a real application of the proposed HCI. For this, we developed an EOG and EMG-based pointing interface system that can be used as a mouse. Through this experiment, we verified if the proposed EOG and EMG-based interface can easily applicable to menu selection command on a computer system as a new interface method.

3.1. Results of Performance Evaluation

The proposed system continuously tracks the movement of the user's eyeball and finds the position where the user is currently gazing on the screen. However, it is a waste of resources to perform such a function continuously even if the user does not need any operation. However, if the system is paused when no operation is needed, the system will lose the position information of the user's current eyeball since the system does not know where to start tracking once the system resumes. Therefore, the performance evaluation program was made by first giving a position to start eye movement tracking and tracking the movement of eye movement started from this position as shown in Figure 12. In this figure, the red circle is the target position to which the eyeball should gaze.

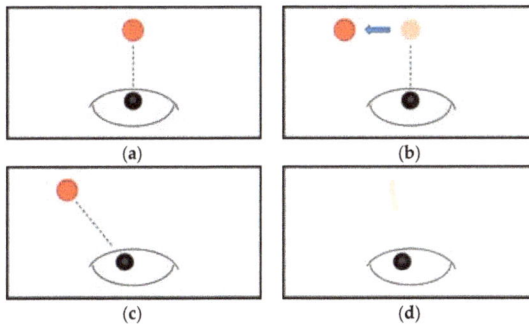

Figure 12. Procedure for command set for performance evaluation. (**a**) Look at the center circle and has a bite action; (**b**) Change the circle position after two seconds. (**c**) Move your eyeball to gazing the circle (**d**) reset by release of biting force.

In the signal acquisition phase for training, the target circles on the screen were designed to reciprocate with a maximum of nine positions. In the performance evaluation stage, ten positions were selected in the same range to confirm the performance of how the SVR can track eye movements when the eye moves to an arbitrary position that has not been trained.

Performance evaluation was performed on 4 out of 7 subjects participating in signal acquisition for training the system. Each subject performed 300 eye movements for testing. The spending time for this experiment was dependent on the subject. We did not give specific time guidelines for repetition of this experiment since each subject showed different levels of eye fatigue. So, we suggested that the subjects re-participate in this experiment after they had fully rested when they felt eye fatigue. In general, they continued the experiments for approximately 30 s to 1 min and then took rest. All subjects were male and aged between 20 and 30 years old. For the performance evaluation, two performance evaluation indices were applied. The first performance evaluation index was obtained by measuring the difference in units of cm on a 51 cm monitor between the position of the target circle and the tracking position of eyeball using the proposed system as shown in Figure 13. As shown in this figure, the significant difference by subject is not significant. In this figure, minimum error was measured at positions −3 and 3 which is the median position of horizontal range. In addition, it can be seen that there is a similar error value at the symmetric position with respect to reference 0 point. This infers that the cause of error may be coming from the approximation of the movement of the eyeball to the linear motion, whereas the human eyeball movement is a rotational motion with varying angle. Therefore, it infers that the error increases as the difference between linear motion and rotational motion becomes larger as the position deviates from the central position where the eyeball motion starts.

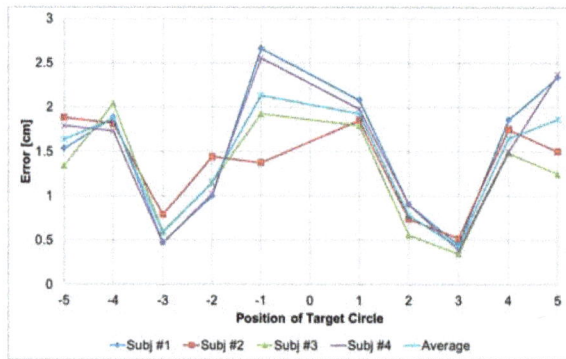

Figure 13. Error plot of eyeball position tracking system.

The second performance index is the time (in units of seconds) required to complete the drag-and-drop operation after the center circle has moved to an arbitrary position as shown in Figure 14. Although the required time varies by subject, it can be confirmed that general trend for each target positions are similar.

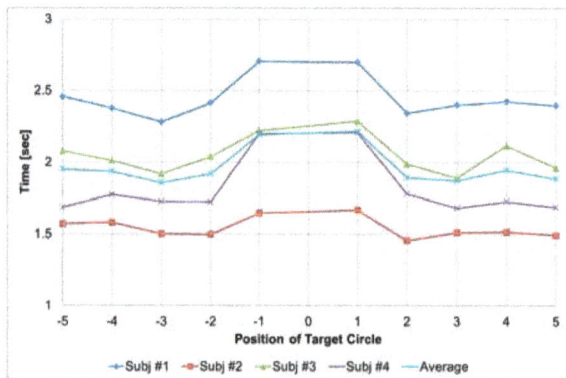

Figure 14. Plot of required time to complete the drag-and-drop operation.

3.2. Result of Mouse Interface Using EOG and EMG

The proposed system can be applied in various ways for HCI or HMI. As one example, we developed EOG and EMG-based pointing interface system that can be used as a mouse. The developed system is similar to the performance evaluation program which is shown in Section 3.1., but it is more user friendly for convenience of use. First, when you run the program, you will only see the tray icon on the Windows operating system without any visible pointing menu on the screen. Right clicking on this tray icon allows you to select your name among the list. If one has finished some set-up procedure in advance, the user can set personalized SF_u value by selecting his or her own name. In the case of a first-time user, a simple set up procedure is required to obtain customized SF_u value for each user. After the user selection is completed, the same tray icon menu can be used to connect the bio-signal acquisition hardware to the PC. After the connection is established, the user's bio-signal is transmitted to the PC and the developed interface system is executed in real time. Figure 15 shows an example of a real demonstration of the developed system. The system initially waits until there is a bite action. When temporal muscle movements are detected, the finger-shaped pointer slowly appeared in the center of the screen and then six selection buttons appeared near the pointer as shown in Figure 15a. After gazing at the central pointer, if the color of the pointer changes to blue, the user can move the pointer to the position of the selection buttons by moving the gaze as shown in Figure 15b. One can click on the button when the eyeball moves over the desired button as shown in Figure 15c and release the teeth biting force as shown in Figure 15d. After all operations are completed by the button click, the pointer and the button that appeared on the screen disappear and the system enters the waiting state until user bites again. This series of operations can run a completed command to the PC or any type of machine including robot or computer through a simple operation. It is similar to the drag-and-drop operation that is similarly used with an existing mouse interface, so a user can adapt this method intuitively. Compare to camera-based HCIs that suffer from high false-positive rate in the idle state because of the unintended ocular movement, proposed system can make start and stop point of our commands and we do not have to worry about error from idle state by adding the EMG signal to EOG Additionally, during the standby state, there is no obstacle to the visual field displayed on the screen and the computational burden is very limited. Therefore, it is possible to execute a desired command through a simple operation for about five seconds without using a conventional mouse.

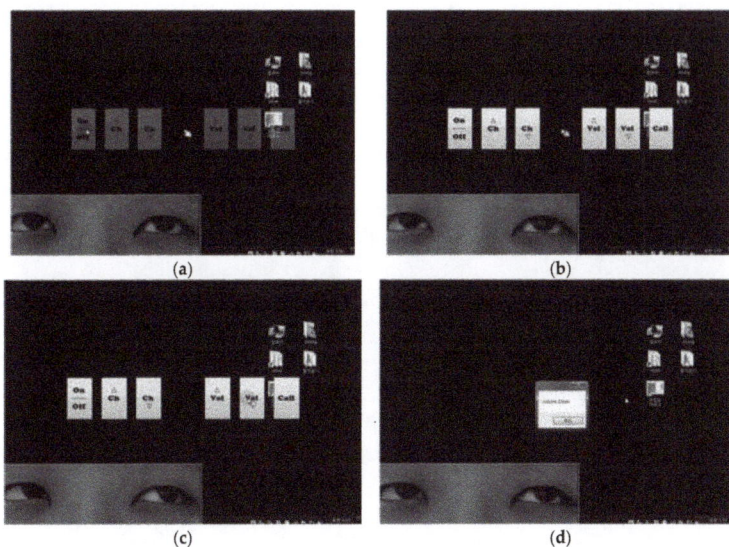

Figure 15. Temporalis EMG signal processing procedure for detection of temporal muscle movement. (**a**) Appearance of pointer and selection menu after detection of bite action; (**b**) Ready to select menu condition; (**c**) Selection of menu by detection of eyeball moving position; (**d**) Click the menu by release of biting force.

4. Discussion

As shown in previous experimental results including performance evaluation test and application test for menu-selection command interface for computer, the proposed system showed acceptable performance in terms of accuracy, speed, and usability. Compared to conventional EOG-based HCI, the proposed interface system has a different way in tracking gaze positions. In general, the EOG-based HCI or EOG mouse detects the up, down, left, and right movements or blinking of eyes using pattern recognition algorithms and uses them as commands. In other words, in conventional ways, only a series of discrete commands can be executed, and it cannot recognize variance of gazing positions since it cannot determine how much the eyeball has moved. In general, the performance of a EOG-based mouse system is measured using accuracy, sensitivity, specificity, or precision since it is in the form of pattern recognition for up, down, left, or right movements. However, we cannot use same performance measures since we continue to track the movement of the eyeball, so we used two kinds of performance measures, traction error and required time to operate as shown in the experimental results. The average difference between position of target circle and position from prediction of the proposed system showed only 1.4 cm difference on 51 cm screen. Since we set the diameter of target circle to 3 cm on a 51 cm screen, 1.4 cm is small enough to use for mouse control. The maximum variation of tracking error according to the subjects is less than 0.3 cm, so we can say that the proposed system is not user-specific. In terms of required time to drag-and-drop command, the system showed 1.97 s of average time, indicating an acceptable level as alternative interface method without using hands. The maximum variation of required time according to the subjects is 0.9 s. Compared to the average of required time to command, 0.9 s is not small, but it cannot be considered as a difficult problem since this difference was based on the familiarity of the subject.

To enhance traction accuracy, we need to consider that human eyeball movement is a rotational motion not linear motion. To find the gazing position on the screen, the proposed system was designed based on an approximation that the eyeball movement is a linear motion. In this case, the error will increase as the difference between linear motion and rotational motion becomes larger as the position deviates from the position where the eyeball motion starts. In addition, for wider application to practical system, it is required to solve the dependency problems of screen size and user position. For successful implementation of a developed system on HCI, it is good interface method for a disabled person with quadriplegia to access computer or mobile device including head mount display (HMD). Also, various applications can be considered for the proposed method. For a direct application, as we show the experimental results, it can be used as a hands-free menu selection interface for systems including TV, smart phone, or computer. By combining eyeball position and EMG signal information, various commands set for mobile robot also can be considered. Also, this can be an alternative way to control the wheelchair for disabled person with quadriplegia. As an indirect application of this method, eyeball position data can be stored as life log and it can be applicable in various ways such as monitoring in healthcare, attention assessment tests and test of human attention. Since the proposed system is designed considering the computational complexity for real-time processing, it can be easily implemented on open source electronics platforms including Arduino and Raspberry-compatible modules. Also, proposed system can be merged into various open-source electronics platforms-based applications including game, IoT, automation and robotics and it enhance the ranges of application for proposed system.

5. Conclusions

In this paper, we developed a single channel bio-signal-based HCI system using PLA and SVR. With only two electrodes, a developed bio-signal sensing system can measure and use EOG and EMG signals independently at the same time by frequency division and signal processing. For real-time processing, a modified sliding window PLA algorithm is developed to reduce the signal complexity. To enhance the eye tracking performance, SVR algorithm is applied and it can enhance the generalization capability. A developed HCI system can perform operations similar to dragging and dropping used in a mouse interface in less than 5 s with only eyeball movement and bite action. Compared to conventional EOG-based HCIs that detects the position of eyeball only in 0 and 1 levels, a developed system can continuously track the eyeball position less than 0.2 s. In addition, compared to conventional EOG-based HCIs, the reduced number of electrodes can enhance interface usability.

6. Patents

The patent of Korea, KR-A-101539923, "Bio-signal-based eye tracking system using dual machine learning structure and eye tracking method using same" invented by Gyeong Woo Gang and Tae Seon Kim is partially resulting from the work reported in this manuscript.

Acknowledgments: This research was supported by the Bio & Medical Technology Development Program of the National Research Foundation of Korea (NRF) funded by the Ministry of Science and ICT (No. 2017M3A9C8064887). This research was supported by Global Research Laboratory (GRL) program through the National Research Foundation of Korea (NRF) funded by the Ministry of Science and ICT (No. 2016K1A1A2912755).

Author Contributions: Jung-Jin Yang and Tae Seon Kim conceived and designed the experiments; Gyeong Woo Gang developed signal acquisition hardware and performed the experiments; All authors participated in the data analysis; Jung-Jin Yang and Tae Seon Kim wrote the paper.

Conflicts of Interest: The authors declare no conflict of interest.

References

1. Sanz-Robinson, J.; Moy, T.; Huang, L.; Rieutort-Louis, W.; Hu, Y.; Wagner, S.; Sturm, J.C.; Verma, N. Large-Area Electronics: A platform for Next-Generation Human-Computer Interface. *IEEE Trans. Emerg. Sel. Top. Circuits Syst.* **2017**, *7*, 38–49. [CrossRef]
2. Jiang, R.; Sadka, A. Multimodal Biometric Human Recognition for Perceptual Human-Computer Interaction. *IEEE Trans. Biomed. Eng.* **2010**, *40*, 676–681. [CrossRef]
3. Mano, L.; Vasconcelos, E. Ueyama, Identifying Emotions in Speech Patters: Adopted Approach and Obtained Results. *IEEE Lat. Am. Trans.* **2016**, *14*, 4775–4780. [CrossRef]
4. Zen, G.; Porzi, L.; Sangineto, E.; Ricci, E.; Sebe, N. Learning Personalized Models for Facial Expression Analysis and Gesture Recognition. *IEEE Trans. Multimedia* **2016**, *4*, 775–788. [CrossRef]
5. Ren, Z.; Yuan, J.; Meng, J.; Zhang, Z. Robust Part-Based Hand Gesture Recognition Using Kinect Sensor. *IEEE Trans. Multimedia* **2013**, *15*, 1110–1120. [CrossRef]
6. Edelman, B.; Baxter, B.; He, B. EEG Source Imaging Enhance the Decoding of Complex Right-Hand Motor Imagery Tasks. *IEEE Trans. Bio-Med. Eng.* **2016**, *63*, 4–14. [CrossRef] [PubMed]
7. Aghaei, A.; Mahanta, M.; Plataniotis, K. Separable Common Spatio-Spectral Patterns for Motor Imagery BCI Systems. *IEEE Trans. Bio-Med. Eng.* **2016**, *63*, 15–29. [CrossRef] [PubMed]
8. Eye Trackers for PC Gaming. Available online: https://tobiigaming.com/products (accessed on 30 January 2018).
9. Wu, S.; Liao, L.; Lu, S.; Jiang, W.; Chen, S.; Lin, C. Controllinf a Human-Computer Interface System with a Novel Classification Method that Uses Electrooculography Signals. *IEEE Trans. Bio-Med. Eng.* **2013**, *60*, 2133–2141.
10. Mulam, H.; Mudigonda, M. A Novel Method for Recognizing Eye Movements Using NN Classifier. In Proceedings of the IEEE International Conference on Sensing, Signal Processing and Security, Chennai, India, 4–5 May 2017; pp. 290–295.
11. Keogh, E.; Chu, S.; Pazzani, M. An Outline Algorithm for Segmenting Time Series. In Proceedings of the IEEE International Conference on Data Mining, San Jose, CA, USA, 29 November–2 December 2001; pp. 289–296.
12. Ang, A.; Zhang, Z.; Huang, Y.; Mak, J. A User-Friendly Wearable Single-Channel EOG-Based Human-Computer Interface for Cursor Control. In Proceedings of the IEEE International EMBS Conference on Neural Engineering, Montpellier, France, 22–24 April 2015; pp. 565–568.
13. Sklansky, J.; Gonzalez, V. Fast Polygonal Approximation of Digitized Curves. *J. Pattern Recognit.* **1980**, *12*, 327–331. [CrossRef]
14. Douglas, D.; Peucker, T. Algorithm for the Reduction of the Number of Points Required to Represent a Digitized Line or Its Caricature. *Can. Cartogr.* **1973**, *10*, 112–122. [CrossRef]

15. Koski, A.; Juhola, M.; Meriste, M. Syntactic Recognition of ECG Signals by Attributed Finite Automata. *Pattern Recognit.* **1995**, *28*, 1927–1940. [CrossRef]

16. Gang, G.; Min, C.; Kim, T. Development of Eye-Tracking System Using Dual Machine Learning Structure. *Trans. Korean Inst. Electr. Eng.* **2017**, *66*, 1111–1116.

17. Chang, C.; Lin, C. *LIBSVM: A Library for Support Vector Machines*; Technical Report; National Taiwan University: Taipei City, Taiwan, 2003; pp. 1–16.

electronics

MDPI

Article

DSCBlocks: An Open-Source Platform for Learning Embedded Systems Based on Algorithm Visualizations and Digital Signal Controllers

Jonathan Álvarez Ariza 🔟

Department of Electronics Technology, Engineering Faculty, Corporación Universitaria Minuto de Dios (UNIMINUTO), 111021 Bogotá, Colombia; jalvarez@uniminuto.edu; Tel.: +57-310-557-9255

Received: 17 January 2019; Accepted: 29 January 2019; Published: 18 February 2019

Abstract: *DSCBlocks* is an open-source platform in hardware and software developed in JavaFX, which is focused on learning embedded systems through Digital Signal Controllers (DSCs). These devices are employed in industrial and educational sectors due to their robustness, number of peripherals, processing speed, scalability and versatility. The platform uses graphical blocks designed in Google's tool *Blockly* that can be used to build different Algorithm Visualizations (AVs). Afterwards, the algorithms are converted in real-time to C language, according to the specifications of the compiler for the DSCs (XC16) and they can be downloaded in one of the two models of development board for the dsPIC 33FJ128GP804 and dsPIC 33FJ128MC802. The main aim of the platform is to provide a flexible environment, drawing on the educational advantages of the AVs with different aspects concerning the embedded systems, such as declaration of variables and functions, configuration of ports and peripherals, handling of Real-Time Operating System (RTOS), interrupts, among others, that are employed in several fields such as robotics, control, instrumentation, etc. In addition, some experiments that were designed in the platform are presented in the manuscript. The educational methodology and the assessment provided by the students ($n = 30$) suggest that the platform is suitable and reliable to learn concepts relating to embedded systems.

Keywords: open-source platform; visual algorithms; digital signal controllers; embedded systems education; dsPIC; Java

1. Introduction

Digital Signal Controllers (DSCs) are hybrid devices that cluster the computing features of a Digital Signal Processor (DSP) with the configuration advantages of a microcontroller [1]. The architecture of these devices is between 16-bit and 32-bit. According to the EE Times survey (2017) [2] that analyzed the opinions of 1234 skilled engineers worldwide in the embedded market, 66% of the survey respondents use both 32-bit and 16-bit processors in their projects. Concerning the 16-bit processors, the company Microchip Technology Inc. provides the best ecosystem for programming and debugging these devices and 45% of the survey respondents that use 16-bit processors plan to employ a Microchip device (dsPIC) in their future designs.

In this context, this manuscript discusses the design and the implementation of DSCBlocks, a novel open-source platform designed in JavaFX [3,4] for learning embedded systems based on Algorithm Visualizations (AVs) and 16-bit DSCs. It has selected the hardware devices dsPIC 33FJ128GP804 and dsPIC 33FJ128MC802 [5,6] that belong to the DSC family from Microchip Technology Inc. known as dsPIC 33 [7]. These devices have interesting features, e.g., remappable peripherals, internal Fast RC Oscillator (FRC) with PLL running at 7.37 MHz, processor speed of 40 Million of Instructions Per Second (MIPS) and multiple peripherals and protocols such as Inter-Integrated Circuit (I2C), Universal

Asynchronous Receiver-Transmitter (UART), Controlled Area Network (CAN), Serial Peripheral Interface (SPI), etc.

At the educational level, the platform takes into account the AVs that have been employed widely to learn and teach algorithmic structures in computer science. The objective of these algorithms is to provide a way in which the students understand the procedures involved in an Algorithm [8]. Shaffer et al. [9] argued that these algorithms allow teaching and exploring the concept of a program as well as promote self-study. Furthermore, the students can debug their algorithms following the different steps involved in them. Some studies have shown that the AVs could improve the understanding of algorithms and data structures, elements that are part of traditional computer science curricula.

However, there exists a current lack of open-source educational tools to learn and explore the different parameters of the Digital Signal Controllers (DSCs). In this sense, this work presents an innovating platform that addresses this issue. *DSCBlocks* combines a Graphical User Interface (GUI) created in JavaFX with Google's tool *Blockly* [10,11]. The GUI includes some elements as a project wizard, a real-time data plotter and a serial port visualizer. Blockly is used to build graphical blocks that could be converted into several programming languages, for instance, Dart, JavaScript, Python, PHP or Lua. Through Blockly, different types of blocks were created and tested to use ports, peripherals, RTOS, interrupts, variables and functions for the dsPICs 33FJ128GP804 and 33FJ128MC802. Furthermore, the blocks designed can work with other dsPICs with similar architecture to these devices.

The blocks are converted in real-time to C programming language according to the structure of the XC16 compiler [12] and they implement a set of functions divided into the categories *oscillator, input–output, peripherals, communications, RTOS, interrupts, delay* and *functions*. Since the main interest in the platform is to contribute to the learning in the embedded systems area, the blocks shape basic functions that students must know and employ to start-up the dsPIC. Afterwards, the programming code produced is downloaded into the device. Then, students or users can debug their algorithms. For this, two models of development board with six analog or digital inputs, four digital outputs and two analog outputs in the voltage range 0–3.3 V with the dsPICs 33FJ128GP804 and 33FJ128MC802 have been proposed.

As support for the educational materials of the platform, the *Xerte* Toolkit Online (XOT) [13,14] was used, which is an open tool suite designed by the University of Nottingham to create interactive learning materials. The different explanations and examples of the platform were produced in this tool and they are accessible in the URL provided in the Supplementary Materials. Additionally, the materials developed in XOT could be embedded in popular e-learning platforms such as Moodle or Integriertes Lern-,Informations- und Arbeitskooperations-System (ILIAS).

The platform has been assessed with 30 students in the course entitled *microcontrollers* that belongs to the embedded systems area of the program of Electronics Technology at UNIMINUTO university. The participants were sophomores and they used the platform in different tasks provided in the classes during 2017-II and 2018-I. Their opinions about the platform were summarized through a survey whose results are shown in this paper. Furthermore, some conclusions derived from several observations of the student interactions with the platform are discussed.

From the elements mentioned, this paper is organized as follows: Section 2 describes the background of this research. Section 3 exposes a general and detailed architecture of the platform. Section 4 explains some experiments with a proposed low-cost development board. Section 5 shows the assessment provided by the students about the platform. Finally, discussion and conclusions are outlined in Sections 6 and 7, respectively.

2. Background

This section discusses the concept of Algorithm Visualization (AV) and related works from an educational perspective.

2.1. The concept of Algorithm Visualization (AV)

Algorithm visualization (AV) is the subclass of software visualization that illustrates the high level mechanisms of computer algorithms in order to help pupils understand in a better way the function and the procedures of an Algorithm [8]. In the same way, Shaffer et al. [9] argued that AVs could be used for the instructor as part of a lecture, a laboratory or an assignment to teach one concept. Additionally, AVs could help explore a concept without explicit direction as well as promote the interactivity to test the comprehension of the students about the algorithmic structures. According to Törley [8], AVs have been employed for the following purposes:

- To illustrate an algorithm by the instructor.
- To understand the mechanism of basic algorithms.
- To debug an algorithm by students.
- To help pupils understand the function and operation of abstract data type.

To find the advantages of AVs in the educational context, Hundhausen et al. [15] conducted a meta-study based on 24 major visualization studies. The research shows how the visualizations influence the learning outcomes and the student's attention.

Rößling et al. [16] considered the impact of AVs on student engagement. In their paper, they mentioned that the members of the group in Innovation and Technology In Computer Science Education (ITICSE) from Association for Computing Machinery (ACM) explored four categories of learning with the AVs, namely responding, changing, constructing and presenting, searching a better impact of the AVs in learning. Regarding the first category, instructors are focused on different questions posed by the students about the visualization of an algorithm. In addition, the algorithm visualization is a source that allows suggesting different types of questions, helping the students with their learning process. As for the second category, the students provide input information for the algorithm, generating a certain visualization response. The students use the visualization to create hypotheses about the behavior of the algorithm. In the constructing category, the students build their own algorithms according to some rules or constraints provided by the instructor. In addition, the students observe the response of their algorithms. In the last category, the students use the visualization to explain an algorithm to the audience in order to interchange ideas in a constructive learning space.

Similarly, Rößling et al. [16] proposed a summary of key-points in different projects that consider the AVs, among which the most important are:

- Providing an input to the algorithm in order to modify its features.
- Including a hypertext that explains the visualization of the algorithm.
- Preferring general-purpose systems over topic-specific systems in virtue of the reuse option.
- Integrating a database for course management.

Pasternak et al. [17] discussed creating visual languages with Blockly. Several features identify Visual Programming Languages (VPL), e.g., drag blocks around the screen, flow diagrams or any mechanism for wiring different blocks and using icons or non-text representations. In addition, every VPL has *grammar* and *vocabulary* that define the behavior of the language that has been created. In any case, the authors marked out some reflection points that could give a horizon for the design of VPLs with Blockly:

- **What** is the audience for your language? **Who** are you building for?
- **What** is the scope of your language? An excessive amount of blocks in the language could overwhelm or to generate distractions in the users.
- **Employ** a natural language in the VLP. With it, the users can build the AVs in an intuitive way.

These elements summarize the key-points for the VLPs and most of them have been taken into account in the design of the XC16 compiler code generator for the platform.

2.2. Related Works

Concerning the related works, the usage of AVs in engineering fields as embedded systems, automatic control or IoT is rather new. Nevertheless, some studies give a guide in this regard.

In [18], a programming language called *Robot Blocky* for ABB's Roberta robot is presented. The authors used a customized language built with Blockly to create a block-based interface for programming a one-armed industrial robot. The authors also presented a small-scale study in which the experiences of students with the developed language are exposed.

Angulo et al. [19] presented a remote laboratory to program robots with AVs called *RoboBlock*. This laboratory is based on a Zumo 32u4 Pololu robot and a Raspberry Pi model 3, and is integrated to WebLab-Deusto Remote Laboratory Management System (RLMS). The authors pointed out the construction of ArduLab, an interface to program and test the mentioned remote robot.

In [20], a visual framework for Wireless Sensors Networks (WSNs) and smart-home applications is exposed. The framework is composed of AVs in which the users can program a smart-home management system. The system is composed of a lightweight tool with an intuitive user interface for commissioning of IP-enabled WSNs. In the research, the authors exposed a prototype to test the visual solution with the WSNs.

In [21], the author proposed an open-source platform called *Controlly* designed with *Blockly* for control systems education. The application has several blocks in order to implement classic controllers, such as Proportional (P), Proportional-Integral (PI), and Proportional-Integral-Derivative (PID). An exposition of the designed controls with the interface for a control plant (DC-DC buck converter) is analyzed in the manuscript.

In [22], a study of the advantages to integrate *Blockly*, Virtual and Remote Laboratories (VRLs) and Easy Java Simulations (EJS) is shown. The research also indicates the design of an API for communication between *Blockly* and a proposed laboratory. In the investigation, the authors worked on a Moodle Plugin that aims to create new experiences in Learning Management Systems (LMS) to foster higher engagement of the students.

The authors of [23] dealt with a language for IoT called *Smart Block* designed specifically for SmartThings that generalizes the Event–Condition–Action (ECA) rules for home automation. The authors also posed a second application to design mobile applications and a development environment to test the language.

In [24], the authors presented an educational mobile robot platform based on MicroPython and *Blockly*. The robot can be programmed easily by the students without wiring for its functioning and it has a STM32F405RG 168 MHz Cortex M4 processor, embedded with MicroPython, an ultrasonic sensor and a Bluetooth module for wireless communication between the interface and the robot. The project was planned for students who are interested in learning programming at a basic level.

From a Computer Science perspective, the research in [25] provides an example of Behavioral Programming (BP), where the individual requirements of certain application are programmed as independent modules. The authors indicated that the BP allows the implementation of interactive technologies such as client-side applications or smartphone customization through *Blockly* and JavaScript. Further work consists of expanding the scope of the developed BP to other areas such as complex robotics or large biological models.

In [26], a study to measure the impact of block-based languages in introductory programming is tackled. The author proposed a method known as Bidirectional Translation System where the students alternate between block-based and textual languages. The authors concluded that block-based language worked to encourage students to focus on high-level algorithm creation.

Finally, the authors of [27] presented the initiative *IOIOAI* that simplifies the programming process through an Android application that can communicate with the IOIO board and the App MIT Inventor platform. The project was tested with adults and children during 2016–2017 in Tunisia. According to the authors, the students enjoyed the laboratory that stimulated their curiosity, making them more open to learn more.

Nonetheless, despite these important studies, there exists a lack of Open Educational Resources (OERs) in the field of embedded systems with DSCs and AVs that the present work intends to tackle. In addition, the platform is novel, as it provides a complete open-source environment to configure and start-up a DSC in which the students can see the different processes to get a determined algorithm in real-time using AVs as learning method.

3. Platform Design and Implementation

This section addresses the aspects entailed in the design and implementation of the platform including software and hardware components.

3.1. Software Component

The software structure in the platform is divided into the layers (Presentation, Application, and Abstraction) depicted in Figure 1. Each layer is explained in the next subsections.

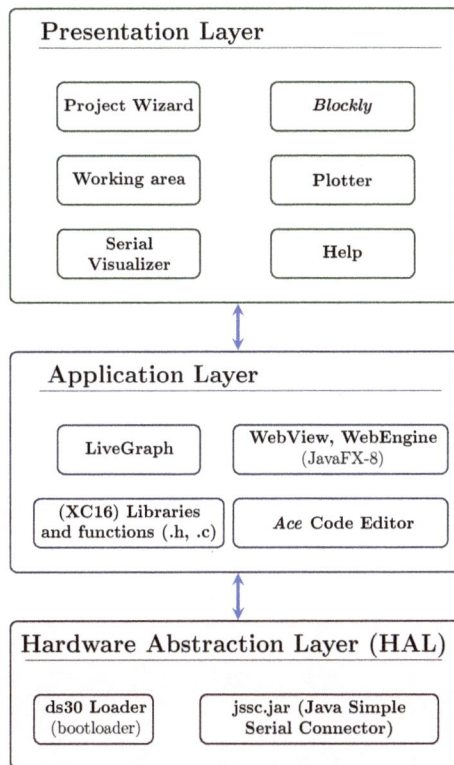

Figure 1. Main software components of the platform by layers.

3.1.1. Presentation Layer

The Presentation Layer is composed of the User Interface (UI) of the application. Firstly, the UI is composed of a project wizard that helps the user with some parameters such as the paths for the compiler and the project, the type of board for the models (dsPIC 33FJ128GP804 and dsPIC 33FJ128MC802) and the communication port to use, e.g., COM1 or COM2, to transfer the created visual algorithm towards the development board. The communication between the development board and the application is managed through a virtual serial port with the USB feature Communication Device Class (CDC) [28].

Secondly, the UI includes a working area in which the students or users can build their algorithms based on the different graphical blocks designed in Blockly and distributed by categories. Thus, a specific C language generator for the XC16 compiler was developed in this platform. The generator converts every block in the C language equivalent and the UI shows this transformation in a tab in real-time. Thereby, students or users can observe the respective code with the configuration of the registers in the DSCs' architecture, helping to learn programming and the implementation related to these devices.

Tables 1 and 2 show a summary that condenses the designed categories with some examples.

Table 1. Summary of the block categories with examples (Part I).

Category	Description	Block Example
Oscillator	Set-up for the (FRC) oscillator. Range ($0 < F_{cy} \leq 40$ MHz) **Parameters:** Values of N1, N2, M (See datasheets dsPIC 33FJ128GP804 and dsPIC 33FJ128MC802).	
Input–Output (I/O)	*High Pin*: Write a logical 1 on selected pin. *Low Pin*: Write a logical 0 on selected pin. *ReadPin*: Read the state of a pin. **Parameters:** Pin number (1–6).	
Peripherals	*ADC Single Channel*: Read an ADC value. **Parameters:** ADC channel number (0–5). *ADC multiple channels*: Sample several ADC channels simultaneously. **Parameters:** Sample time in (ms). *PWM Output*: Configure the selected PWM Channel. **Parameters:** Duty Cycle; PWM frequency(Hz); Pre-scale (1,8,64,256); PWM Channel (0-4). *DAC Channel A,B*: Configure the 12-bit Digital to Analog Converter (DAC) MCP4822 [29] through SPI protocol. **Parameters:** Channel (A,B); Digital value to convert in analog (scale 0 to 3.3 V).	
Communications	*UART Send Integer*: Send an integer value to UART peripheral. **Parameters: Integer number.** *UART Write Text*: Write a string to UART peripheral. **Parameters:** Text String. *UART Write Float*: Write a float number to UART peripheral. **Parameters:** Float number.	

Table 2. Summary of the block categories with examples (Part II).

Category	Description	Block Example
Communications	*ReadUart*: Read a byte from UART peripheral. **Parameters:** None. *DataRdyUART*: Check if a byte is available in the UART peripheral. **Parameters:** None.	ReadUart() DataRdyUART1()
RTOS	*OS Init*: Start the RTOS. **Parameters:** None. *OS Run*: Start the created tasks in the RTOS. **Parameters:** None. *OS Yield*: Free the current task executed in the RTOS. **Parameters:** None. *OS Timer*: Start the Timer for the RTOS. **Parameters:** None. *OS Task Create*: Create a task with priority. **Parameters:** Priority (0–7); Task name. *OS Delay*: Create a configurable delay for a task. **Parameters:** Ticks or cycles for the delay.	OS_init() OS_Run() OS_Yield() OS_Timer() Priority Task name OSTaskCreate OS_Delay() delay in ticks (cycles)
Interrupts	*Timer interrupt*: Configure a Timer interrupt with priority and pre-scale. **Parameters:** Timer number (0–5); Pre-scale (1,8,64,256); Elapsed time in milliseconds (ms); Priority (0–7).	Timer interrupt (timer) 1 Pre-scale 1 Time interrupt (ms) Priority (0...7) 0
Delay	*Delay (ms)*: Delay in ms. **Parameters:** Time in ms. *Delay (µs)*: Delay in microseconds (µs). **Parameters:** Time in (µs). *NOP*: NOP instruction. **Parameters:** None. *Delay in cycle*: Delay in instruction cycles. **Parameters:** Number of instruction cycles for the delay.	Delay(ms) Delay(us) NOP Delay(Cycle)
Functions	*Function without parameter to return*: Create a function without a return parameter. **Parameters: Invoking variables.** *Function with parameter to return*: Create a function with a return parameter. **Parameters: Return variable and invoking parameters.**	to do something to do something2 return

The elements in Tables 1 and 2 are explained below.

- *Oscillator*: In this category, the user can configure the oscillator of the DSC. For the dsPICs 33FJ128GP804 and 33FJ128MC802, the equation that defines their operation frequency in Hz is the following:

$$Fcy = \frac{F_{osc}}{2} = F_{in}\left(\frac{M}{N_1 \cdot N_2}\right) \tag{1}$$

where F_{in} is the frequency of the internal FRC oscillator (7.37 MHz), F_{osc} is the output frequency of the Phase-Locked Loop (PLL), M is the PLL multiplier and N_1 and N_2 compound the PLL postscale. The values of N_1, N_2, and M are established by the user. For instance, when $N_1 = 2$,

$N_2 = 2$, $M = 40$, these values yield a frequency of 36.85 MHz (36.85 MIPS). The graphical block converts the previous values in the respective code for the registers involved in the configuration of the oscillator, in this case CLKDIV and PLLFBD.

- *Input–Output*: In this category are located the blocks for reading and writing the logical state of the DSC pins. Every block sets up the specific configuration register (TRISx) and assigns the logical state specified by the user (1 or 0). The pins that a user can employ are mapped in a C header file called <HardwareProfile.h>. This file contains the names of every pin provided by the manufacturer with a macro, e.g., the Pin1 in the application is RB12 in the DSC.

- *Peripherals*: In this category, the user can configure the Analog to Digital Converter (ADC), Pulse Width Modulation (PWM) or Digital Analog Converter (DAC) peripherals. With respect to ADC, it was configured in 12-bit mode with an input voltage in the scale 0–3.3 V. The resolution for the ADC is given by Equation (2).

$$\frac{3.3 \text{ V}}{2^{12} \text{ bits}} \approx \frac{0.81 \text{ mV}}{\text{bit}} \tag{2}$$

The block ADC Single reads the ADC channel selected by the user and returns an integer variable with the respective value in 12-bit. The block ADC Multiple Channels reads simultaneously up to four channels and returns the values in a set of preloaded variables (L0, L1, L2, and L3). The user must declare them, employing the category variables available in *Blockly*. In addition, this block uses the Timer 3 of the DSC to sample the indicated channels. PWM was configured in 10-bit mode with the XC16 library <pwm.h>. Four channels could be selected with the designed block. Moreover, the user can adjust the frequency in Hz and the pre-scale when a value for the register OCxR (Duty cycle register) is outside the maximum value in a 16-bit architecture. The pre-scale decreases this value.

Regarding DAC, the block configures the device MCP4822 [29], which is a 12-bit DAC in a range from 0 to 3.3 V. This device operates with SPI protocol and it contains a single register to write the digital value to convert. The user must indicate an integer variable and one channel (A or B) to write a voltage value in the mentioned scale. For instance, when the user writes the value 2048, it will correspond to the analog value of 1.65 V.

- *Communications*: This category contains several blocks to handle the UART peripheral with a default baud rate of 57,600. The blocks were designed utilizing the XC16 library <UART.h>. The library starts up the UART peripheral with the parameters specified by the user (interrupts in transmission or reception, addressing mode in eight or nine bits, interruption priority, etc.). Additionally, several functions were developed to transform either an integer or float variable in a string ready to transmit. The category has several blocks to writing and reading data: *UART write text*, *UART write integer*, *UART write float* and *UART read data*.

- *RTOS*: For the platform, OSA was selected, which is a small RTOS compatible with dsPIC. OSA [30] is a cooperative multitasking RTOS that uses, in this case, Timer 1 to generate the *Tslice* for the different assigned tasks. In a cooperative RTOS, each tasks is executed periodically with a time provided by the system scheduler [31]. In the category, some blocks were designed to create and run tasks with priority in the range 0–7, where 0 is the highest level of priority and so on. A Java class copies the contents (folders and subfolders) of this RTOS into the user's folder to compile with XC16. An example of code with the blocks is shown in Algorithm 1.

- *Interrupt*: In this category, a timer interrupt was designed with the associated Interruption Service Routine (ISR). The graphical block contains several inputs such as pre-scale, clock tick between interrupts in ms and priority in the range 0–7. These elements serve to open and configure the timer selected by the user. The block operates with the frequency provided by the oscillator block that the student must configure previously. An "Interrupt" is a key concept because it allows understanding the architecture of any embedded system, in this case, concerning the DSCs. As concept, Di Jasio [32] defined an interrupt as an external or internal event that requires a quick

CPU intervention. A code example with this block is shown in Algorithm 2 with a time between interrupts of 1 ms, a DSC frequency of 36.86 MHz and 1:1 pre-scale.

- *Delay*: In this category, several blocks for delays in ms and μs, and instruction cycles were designed. The XC16 library <libpic30.h> was employed to create the delays based on instruction cycles according to the frequency specified by the user. For example, for a frequency of 36.86 MHz and a delay of 10 ms, the block delay(ms) will return the statement _delay32(368500).
- *Functions*: In this category, the user can create C functions either with or without return variable. In addition, the names and invoke parameters of each function must be defined by the user as global variables. The functions make the code more readable and organized for the students.

Algorithm 1 Example of generated code for the RTOS (OSA).

```
//Task (Pin oscillator)
void Task2(void){
        while(1){//Task in infinite loop. Feature of cooperative RTOS.
                PIN1=1;//Write logical 1 on PIN1.
                OS_Delay(10000);//Task Delay (10000 clock ticks).
                PIN1=0;//Write logical 0 on PIN1.
                OS_Delay(10000);//Task Delay (10000 clock ticks).
                OS_Yield();//Return to scheduler.
        }
}
```

Algorithm 2 Example of generated code for the block (*Timer interrupt*).

```
#include <timer.h>//XC16 Timer Library
//Interruption Service Routine (ISR) for Timer1
void __attribute__((interrupt,no_auto_psv)) _T1Interrupt( void )
{
        IFS0bits.T1IF=0;//Clear Timer flag
        WriteTimer1(0);//Restart Timer
}
//Routine to configure the selected timer
void ConfigTimer1(void){
        T2CONbits.T32=0;//Timer register in 16-bit mode
        T4CONbits.T32=0;
        ConfigIntTimer1(T1_INT_PRIOR_0 & T1_INT_ON ); //Set-up of Timer1.
        WriteTimer1(0);//Write 0 to the Timer
        OpenTimer1(T1_ON & T1_GATE_OFF & T1_PS_1_1
        & T1_SYNC_EXT_OFF &T1_SOURCE_INT,36850);//Clock tick of 1ms.
}
```

Thirdly, a plotter and a serial port visualizer were added to the application. For the plotter, the Java library known as LiveGraph was utilized [33,34]. LiveGraph allows plotting up to 1000 data series simultaneously, with an update between 0.1 s and 1 h. Data are saved in the application in CSV format and the user should build a AV to plot the different data from the development board. A Java class embedded in the application captures the data incoming from the serial port and sends them to the framework of LiveGraph. With the CSV file, the users can export the data towards different mathematical environments such as MATLAB, Octave, etc. for further processing. When a user wants to plot data, the application will request the number of variables to plot, the update frequency and one option to buffer data. Data sampling can be stopped, closing the serial port from the application.

The serial port visualizer shows the user's data, using the same COM port indicated in the project wizard. The serial port visualizer could allow the users to check whether their algorithm is correct.

As mentioned above, the e-learning tool *Xerte* was employed as support for the construction of the educational materials of the platform that begin with the concepts of the DSCs and the embedded systems to end in the features of the application. *Xerte* provides a complete environment to develop materials with resources such as videos, quizzes, games or interactive web pages that accompany the learning process of the student.

Finally, an overall perspective of the UI with its main components and the educational materials are depicted in Figures 2 and 3, respectively.

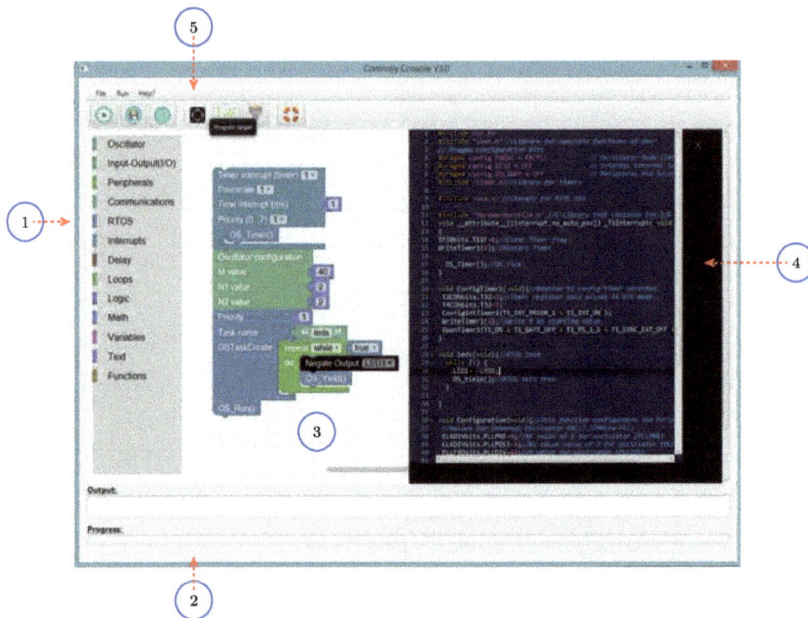

Figure 2. Overall components of the UI: (1) blocks palette; (2) console output; (3) working area; (4) real-time code tab; and (5) toolbar.

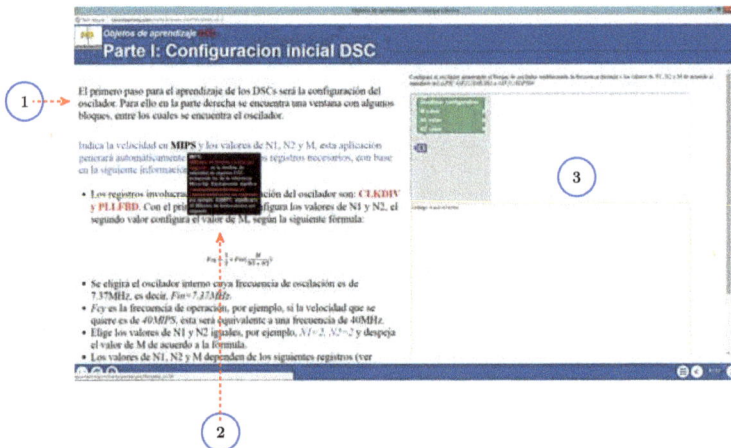

Figure 3. Educational materials in *Xerte* (Spanish version). (1) example of educational material; (2) glossary option of *Xerte*; and (3) embedded resource (interactive web page) with AVs.

3.1.2. Application Layer

The core of the software component of the platform is composed of the objects WebView [35] and WebEngine [36] available in JavaFX, which interact with *Blockly* to get the respective code for the DSCs. A WebEngine is a non-visual component to manage a specific webpage, which allows invoking the different functions made in JavaScript into *Blockly*. Because *Blockly* is web-oriented, a WebEngine is a useful component to exchange data with this application. The WebView displays adequately the content of a webpage into a *scene* in JavaFx. The scenes posed in this section were created with the software JavaFX Scene Builder 2.0 [37] and their respective controllers with NetBeans IDE 8.2. The Application Layer is composed of the elements: WebView, WebEngine, LiveGraph, Ace code editor [38] and some XC16 libraries. An *actor* can select seven options in the UI: *Project Wizard, Program, Plot, Serial Port Visualizer, Help, Open* and *Save*. The Unified Modeling Language (UML) sequence diagram for the option Project Wizard is depicted in Figure 4.

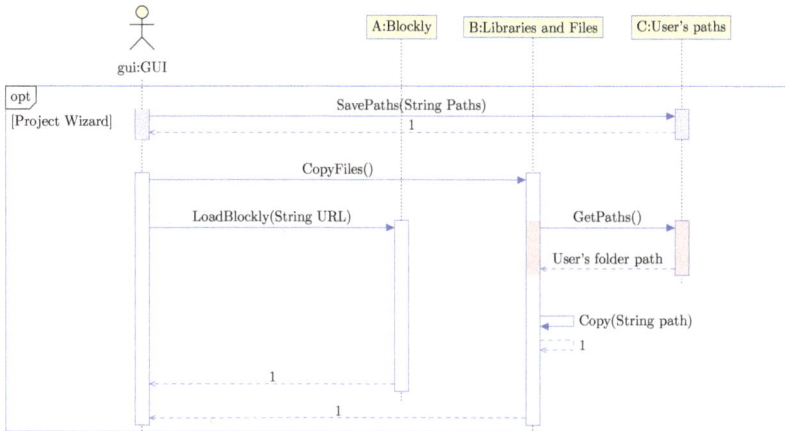

Figure 4. UML sequence diagram for the option Project Wizard.

When a user clicks on the button *Project Wizard*, this event opens a scene that asks for some parameters such as the user's folder path, the compiler's path and the serial port for the communication with the development board. These elements are saved as constructors in a Java Class to be used with other methods. Posteriorly, another class copies all files and folders needed into the user's folder. Table 3 shows a description of these files.

Table 3. List and description of copied files and folders into the user's folder.

Name	Type (File or Folder)	Description
OSA	Folder	It contains the files (.c, .h) for the RTOS (OSA).
User.h	File	Header file to configure, read and write the UART peripheral.
User.c	File	It contains the implementation of the functions to configure, read and write the UART peripheral.
Data.csv	File	File with the user's plotter data.
Hardwareprofile.h	File	Header file with the pin-out definitions for the development board.

Last in the sequence, *Blockly* is loaded into the WebView, employing the JavaFX constructor *WebEngine()* with the method *load(String URL)*. The URL contains the local path for *Blockly*, which is found in the resources' folder of the application. A tab in the workspace deploys in real-time the code for the elaborated AV, as illustrated in Figure 2. Moreover, the generated code is highlighted with the Ace code editor [38], which is a syntax highlighter for diverse statements in over 110 programming languages, including C.

Each block inside *Blockly* has two associated files: (1) a shape file with the definition of the graphical attributes; and (2) a file that describes the behavior's block, that is, the returned C language code by the block. For instance, the block delay(ms) depicted in Table 2 is associated with the JavaScript function shown in Algorithm 3 for its behavior. In this algorithm, the value with the time in milliseconds (ms) is assigned to the variable *OSCVal*. Then, the code for the block is returned and injected by *Blockly* to be represented in the code tab for the user. As mentioned, this operation is made in real-time with the component WebView in JavaFX that allows to invoke the JavaScript functions nested in *Blockly*.

Algorithm 3 JavaScript Behavior function for the Block *Delay (ms)*.

```
Blockly.Dart.delay=function() {//Delay function
        var OSCVal=Blockly.Dart.valueToCode(this,'Time',
        Blockly.Dart.ORDER_ATOMIC);
        var code=__delay32('+OSCVal+');//Returned code for the block.
        return code;
};
```

The second option available in the UI is to *program*. The respective sequence diagram for this function is depicted in Figure 5.

The method GetUserCode() calls the JavaScript function (content()) that returns through a JavaScript *alert* the programming code for the blocks. A callback registered with the constructor (WebEngine) detects this alert and it proceeds to get the code as a string. Subsequently, the content of this string is saved in a C file to be compiled with the compiler XC16, employing its command line. The previous operation is executed by the class *Runtime* in JavaFX. The command line option provides the different instructions to compile and generate the Hex file to program the DSC and a report with errors, memory distribution (RAM, Flash), etc.

With the Hex file, the method CallBootloader() invokes the Bootloader ds30 Loader (free edition) [39]. By definition, a bootloader is a small piece of code inside the DSC that manages its programming mode through some peripheral such as UART, CAN or USB. For the bootloader, an assembler file provided by the developer was loaded into the program memory of the DSC with a default baud rate of 57,600. Some modifications were made to this file according to the remappable pins of the UART peripheral of the DSC. The bootloader is executed by the application in the command line mode, in which a string command with the path for the Hex file, the name of the COM port, the reset command and the device to program is provided.

When a reset occurs by the mentioned command, the DSC starts with the bootloader during 5 s, waiting for a new programming request from the application. When no programming request is made, the bootloader gives the control to the code loaded previously in the flash memory of the DSC. Independently, the bootloader opens and closes the serial port and it sends the Hex file towards the DSC, managing all the process.

Concerning the option *plot*, the UML sequence diagram for this option is represented in Figure 6.

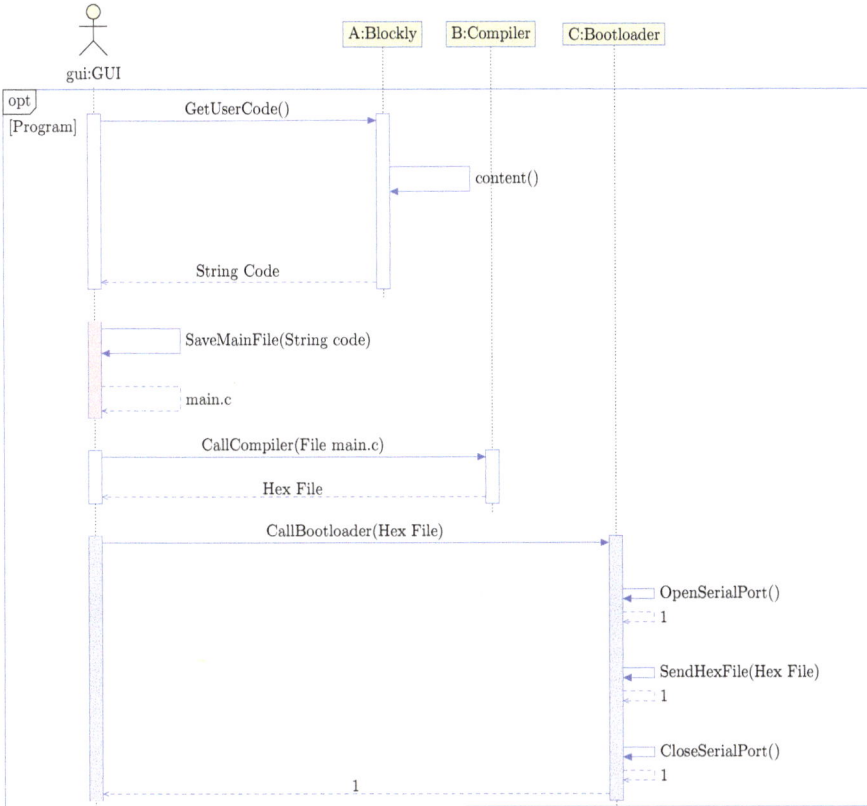

Figure 5. UML sequence diagram for the option Program.

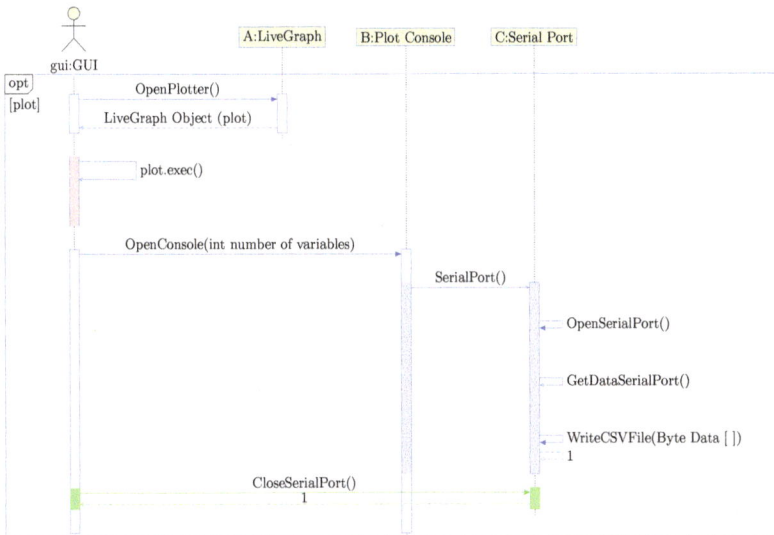

Figure 6. UML sequence diagram for the option Plot.

To create an instance for the plotter, the method OpenPlotter() is invoked and an object is returned by it. This object opens the different elements of the class LiveGraph. A new scene is deployed to the user that asks the number of variables to plot. Hence, the result of this transaction is a new window in which the user can see and update the data, as shown in Figure 7.

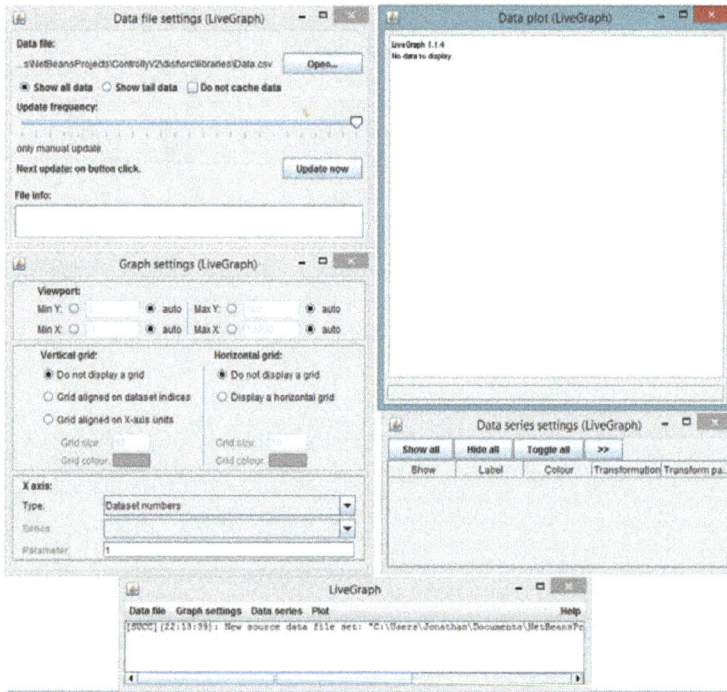

Figure 7. LiveGraph components composed of windows (plotter and update data).

Afterwards, the serial port (COM) is opened and it gets the data incoming from the development board in a buffer. Meanwhile, a method in the application saves the data in a CSV file.

The user also can save or open the designed AV and to request help from the application. The programming structure of these functions is summarized in the sequence diagrams in Figure 8.

Whether a user wants to open or save the AV created, *Blockly* contains the JavaScript function *Blockly.Xml.domToPrettyText* that converts the AVs into an XML file, which is saved in the user's folder. The Java class *WebEngine* calls the mentioned JavaScript function and it returns a text with the XML code in a JavaScript alert. The code is retrieved by means of a callback event in the application that detects this alert. In the same way, the XML file in the user's folder can be upload to the application, recovering the AV made through the functions *Blockly.Xml.textToDom* and *Blockly.Xml.domToWorkspace*.

With the help of the application, it releases a new browser window with the URL to the *Xerte* materials, as depicted in Figure 3. To open this window, the application uses the Java class *Desktop* with the method *browse*.

Finally, a *serial port visualizer* was created through library, Java Simple Serial Connector (jSSC) [40], which implements the necessary methods to write and read data from the serial port. The library provides an asynchronous event to get data in a byte buffer that are converted to string and displayed in a JavaFx TextArea. Regarding the asynchronous event, it allows the user to execute other functions in the application with the serial port receiving data of the development board at the same time. Figure 9 shows the sequence diagram for this function. Below, it is shown how the hardware component of the platform interacts with the software application.

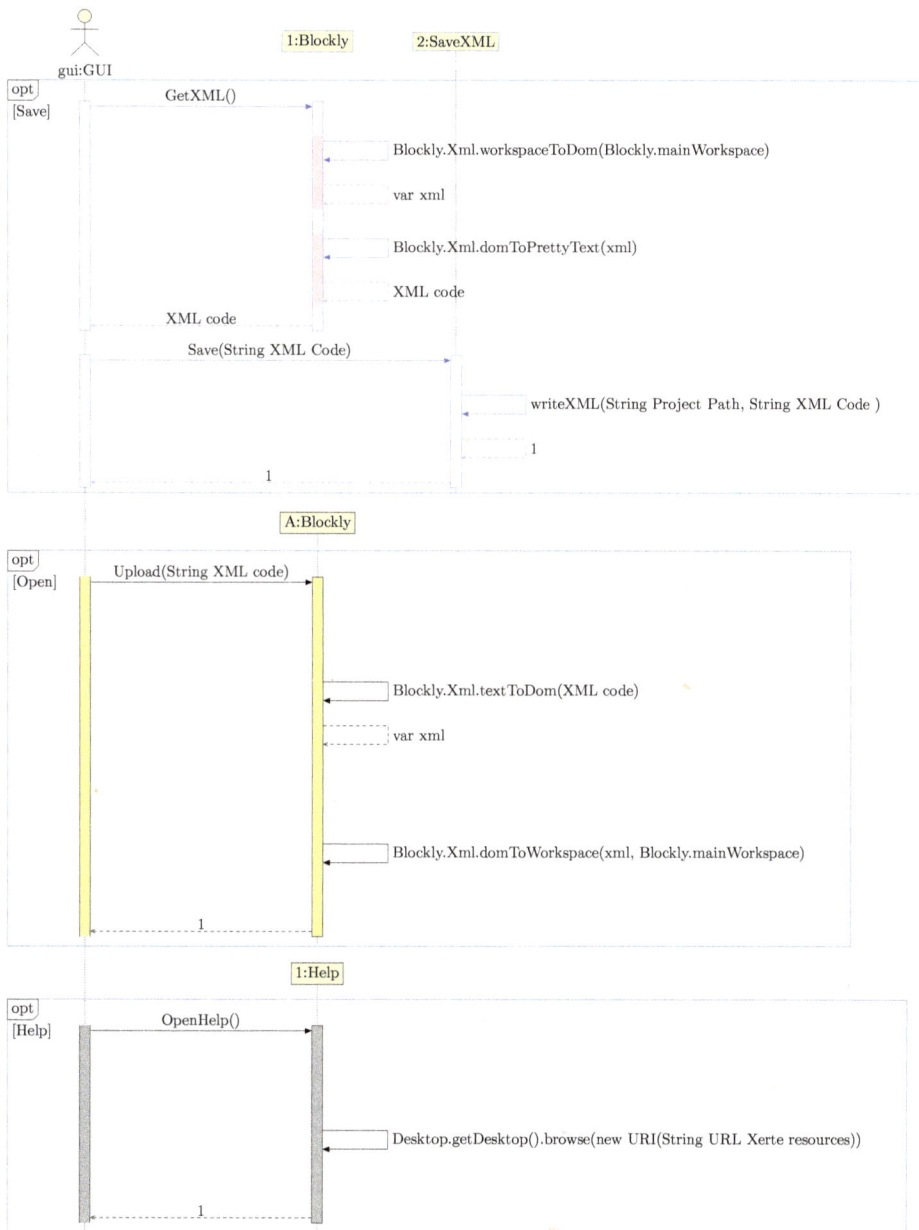

Figure 8. UML sequence diagrams for the functions Open, Save and Help.

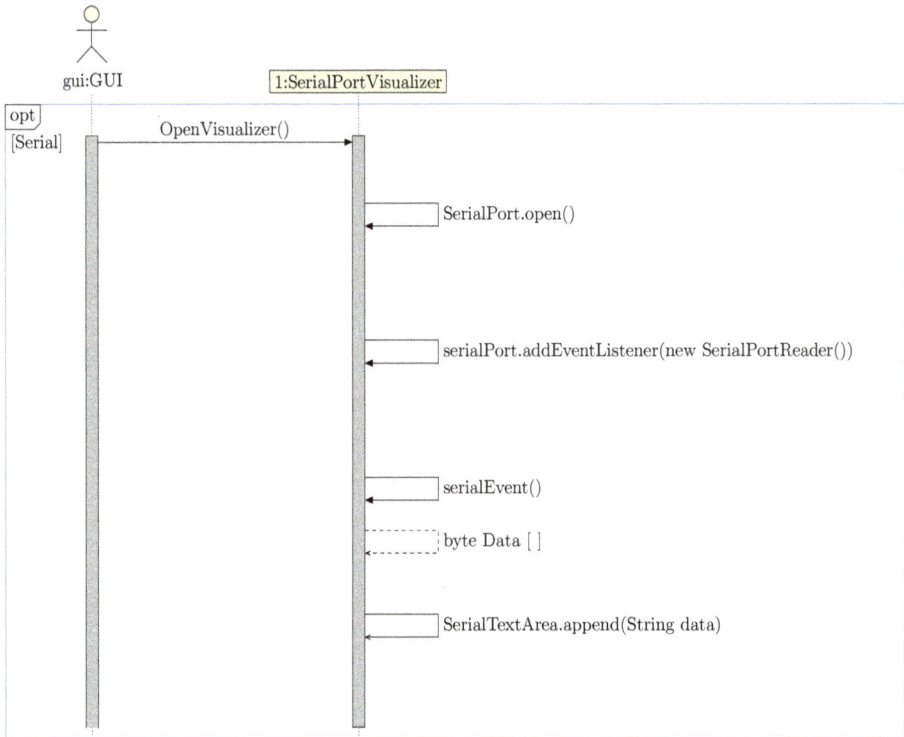

Figure 9. UML sequence diagram for the function Serial Port Visualizer.

3.1.3. Hardware Abstraction Layer

The communication between the software and the hardware components in the application is managed by the Hardware Abstraction Layer (HAL), which is composed of the bootloader (ds30 Loader) and the Java library (jSSC). These elements form the HAL because they interchange information between the software and hardware elements in the application. As mentioned, the bootloader allows programming the DSC with the Hex file provided by the compiler and it is composed of two overall files, namely, the firmware and a console that is invoked through a command line, sending the Hex file towards the DSC. Table 4 shows a summary of the files provided in the firmware with their description.

Table 4. List and description of files for the bootloader (ds30Loader).

File Name	Description
ds30loader.s	It contains the assembler implementation for the bootloader.
devices.inc	It defines the memory size for the bootloader in *words and pages* according to the selected device.
settings.inc	It indicates the DSC's reference, the configuration bits (fuses) and the UART's baud rate.
uart.inc	It describes the configuration for the UART's registers.
user code.inc	It configures the remappable pins for the UART and the oscillator settings so as to start-up the DSC.

The firmware should be modified according to the specifications of the DSC in assembler (asm30). For instance, in the hardware, pins RB9 and RC6 of the dsPIC 33FJ128GP804 serve for reception and transmission of data, respectively. Thus, the RPINR19 and RPOR11 registers were configured for these functions and the FRC oscillator was enabled with the default frequency of 7.37 MHz. With these parameters, the bootloader is programmed in the DSC, utilizing a programmer such as PICkit 3 [41] or ICD3 [42].

3.1.4. Application Summary

Finally, a summary with the software features of the platform is described in Table 5. The application was tested in a PC with the following parameters:

- Processor: Intel(R) Core (TM) i5-4460T @ 1.9 GHz.
- Installed memory RAM: 8 GB
- System type: 64-bit operating system. (Windows 8).
- Local disc capability: 1.5 TB
- Java version: 1.8.0.121

Table 5. Summary of main features of the software component in the platform.

Feature	Description
Application size in disc	161 (MB)
Average Heap size (Java)	22 (MB) of 100 (MB) assigned.
Peak CPU usage	31%
Programming language	Java (JavaFX 8)

The application was also tested in a PC based on Windows 7 with similar features. To complement the previous information, VisualVM 1.4.2 [43] was used, which is a software that monitors and troubleshoots applications running on Java. The analysis information provided by this software is shown in Figure 10.

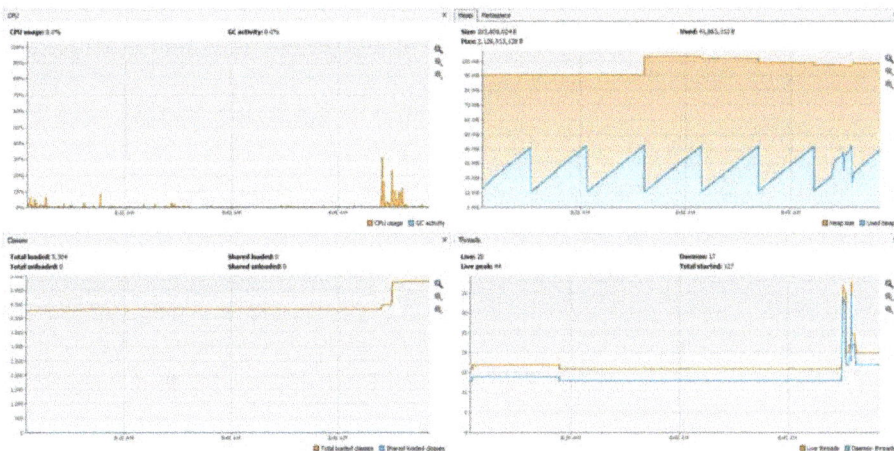

Figure 10. Benchmark information for the application provided by VisualVM 1.4.2.

The information in Figure 10 was retrieved when the AV represented in the Example 1 in Section 4 was built. In this case, the peak CPU usage in percentage was 31% with an average heap size for Java of 22 MB.

3.2. Hardware Component

Two models of low-cost development boards were built as hardware components in the platform. They are composed of the DSCs (dsPIC 33FJ128GP804 or dsPIC 33FJ128MC802) with the respective conditioning for their inputs, a 3.3 V power supply and a UART to USB bridge (FT232RL). Each development board contains six analog or digital inputs (depending on the function assigned by the user), four digital outputs and two DAC outputs (channels A and B). A scheme with the main components of the developments boards is depicted in Figure 11.

The analog inputs are conditioned by voltage followers built with Operational Amplifiers (Op-Amps) (MCP6004) [44] that limit the voltage of the input signals to a range of 0–3.3 V and reduce their noise. In addition, the selected Op-Amps are rail-to-rail, that is, the output voltage (V_{out}) of each when the Op-Amp is in saturation mode is approximately $V_{out} = V_{sat} - 150\,\text{mV}$ with $V_{sat} \approx 3.3\,\text{V}$. This feature allows the user to have a full span of the applied analog signals.

For the inputs, an example of schematic with the voltage followers is presented in Figure 12.

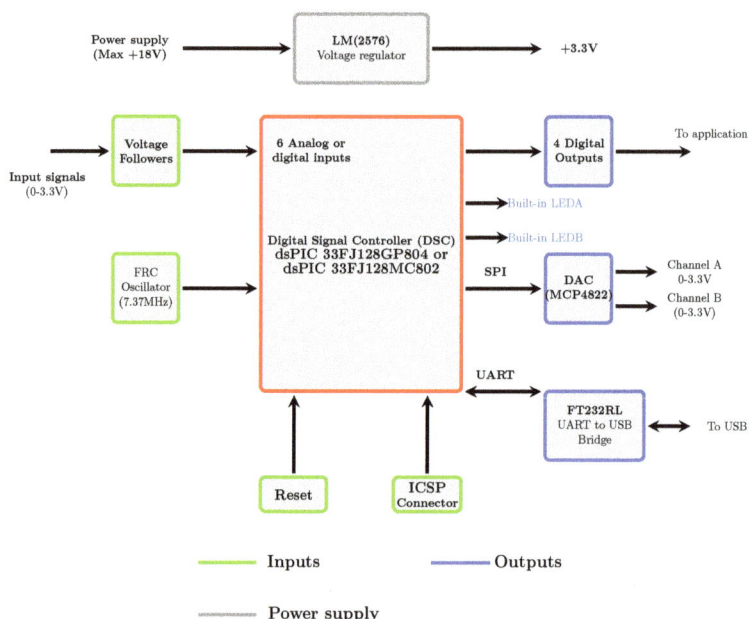

Figure 11. Overall scheme of the development board by blocks *inputs, outputs* and *power supply*.

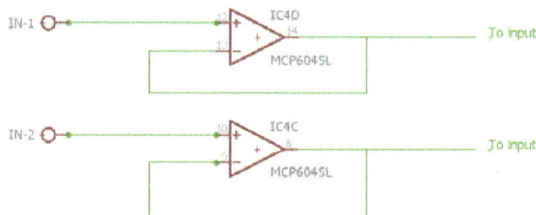

Figure 12. Schematic of the inputs' conditioning through voltage followers for the development boards.

The user connects their signal to the terminal blocks (IN-1 to IN-6) that are wired to the Op-Amps (MCP6004). This hardware device has four Op-Amps with a 1 MHz Gain Bandwidth Product (GBWP),

90° phase margin, a power supply voltage in the range of 1.8–6 V and a quiescent current of 100 μA. The device is suitable for the hardware requirements of the platform. The development boards also have a reset push button and one In-Circuit Serial Programming (ICSP) header for a programmer such as PICkit 3 or ICD3. For the design of the application, the low-cost programmer (PICkit 3) from Microchip Inc. was used.

As regards to the power supply of 3.3 V, it was designed through a step-down voltage regulator (LM2576) [45] with a maximum load current of 3 A for the model (dsPIC 33FJ128GP804). The current consumption for a frequency of 40 MIPS in the DSC is roughly 0.17 A. The regulator provides thermal shutdown and current limit protection with a efficiency (η) around 88% for the parameters V_{in} = 18 V and I_{Load} = 3 A. Figure 13 illustrates the schematic for the power supply.

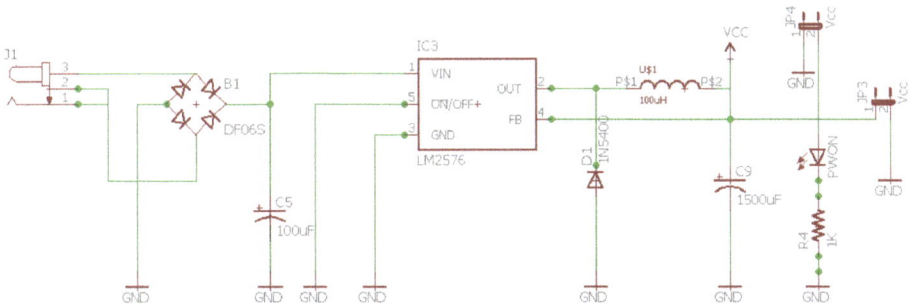

Figure 13. Schematic of the Power Supply with a step-down voltage regulator (LM2576). The connectors JP3 and JP4 provide voltage test points for the user designs.

For the model dsPIC 33FJ128MC802, a voltage regulator (LM1117-3.3) with a maximum load current of 800 mA was used. Concerning the UART to USB bridge, it was built with the chip FT232RL [46] from FTDI that has a CDC emulation and it transfers the USB data to the UART peripheral for the DSC. Figure 14 shows the followed schematic for this device in the two models of development board.

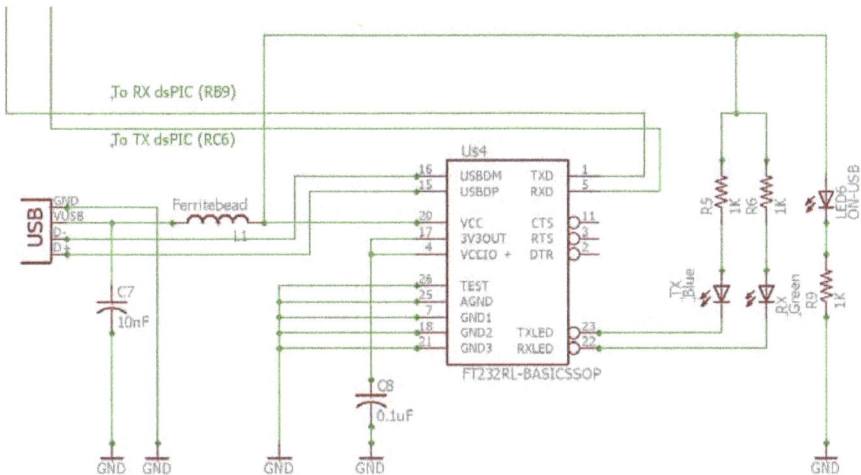

Figure 14. Schematic of the USB to UART bridge (FT323RL). The points RXD (Data reception) and TXD (Data transmission) are connected to dsPIC pins RB9 and RC6, respectively, for the dsPIC 33FJ128GP804.

The previous elements compose the development boards depicted in Figures 15–17 for the dsPIC 33FJ128GP804 and dsPIC 33FJ128MC802. The design of each development board can be modified effortlessly to use other type of dsPICs, e.g., the dsPIC 33FJ128GP802. Each development board has an average cost of US$30.

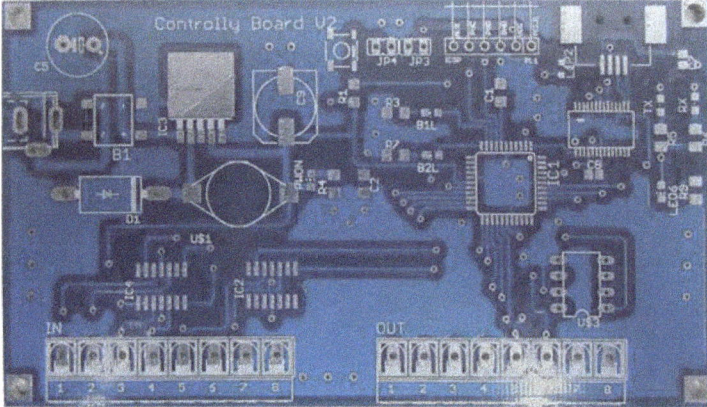

Figure 15. Development board's solder screen.

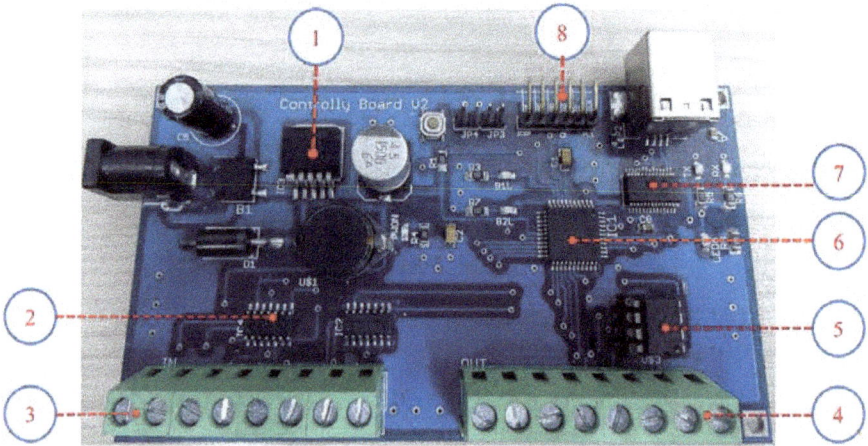

Figure 16. Overall appearance of the development board with dsPIC 33FJ128GP804: (1) power supply unit; (2) voltage followers; (3) inputs; (4) outputs; (5) DAC (MCP4822); (6) dsPIC 33FJ128GP804; (7) FT232RL (UART to USB bridge); and (8) ICSP header connector.

Figure 17. Overall appearance of the development board with dsPIC 33FJ128MC802: (1) power supply unit; (2) input–output connectors; (3) dsPIC 33FJ128MC802; (4) FT232RL (UART to USB bridge); and (5) ICSP header connector.

4. Experiments

This section exposes several examples with the platform, employing the mentioned elements of software and hardware. The main purpose of these examples is to provide a description of some usage cases that can be built with the platform.

4.1. Example 1: ADC plotting

In this example, the analog data on input (1) of the development board is plotted. The steps involved are as follows:

1. Plug-in the development board to a PC. Configure the project in the application using the project wizard.
2. Create the AV according to the specifications of a design. Use the palette to get the different graphical blocks that are needed in the AV. Program the development board with this.
3. Debug the AV. For this example, connect a potentiometer on *input 1*.
4. Open the plotter, specifying the number of variables to plot, in this case, 1.
5. Click on the run button to start the plotter.

Taking into account the previous steps, Figures 18 and 19 show the procedure starting with the flow diagram, Figure 20 illustrates the plot with the data from ADC peripheral and Figure 21 depicts the testing environment for this example:

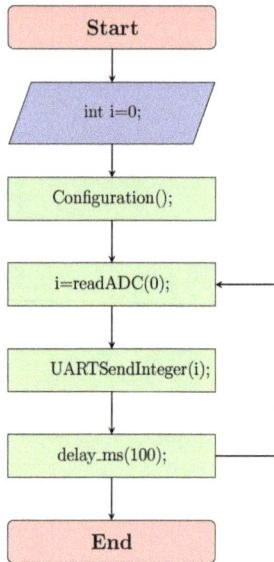

Figure 18. Flow diagram for Example 1.

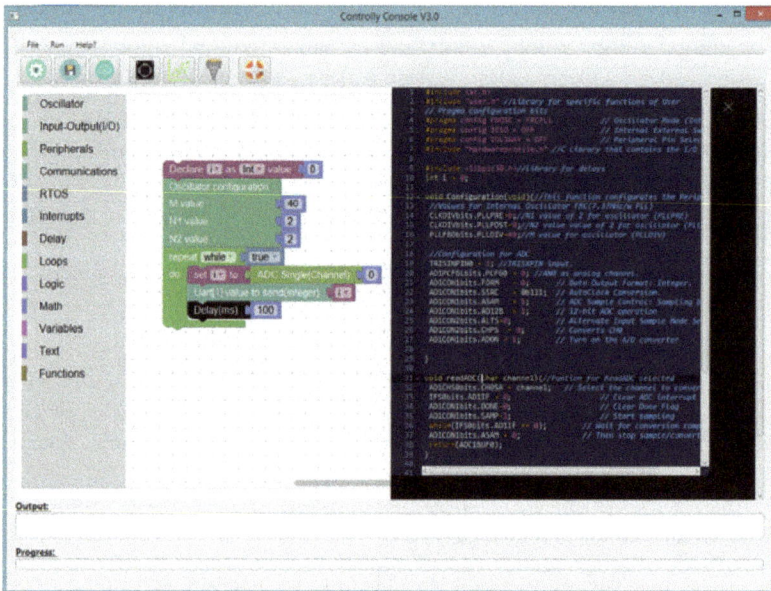

Figure 19. Example of the created AV for the function ADC plotting.

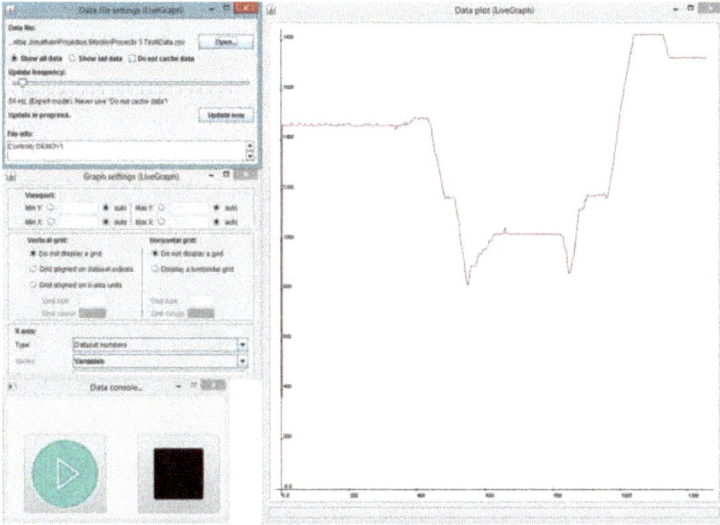

Figure 20. Data plot of the ADC channel (1) for Example 1.

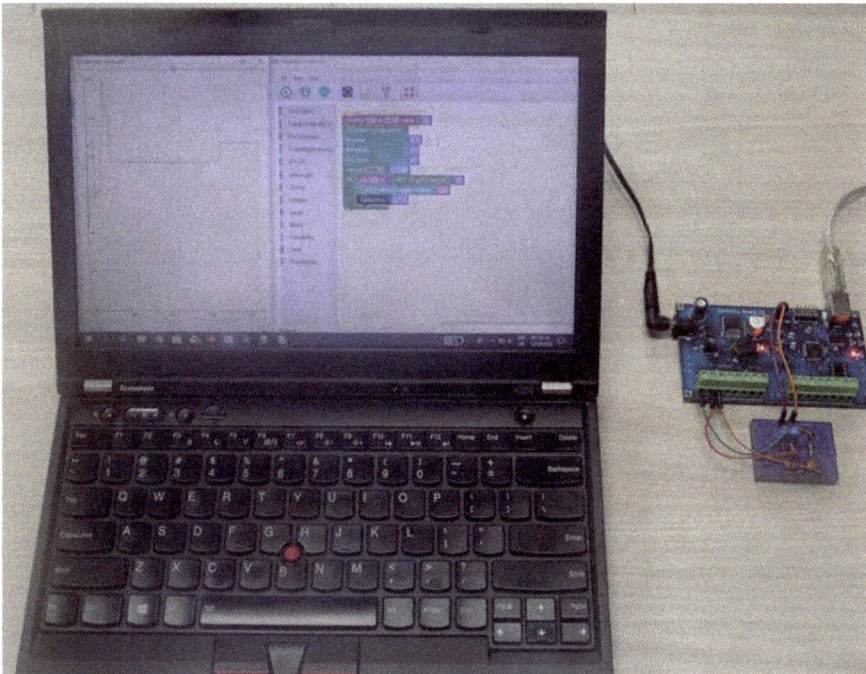

Figure 21. Testing environment for Example 1.

The user declares the variable i that reads the value of the ADC. This value is sent to the UART peripheral, employing the block *UART send (integer)*. The program runs continuously in an infinite loop.

4.2. Example 2: RTOS

This example uses the RTOS (OSA) to start and stop an AC Induction Machine (ACIM). The RTOS reads two switches (start and stop) and also sends the texts *Motor On, Motor Off* to the serial port, relying on the state of the ACIM. For this example, the steps are as follows:

1. Plug-in the development board to a PC. Configure the project in the application using the project wizard.
2. Create the AV according to the specifications of a design. Indicate the name of the tasks for the RTOS (*TurnOn, Turnoff*). To use the RTOS, a Timer (1) interrupt must be configured.
3. Program the development board with the AV built in Step 2.
4. Debug the AV. Connect two switches with their respective pull-up resistors to inputs 2 and 3 of the development board. In addition, connect a relay and an AC contactor for the ACIM with their protections.
5. Start and stop the ACIM. Test the message transmission on the serial port visor in the GUI of the platform.

Figures 22–24 show the flow diagrams, the AV and the testing environment for the example.

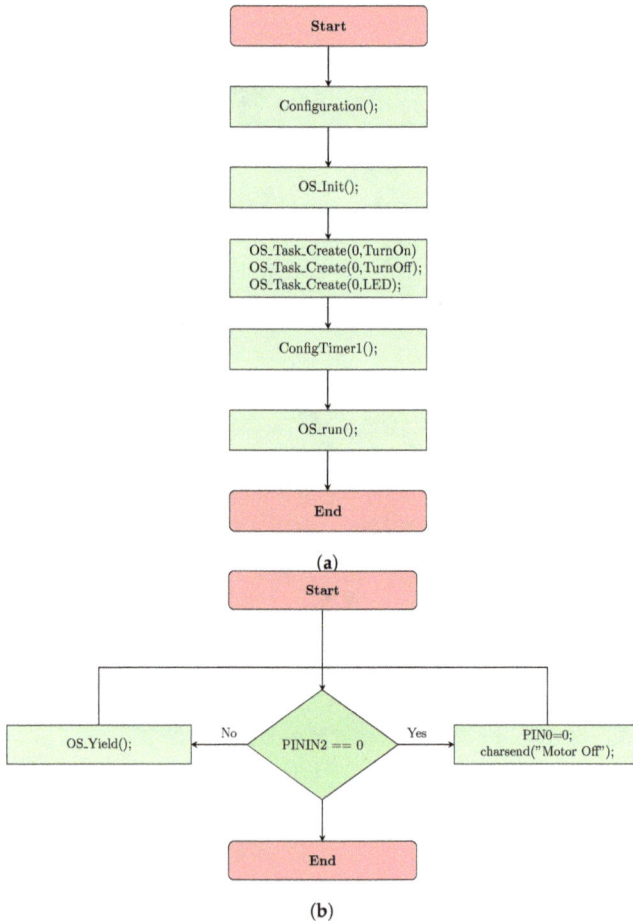

Figure 22. Flowcharts for Example 2: (a) flow diagram to configure the tasks of the RTOS: and (b) flow diagram for the task *Turn off*.

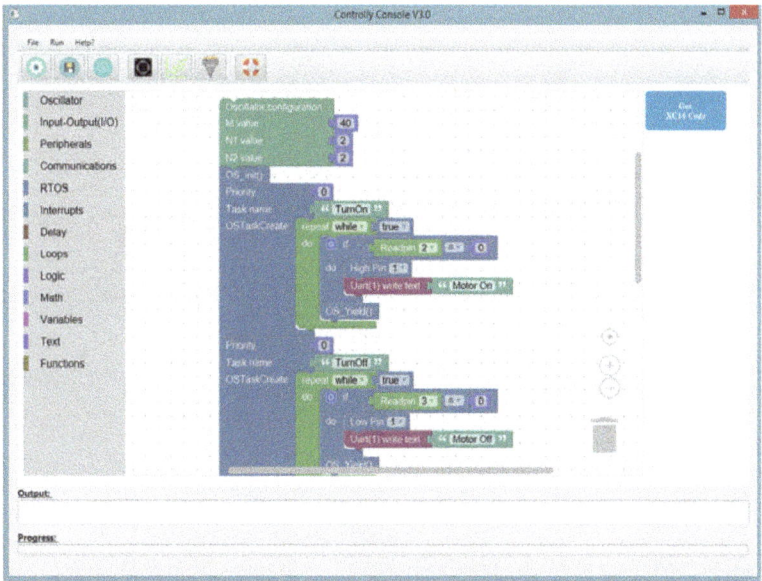

Figure 23. Designed AV for Example 2.

Figure 24. Testing environment for Example 2.

5. Assessment

To assess the platform, a survey was applied to 30 sophomores ($n = 30$) in the periods 2017-II and 2018-I. The students belonged to the program of Electronic Technology at UNIMINUTO university and they were informed and invited to the study conducted with the platform. All students decided to participate in this study giving their approval. This assessment provides the standpoint of the students that used the platform in aspects such technical as educational, which could give a perspective on its use. Table 6 sums-up the questions and answers of the survey.

Table 6. Survey's questions and answers ($n = 30$).

Question	Answers
1. In a scale of 1 to 10, assess the technical functioning of the platform.	$\bar{x} = 8.5$
2. With the graphical blocks, did you understand in a better way the concepts involved in the DSCs?	Yes (100%), No (0%)
3. What element do you consider most relevant in the platform?	• Designed Blocks to configure the peripherals and ports of the DSC. **(12)** • Algorithm Visualizations (AVs). **(9)** • Development board to implement the designed algorithms. **(6)** • Interface (Robust and User-friendly). **(3)**
4. Did the platform help you in the way that you understand an algorithm?	Yes (100%), No (0%)
5. Were learning materials pertinent to your needs in the embedded systems area?	Yes (100%), No (0%)
6. Would you employ the platform in your own designs into academic and work contexts?	Yes (100%), No (0%)

Questions 1 and 3 evaluated the platform at the technical level. Due to some presented bugs that were fixed, the average grade provided by the students was 8.5. In the same way, Question 3 indicated what elements the students considered relevant about the platform. The answer to Question 2 showed the learning that the students had with the platform in comparison with the traditional educational method of embedded systems. The traditional method to teach embedded systems at Uniminuto University typically consists of employing a text-based structured programming, flow diagrams and microcontrollers (PIC) in projects, instead of the AVs and DSCs. The difference between this traditional method and the learning with the *DSCBlocks* platform lies, as mentioned above, in the possibility to design and implement in real-time an algorithm through AVs, in which the students can see different algorithmic procedures and the registers involved in the DSC's architecture. Nevertheless, a dedicated questionnaire to assess the learning with the platform is a research question still needs to be addressed.

In addition to the previous questions, the students were inquired in the same survey about the elements that they would improve in the platform. The students agreed with the need to have more graphical blocks to deploy protocols as I2C, CAN or SPI. Moreover, the students indicated that the methodology with the platform needs to be longer to fulfill the proposed learning outcomes.

The concept of the platform in the educational context arises on the one hand due to the lack of open-source resources pertaining to embedded systems area and on the other hand the problems detected in the algorithmic thinking that limit somehow the learning of programming in the students. In this way, the platform is an educational alternative that could address the described issues.

During 2017-II and 2018-I, the students employed the platform in different types of tasks into the curriculum of the microcontrollers with an average class time of 2 h per week during 15 weeks. The class sessions varied between the usage of the developed block-based language with the platform and text-based programming with MPLABX IDE [47] in a similar process mentioned by the authors of [15,26].

This method is suitable because the students must understand the DSC architecture with the different configuration registers involved in it and they must familiarize with the development environment (compiler and IDE) in order to program the DSCs. Thus, the students used the text-based programming in the first classes of the subject. Then, the students employed the platform in different laboratories with the visual language, considering the different designed AVs. In addition, the students debugged their algorithms in real practices, which represents an advantage over the simulation.

The educational process raised with the platform took into account four moments: abstraction, design, practice and arguing. These moments are described as follows:

- *Abstraction*: In this stage, the class session was focused on theoretical aspects regarding the DSC's architecture (registers, ports, peripherals, data bus, etc.) and the parameters such of the compiler XC16 as of MPLABX IDE.
- *Design*: The students developed an algorithm for a proposed problem, alternating between the text coding and the AVs. The algorithms required the usage of peripherals, ports, variables and loops, which are elements commonly used in the embedded systems area.
- *Practice*: The students implemented the designed algorithm in several proposed laboratories, typically, clustering industrial devices such as AC motors, AC contactors, relays and sensors.
- *Arguing*: The students explained the developed algorithms in their structure. For example, when the students configured a port or peripheral for certain design, they explained the configuration of the registers, loops, variables or functions involved in this operation.

According to the observations made in the class sessions and the indicated poll, it is possible to say that the platform is suitable and reliable to learn the concepts of the embedded systems with the DSCs, although it needs to be alternated with the text-based programming. In addition, the arguing concerning the developed AVs is a key process because the students explain the different mechanisms that structure an algorithm from their learning perspective.

6. Discussion

In agreement to the described technical and educational aspects and the assessment provided by the students, it can be concluded that the platform is suitable to learn the different concepts relating to the DSCs through AVs as educational way in the construction and understanding of an algorithm that requires the usage of these devices. In addition, the platform allows the user to design and test applications in the embedded systems area due to the number of graphical blocks to handle diverse peripherals, ports, RTOS and its user-friendly interface. With the platform, the students can observe the configuration for the registers, loops, functions and variables involved in the functioning of the DSCs, following an algorithmic structure.

DSCBlocks was tested with industrial devices (ACIMs, AC contactors, relays, etc.) to evaluate its robustness, as depicted in Example 2. Furthermore, the graphical blocks that compound the application were divided into 28 functions, distributed in different categories and 120 codes were tested.

Conceptually, open-source hardware [48] is composed of physical technological artifacts, e.g., PCB layouts, schematics, HDL source codes, mechanical drawings, etc., that could be used in other projects or designs in some cases with the respective General Public License (GPL). This work is entirely open-source in hardware and software, that is to say, a user can modify the structure of hardware and software according to the requirements of a design only with an attribution requirement. Finally, the importance of this work lies in that the platform addresses the lack of open-source educational resources for the embedded systems regarding the DSCs. Although the hardware devices employed in the application are dsPICs, the followed schema could adapt to other DSC's architectures, e.g., ARM, Texas Instruments or Atmel.

One important difficulty that deserves to be mentioned in the design of the interface is the modification of *Blockly*, because the documentation concerned was not totally accessible at the time to create the XC16 language generator and the graphical blocks. *Blockly* combines many JavaScript functions that could confuse to a developer. The recommendation to overcome this issue is to adapt a preloaded language converter in *Blockly*. For example, to build the platform, a Dart language converter was adapted to its design requirements.

It is important to create open-source resources in software and hardware that contribute to the learning of the students or users in engineering areas such as control, embedded systems, power electronics or robotics. This type of resources could become a source of knowledge that deserves to be shared and can be used as reference guides in the construction of different types of applications and designs in the academic and industrial sectors. Besides, the users can modify the structure of these resources, generating new developments that could help to solve diverse problems in the engineering context.

7. Conclusions and Further Work

In this paper, an open-source platform for embedded systems called *DSCBlocks* is presented. The main interest of the platform was to provide an open, flexible, suitable and efficient environment to program DSCs through Algorithm Visualizations (AVs). Furthermore, the platform aims to cope with the lack of open-source resources in the area of the embedded systems, specifically associated with the DSCs.

As described above, the AVs have important advantages for the students or users that want to design any type of algorithms in order to learn embedded systems. With the AVs, the students can observe the different configurations of the registers of the DSCs for different peripherals and ports; create variables, loops, and functions; and can see the algorithmic procedures in real-time, helping to understand the architecture of these devices.

The assessment provided by the students suggests that the platform is reliable for the design and the implementation of different types of algorithms needed in embedded systems, and it allows the learning of the concepts concerning this area.

Although the platform has been conceived for academic work, the software and hardware components can be adapted effortlessly for any kind of project that employs DSCs. Further research of the platform will consist, on the one hand, in the design of new blocks for communication peripherals such as as I2C or CAN. In addition, the development board will be expanded to give a better functionality to the users for applications that utilize, e.g., Liquid Crystal Displays (LCDs), sensors and actuators that require a greater number of inputs and outputs. On the other hand, it will make an educational study of the platform to know the implications of it in the learning of the students.

Supplementary Materials: Complete version of *DSCBlocks* for PC or laptop is available online at http://www.seconlearning.com/DSCBlockV2/DSCBlocksFull.zip. GitHub repository of the application is available online at https://github.com/Uniminutoarduino/DSCBlocksV2. Xerte materials are available online at http://seconlearning.com/xerte/play.php?template_id=5. *DSCBlocks* web application is available online at http://seconlearning.com/DSCBlockV2/BlocklyOPt/demos/code/index.html.

Author Contributions: The author carried out the conceptualization, design, implementation of the platform and writing of the paper.

Acknowledgments: This work was supported by the Control Research Incubator (SeCon) funded by the Corporación Universitaria Minuto de Dios (UNIMINUTO).

Conflicts of Interest: The author declares no conflict of interest.

References

1. Microchip Technology Inc. dsPIC ® Digital Signal Controllers The Best of Both Worlds. Available online: http://www.farnell.com/datasheets/133476.pdf (accessed on 24 July 2018).
2. Aspencore. 2017 Embedded Markets Study. Available online: https://m.eet.com/media/1246048/2017-embedded-market-study.pdf (accessed on 28 July 2018).
3. Clarke, J.; Connors, J.; Bruno, E.J. *JavaFX: Developing Rich Internet Applications*; Pearson Education: London, UK, 2009.
4. Heckler, M.; Grunwald, G.; Pereda, J.; Phillips, S.; Dea, C. *Javafx 8: Introduction by Example*; Apress: New York, NY, USA, 2014.
5. Microchip Technology Inc. dsPIC 33FJ128GP804 Datasheet. Available online: https://www.microchip.com/wwwproducts/en/dsPIC33FJ128GP804 (accessed on 31 July 2018).
6. Microchip Technology Inc. dsPIC 33FJ128MC802 Datasheet. Available online: https://www.microchip.com/wwwproducts/en/dsPIC33FJ128MC802 (accessed on 31 July 2018).

7.   Microchip Technology Inc. dsPIC 33F Product Overview. Available online: https://cdn.sos.sk/productdata/fb/55/a9c85743/dspic33fj256gp710-i-pf.pdf (accessed on 31 July 2018).
8.   Törley, G. Algorithm visualization in teaching practice. *Acta Didact. Napoc.* **2014**, *7*, 1–17.
9.   Shaffer, C.A.; Cooper, M.L.; Alon, A.J.D.; Akbar, M.; Stewart, M.; Ponce, S.; Edwards, S.H. Algorithm visualization: The state of the field. *ACM Trans. Comput. Educ. (TOCE)* **2010**, *10*, 9.
10. Google LLC. Blockly Demo: Code. Available online: https://developers.google.com/blockly/ (accessed on 31 July 2018).
11. Fraser, N. Ten things we've learned from Blockly. In *Blocks and Beyond Workshop (Blocks and Beyond)*; IEEE Computer Society: Washington, DC, USA, 2015; pp. 49–50. [CrossRef].
12. Microchip Technology Inc. MPLAB ® XC16 C Compiler User's Guide. Available online: http://ww1.microchip.com/downloads/en/DeviceDoc/MPLAB%20XC16%20C%20Compiler%20Users%20Guide%20DS50002071.pdf (accessed on 28 July 2018).
13. Ball, S.; Tenney, J. Xerte—A User-Friendly Tool for Creating Accessible Learning Objects. In *International Conference on Computers for Handicapped Persons*; Springer: Berlin/Heidelberg, Germany, 2008; pp. 291–294. [CrossRef].
14. González, G.G. Xerte Online Toolklts Y El Diseño De Actividades Interactivas Para Fomentar La Autonomía De Aprendizaje en Ele. La Red Y Sus Aplicaciones en La Enseñanza-Aprendizaje Del Español Como Lengua Extranjera. Asociación Para La Enseñanza Del Español Como Lengua Extranjera, 2011; pp. 653–662. Available online: https://cvc.cervantes.es/ensenanza/biblioteca_ele/asele/pdf/22/22_0063.pdf (accessed on 17 August 2018).
15. Hundhausen, C.D.; Douglas, S.A.; Stasko, J.T. A meta-study of algorithm visualization effectiveness. *J. Vis. Lang. Comput.* **2002**, *13*, 259–290.
16. Rößling, G.; Naps, T.L. Towards Improved Individual Support in Algorithm Visualization. In Proceedings of the Second International Program Visualization Workshop, Århus, Denmark, 22–24 November 2002; pp. 125–130.
17. Pasternak, E.; Fenichel, R.; Marshall, A.N. Tips for creating a block language with blockly. In Proceedings of the Blocks and Beyond Workshop (B&B), Raleigh, NC, USA, 9–10 October 2017; pp. 21–24. [CrossRef].
18. Weintrop, D.; Shepherd, D.C.; Francis, P.; Franklin, D. Blockly goes to work: Block-based programming for industrial robots. In Proceedings of the Blocks and Beyond Workshop (B&B), Raleigh, NC, USA, 9–10 October 2017; pp. 21–24. [CrossRef].
19. Angulo, I.; García-Zubía, J.; Hernández-Jayo, U.; Uriarte, I.; Rodríguez-Gil, L.; Orduña, P.; Pieper, G.M. RoboBlock: A remote lab for robotics and visual programming. In Proceedings of the Experiment@ International Conference (exp. at'17), Faro, Portugal, 6–8 June 2017; pp. 109–110. [CrossRef].
20. Serna, M.; Sreenan, C.J.; Fedor, S. A visual programming framework for wireless sensor networks in smart home applications. In Proceedings of the International Conference on Intelligent Sensors, Sensor Networks and Information Processing, Singapore, 7–9 April 2015. [CrossRef].
21. Ariza, J.A. Controlly: Open source platform for learning and teaching control systems. In Proceedings of the 2015 IEEE 2nd Colombian Conference on Automatic Control (CCAC), Manizales, Colombia, 14–16 October 2015; pp. 1–6. [CrossRef].
22. Galán, D.; de la Torre, L.; Dormido, S.; Heradio, R.; Esquembre, F. Blockly experiments for EjsS laboratories. In Proceedings of the Experiment@ International Conference (exp. at'17), Faro, Portugal, 6–8 June 2017; pp. 139–140. [CrossRef].
23. Bak, N.; Chang, B.; Choi, K. Smart Block: A Visual Programming Environment for SmartThings. In Proceedings of the 2018 IEEE 42nd Annual Computer Software and Applications Conference (COMPSAC), Tokyo, Japan, 23–27 July 2018; Volume 2, pp. 32–37. [CrossRef],
24. Khamphroo, M.; Kwankeo, N.; Kaemarungsi, K.; Fukawa, K. MicroPython-based educational mobile robot for computer coding learning. In Proceedings of the 2017 8th International Conference of Information and Communication Technology for Embedded Systems (IC-ICTES), Chonburi, Thailand, 7–9 May 2017; pp. 1–6. [CrossRef],
25. Marron, A.; Weiss, G.; Wiener, G. A decentralized approach for programming interactive applications with javascript and blockly. In Proceedings of the 2nd Edition on Programming Systems, Languages and Applications Based on Actors, Agents, and Decentralized Control Abstractions, Tucson, Arizona, USA, 21–22 October 2012; pp. 59–70.

26. Matsuzawa, Y.; Tanaka, Y.; Sakai, S. Measuring an impact of block-based language in introductory programming. In *International Conference on Stakeholders and Information Technology in Education*; Springer: Berlin/Heidelberg, Germany, 2016; pp. 16–25. [CrossRef].

27. Chtourou, S.; Kharrat, M.; Ben Amor, N.; Jallouli, M.; Abid, M. Using IOIOAI in introductory courses to embedded systems for engineering students: A case study. *Int. J. Electr. Eng. Educ.* **2018**, *55*, 62–78. [CrossRef].

28. Microchip Technology Inc. USB CDC Class on an Embedded Device. Available online: http://www.t-es-t.hu/download/microchip/an1164a.pdf (accessed on 2 August 2018).

29. Microchip Technology Inc. MCP4822 datasheet. Available online: https://people.ece.cornell.edu/land/courses/ece4760/labs/f2015/lab2_mcp4822.pdf (accessed on 3 August 2018).

30. RTOS, O. What Is OSA? Available online: http://wiki.pic24.ru/doku.php/en/osa/ref/introduction/intro (accessed on 7 August 2018).

31. Heath, S. *Embedded Systems Design*; Elsevier: Amsterdam, The Netherlands, 2002.

32. Di Jasio, L. *Programming 16-Bit PIC Microcontrollers in C: Learning to Fly the PIC 24*; Elsevier: Amsterdam, The Netherlands, 2007.

33. Sain, M.; Lee, H.; Chung, W. MUHIS: A Middleware approach Using LiveGraph. In Proceedings of the 2009 International Multimedia, Signal Processing and Communication Technologies, Aligarh, India, 14–16 March 2009; pp. 197–200. [CrossRef],

34. Paperin, G. LiveGraph Summary. Available online: https://sourceforge.net/projects/live-graph/ (accessed on 9 August 2018).

35. Oracle Corporation. WebView JavaDoc. Class WebView. Available online: https://docs.oracle.com/javase/8/javafx/api/javafx/scene/web/WebView.html (accessed on 9 September 2018).

36. Oracle Corporation. WebEngine JavaDoc. Class WebEngine. Available online: https://docs.oracle.com/javase/8/javafx/api/javafx/scene/web/WebEngine.html (accessed on 9 September 2018).

37. Oracle Corporation. JavaFX scene builder 2.0. Available online: https://www.oracle.com/technetwork/java/javase/downloads/sb2download-2177776.html (accessed on 28 July 2018).

38. Mozilla Foundation. Ace Code Editor. Available online: https://ace.c9.io/ (accessed on 9 September 2018).

39. Gustavsson, M. ds30 Loader. Available online: https://www.ds30loader.com/ (accessed on 28 July 2018).

40. jSSC (Java Simple Serial Connector). Available online: https://code.google.com/archive/p/java-simple-serial-connector/ (accessed on 28 August 2018).

41. Microchip Technology Inc. Pickit3 Datasheet. Available online: https://ww1.microchip.com/downloads/en/DeviceDoc/51795B.pdf (accessed on 12 October 2018).

42. Microchip Technology Inc. ICD3 Datasheet. Available online: http://ww1.microchip.com/downloads/en/DeviceDoc/50002081B.pdf (accessed on 12 October 2018).

43. Jiri Sedlacek, T.H. VisualVM: All-in-One Java Troubleshooting Tool. Available online: https://visualvm.github.io/index.html (accessed on 10 October 2018).

44. Microchip Technology Inc. MCP6004 Datasheet. Available online: http://ww1.microchip.com/downloads/en/DeviceDoc/21733j.pdf (accessed on 28 August 2018).

45. Instruments, T. LM2576 Voltage Regulator Datasheet. Available online: http://www.ti.com/lit/ds/symlink/lm2576.pdf (accessed on 28 August 2018).

46. FTDI Chip. FT232RL Datasheet. Available online: https://www.ftdichip.com/Support/Documents/DataSheets/ICs/DS_FT232R.pdf (accessed on 28 July 2018).

47. Microchip Technology Inc. MPLABX IDE. Available online: https://www.microchip.com/mplab/mplab-x-ide (accessed on 10 October 2018).

48. Ray, P.P.; Rai, R. *Open Source Hardware: An Introductory Approach*; Lap Lambert: Saarbrücken, Germany, 2013.

electronics

MDPI

Article

Design of an Open Platform for Multi-Disciplinary Approach in Project-Based Learning of an EPICS Class

Ha Quang Thinh Ngo [1] and **Mai-Ha Phan** [2,*]

1   Department of Mechatronics Engineering, HCMC University of Technology, Vietnam National
     University-Ho Chi Minh City (VNU-HCM), Ho Chi Minh City 700000, Vietnam; nhqthinh@hcmut.edu.vn
2   Department of Industrial Systems Engineering, HCMC University of Technology, Vietnam National
     University-Ho Chi Minh City (VNU-HCM), Ho Chi Minh City 700000, Vietnam
*   Correspondence: ptmaiha@hcmut.edu.vn

Received: 31 December 2018; Accepted: 3 February 2019; Published: 10 February 2019

Abstract: Nowadays, global engineers need to be equipped with professional skills and knowledge to solve 21st century problems. The educational program, created in digital learning rooms of the Higher Engineering Education Alliance Program (HEEAP) program supported by Arizona State University, became a pioneer in teaching learners to work within the community. First, the combination of a novel instructional strategy and an integrated education in which project-based approach is employed to apply the technical knowledge. During this, students in mechatronics, computer science, and mechanics must collaborate with peers from industrial systems engineering. Second, in this paper, the design of an open structure connecting multi-disciplinary major is illustrated with a new teaching approach. It is proved to be better by combining specialized understandings of various types in a wide range of applications. From this basis support, participants could implement additional components quickly while keeping the cost low, making the products feasible and user-friendly. Last but not least, students are facilitated with a free library that helps to control simply despite lacking experience in robotics or automation. Several examples show that students are capable of developing things by themselves on open design. In brief, this platform might be an excellent tool to teach and visualize the practical scenario in a multi-disciplinary field.

Keywords: service learning; robotics; open platform; automated vehicle; EPICS

1. Introduction

Currently, universities are improving their undergraduate programs to apply the teaching trend based on science, technology, engineering, and math (STEM). Their purpose is to develop both students' skills and attitudes simultaneously with knowledge obtained before. The integrated subjects of these skills are often focused. A project-based learning method is used to teach students skills and attitudes towards team work and self-studying. For many programs, the coursework project is basically a specialized background to apply a project-based learning method (PBL). However, in order to develop multi-disciplinary skills, projects should be organized to allow students from different majors to work in groups to solve certain practical problems. With that idea, the Engineering Projects In Community Service (EPICS) course is a good way to effectively implement skills training for students. Founded at Purdue University in Fall 1995, EPICS courses are taught in 35 universities so far [1]. EPICS is a service-learning design course where teams of students collaborate on long-term projects that contribute to the community [2]. Project work concentrates on engineering and technology needs of a community partner while interdisciplinary team interaction is an essential element for project success.

EPICS is a model that has been recognized within engineering education globally. Teaching EPICS according to PBL method helps the teams design, build, and deploy systems to solve engineering-based problems. The EPICS course is organized for students from different backgrounds to work on multi-disciplinary practical problems from life or industries. Occasionally, lecturers, as an instructor role in the working process, do not have sufficient in-depth knowledge in all areas, which is a big challenge for EPICS classes. This paper suggests a feasible solution: to create open learning resources for in-depth knowledge and deploy a flipped classroom to introduce basic knowledge to relevant STEM issues for EPICS course. The application of open learning resources for an EPICS course is implemented for the Faculty of Mechanical Engineering, comprised of Mechanical Engineering, Mechatronics Engineering, Industrial Systems Engineering, Logistics and Supply Chain Management, and so on.

. The more projects that are associated with the development trend, the more they support students to quickly meet employers' requirements. With the goal of deploying EPICS course into undergraduate programs at the Faculty of Mechanical Engineering, combined with the development trend of society, projects that match are analyzed. One easily recognized trend in the industry 4.0 is the need for automation in global logistics operations.

In that trend, logistics is concerned with the arrangements and movement of material and people. While procurement, inventory, transportation, warehouse, and distribution management are all important components, logistics is concerned with the integration of these and they are related to create time and space value for organizations. Logistics providers are required to do more transactions, in smaller quantities, in a shorter time, at minimum cost, and with greater accuracy. New trends, such as mass customization and e-commerce, are intensifying these demands. To maximize the value in a logistics system, various planning decisions need to be made, from task-level choice of which items to pick next for order fulfillment to strategic level decisions of investing an automatic handling machine system or not.

Furthermore, e-commerce development leads to increasing demand for transporting goods. Businesses that invest in automatic handling systems will have the advantage due to the precise positioning of the desired goods despite the small quantity. One of the pieces of handling equipment that is commonly used in warehouses that meets automation needs is an automatic guided vehicle (AGV). Therefore, designing AGV to cope with the logistics system needs is a promising topic at the moment. After analyzing this trend in the industry, the university proposed the EPICS course for students in mechanical engineering, mechatronics, computer science, and industrial engineering to work in groups with the direction on an AGV design.

2. Literature Review

2.1. Project-Based Learning in a Specilized Subject

For applying a PBL approach to a specialized subject, there are previous studies reporting the teaching experience and authors' reflection. Nasser et al. [3] spent their time in a power electrical system engineering subject via technical projects for students. They recognized that PBL helps with providing both professional attitude and skills. However, they lacked criteria to choose suitable projects in the first stage. This resulted in there being impractical or obsolete knowledge discussed. In a similar approach with a different manner, Aaron et al. [4] presented a process to develop, implement, and assess a project-based bioinstrumentation course. The learners must apply principles or theories of the course to design their devices. Authors innovated two assessments: direct assessment, which is evaluated via exercises related to constructing the benchmarks, and an indirect one through surveys. The direct rating is done at the end of the course to avoid bias in grading while the indirect rating is collected before and after the course. The indirect and direct evaluation with respect to learning outcome provides evidence to determine whether there was significant improvement in student-perceived skills. There were no prerequisite subjects or skills for enrolling in the course. A one-term period is insufficient to build up a bioinstrumentation device, so teachers in [5] focused on a semi-autonomous vehicle project divided into several stages. Design of a project requires detailed planning, an extensive literature survey, and comprehensive hardware and software design. They delivered basic control system knowledge to students such as circuit design, practical control theory, and hardware implementation. In each stage, understanding in analog and digital electronics, sensors and actuators, control, and power were detailed. In addition, extensive experiments were conducted to emphasize the lateral distance, speed control performance, and obstacle avoidance ability. Via this project, students learnt how to solve control problems, planning, survey, reading technical datasheets, and validation. Although mentors can be an expert in the control field, the process of studying was not open to feedback from participants, along with there being the absence of learning outcomes.

Regarding the other concern of PBL, the cooperative learning in a project becomes one of the key topics. In [6,7], Laio et al. introduced an agent-based simulation that focuses on students allows testing of various situations for project-based learning applications. In this test, the assessment methodology for student's perception in PBL activities is verified on a partial least squares path modeling. The theory assumed that the learning rate of a team of agents with different educational levels and amount of instruction given is different in a proposed simulation environment. However, the results of simulation were only validated in soft skills, not in technical courses.

2.2. PBL in a Flipped Classroom

PBL is a comprehensive approach to classroom activities that is designed to engage students in working on authentic problems [8]. Barron et al. in [9] remains valid until today. They shared a perspective to design, implement, and evaluate problems including project-based curricula that emerged from a long-term collaboration with teachers. Their four design principles are: (1) defining learning-appropriate goals towards deep understanding; (2) providing scaffolds such as "embedded teaching," "teaching tools," set of "contrasting cases," and an overview of problem-based learning activities before initiating projects; (3) ensuring multiple opportunities for informative self-assessment and revision; and (4) developing social structures that promote participation and a sense of agency. Similarly, this paper also designed the goal for the coursework project, which is the application of EPICS and an automation handling system that serves in logistics by applying the PBL method (principle 1). The support tool for deployment is a model that proposed an open platform along with a flipped classroom model (principle 2). Some issues related to principles 3 and 4 are also discussed in the previous section.

Thomas J.W. [10] reviews all studies on PBL in the past ten years and divided them into eight topics: definition, underpinning, effectiveness, role of student characteristics, implementation, intervention, conclusions and future. There is evidence that: (1) students and teachers believe PBL is beneficial and effective as an instructional method; (2) PBL, in comparison to other instructional methods, has value for enhancing the quality of students' learning in the concerned areas; (3) PBL is an effective method for teaching students complex processes and procedures such as planning, communicating, problem solving, and decision making; (4) PBL is relatively challenging to plan and enact; and (5) students have difficulties benefiting from self-directed situations, especially in complex projects.

Stephanie B. [11] also acknowledges the innovative approach of PBL for success in the twenty-first century. Student should drive their own learning through inquiry, as well as co-operate to research and join profession-related projects. Integrating problems and PBL into sustainability programs are applied in some universities, such as Arizona State University [12] and University of Saskatchewan [13].

Students benefit greatly from participating in PBL courses. Students engage and learn critical thinking and problem-solving skills throughout each step of the project. Moreover, soft skills could be developed as well [14]. PBL also has a great impact on English teaching, as students that learn English through PBL are more successful and have a higher motivation level than the students who are educated by the traditional instructional methods [15]. Even with the tendency to increase English communication skills for students, PBL can still be applied in content and language integrated learning (CLIL) or English as a medium of instruction (EMI). CLIL is an approach for learning content through an additional language (foreign or second), thus teaching both the subject and the language. EMI is the use of English to teach academic subjects in countries or areas where English is not the first language. Some undergraduate programs in Vietnam National University (VNU) are also taught under EMI and CLIL. Ai Ohmori [16] broadly indicates positive results in terms of student attitudes towards English language classes using a CLIL approach with PBL. Lastra-Mercado [17] outlines the connections between PBL and CLIL to suggest some theoretical implementations in the education framework.

In PBL, students should search more in the problem-space of authentic problems. By far, various technology scaffolds have been researched to understand how these tools support students. One recent strategy includes a "flipped classroom," when lectures are online and class time is spent engaging in active forms of learning. A study qualitatively investigated the implementation of a flipped classroom with PBL [18]. A flipped classroom, which is created from a blended learning model, emphasizes student-centered learning activities. Flipped classrooms, through the use of technology, have been proven empirically to improve student achievement [19]. The combination of PBL and a flipped classroom can be applied to a wide variety of subjects, from subjects in elementary or high school to graduate programs [20–23]. In these studies, this combination is likely to improve students' learning performance, and was significantly better than other teaching methods mentioned.

2.3. From PBL to EPICS

EPICS, an engineering design-based service learning and social entrepreneurship program, incorporates the engineering and human-centered design processes in providing solutions to real-world problems. With the same orientation, Cheryl et al. [24] desire to discover the relationship between the Interaction Design program and EPICS program at Purdue university. By focusing on students' experience, they identified several characteristics of a collaboration process via models. Different aspects of designers and engineers are useful to find differences between the two programs. The drawbacks of the research are that the product of an EPICS program is an industrial hardware, social facility, or practical result, while the output of design program is only a draft or sketch. They did not carry out a proposed model of interdisciplinary collaboration between the two programs. Jonathan et al. [25] discussed a brief story of an EPICS program in a computer science major and software engineering major of Butler university. Reseachers summarized the teaching experience, how to organize the class, and some theoretical projects that learners carried out. However, there are no conclusions, evaluating model, or technical achievements from this work. In References [26,27], teachers provide other views from high school about the EPICS program. They took advantage of acquirements in social activities to help communities. The results from this report is not validated in higher education, such as college or university, and is inappropriate for engineering.

In exploring the implication of this approach, Carla et al. [28] applied the EPICS model to develop learning through service-learning and community-based engineering. They found that this model effectively prepares learners for a wide range of careers more than teaching conventional engineering methods. More specially, it established and sustained a life-long relationship in the community by providing mutual advantages and considerable community impact. However, its contents must be updated and evolved continuously. Also, its scope should be expanded to pre-engineer learner or graduate student compared to the college or pre-college level of current research. On the research of added value from service-learning, References [29–31] introduced possible methods to measure when emerging project-based strategy and service learning took place in engineering education. It was recognized that the impact of the proposed approach of mixing quantitative and qualitative research on the knowledge, skills, attitudes, and attendance of participants are attractive to students, faculty, and employers. Via analyzing evidence, undergraduate students achieved superior cognitive levels in professional skills, industrial attitudes, and social outcomes. One of EPICS' characteristics is that it requires a multi-semester project. In this context, team members could be altered depending on unpredicted reasons. To appreciate the discretion of EPICS alumni, James et al. [32] investigated

EPICS based on sequential mixed-methods study. They interviewed EPICS participants and conducted a thematic analysis of transcripts to explain how members related their EPICS experiences to the workplace.

In Vietnam, to connect between theories and practice, a technical festival named the Tech Show (http://www.hcmut.edu.vn/vi/newsletter/view/tin-tuc/3445-tung-bung-ngay-hoi-ky-thuat-bach-khoa-) is organized every year in Ho Chi Minh city University of Technology (HCMUT), Vietnam National University Ho Chi Minh city (VNU-HCM). On this occasion, students in many fields gather on campus to display their product range from technical solutions, useful devices, and applications in engineering. Professional practitioners, foreign enterprises, and research institutes also join the Tech Show. Positive feedback are given by employers to the school. Additionally, there are numerous challenges for first year students to fourth year students. In the first year, students must accomplish the mouse trap car in a mini-project of an introduction to engineering course, as in Figure 1. During this time, learners become accustomed to the technical design process, i.e., planning, drawing, reviewing, assembling, and testing. For last year students, they are able to join in the racing of a line-following car (http://www.hcmut.edu.vn/vi/event/view/noi-san-bk/2972-chung-ket-cuoc-thi-lap-trinh-xe-dua-tu-dong-%E2%80%93-bkit-car-rally-nam-2015). In this contest, they need to program a micro-controller to drive car that tracks the reference line. Theoretically, students have fundamental skills and knowledge to complete the tasks. Students' attitudes, interests, and active learning spirit remain unclear if they are taught in the conventional education approach.

(a)   (b)

Figure 1. Mini-project of introduction to engineering class: (**a**) at HCMUT, and (**b**) at HUTECH.

2.4. Logistics System and Handling Equipment

Logistics deals with the planning and control of material flows and related information in organizations [33]. Logistics is one of the most important activities in modern societies. Logistics systems are made up of a set of facilities linked by transportation services. Facilities are sites where materials are processed, such as stored, sorted, etc. Logistics systems are made up of three main activities: order processing, inventory management, and transportation. Order processing is related to information flows in the logistics system and includes a number of operations. First, the orders requested by customers are transmitted and checked. After verifying the availability, the items in orders are retrieved from the stock, packed, and delivered along with their shipping documentation. Traditionally, it takes around 70% of the total order-cycle time for order processing activities. In recent years, it has gained much support from advances in electronics and information technology. Barcode scanning allows retailers to rapidly identify the required products and update inventory level records. As a result, the items' handling process is more accurate.

There is a lot of research about these handling systems. Whelan P.L. [34] suggested a material handling system that can be utilized for electronic component manufacturing, as well as crossdocking warehouse (for just-in-time distribution). A method and system for processing packages that is designated for special handling and notifying an appropriate party was introduced in the patent of Kadaba N. [35]. The method of automatically picking items to fill a purchase order supported a clear process from receiving an order to releasing the additional inventory [36]. References [37,38] introduced an inventory system including multiple mobile inventory trays with a positioning system that enables the mobile inventory trays to determine their three-dimensional coordinates and thereby navigate on factory floors. These inventions are the beginning studying and designing automation machines for material handling such as AGVs.

In 2002, Ito et al. [39] proposed an agent-based model for a warehouse system particularly for designing and implementing warehouse systems simulation with agents. Babiceanu et al. [40] focuses on the integration of the holonic-based control concept in the design of an automated material-handling control system. An important factor for implementing automation systems is to accurately identify the goods to be loaded and unloaded. Traditionally, printed labels that are easy to read are used. With e-commerce trends and automated handling system, RFID systems are being applied. A RFID-based resource management system (RFID-RMS) is designed to help users select the most suitable resource usage packages for handling warehouse operation orders by retrieving and analyzing useful information from a case-based data warehouse, which results in both time saved and cost-effectiveness [41]. They also developed and embedded a pure integral-linear programming model using a branch and bound algorithm to define the optimum travel distance for forklifts, which can be used for AGVs in warehouses.

2.5. An Open Platform for Multidisciplinary Collaborations

Mechatronics, which is composed of mechanics, electronics, and informatics, is one fundamental course for learners majoring in machinery. With the fast development of intelligent technologies and knowledge management, mechatronics needs to integrate with other fields, namely industrial management, to satisfy practical needs. The products of mechatronics must be combined with the others. One typical example of product in mechatronics is the autonomous vehicle. Ting et al. [42] launched the autonomous car as a topic in mechatronics education to cultivate successful engineers. Nonetheless, in this context, the global engineers require professional skills rather than the theoretical knowledge taught in mechatronics courses. They are compelled to co-work with other engineers. Team-work skills, practical techniques, and industrial application, which are not related to this course, becomes a must. With the same approach but different purposes, Hector et al. [43,44] want to apply a methodology using practical activities to determine the integration of different students whilst maintaining their interest in the learning process. In contrast, one team mentioned in this research includes students only in mechatronics. Over and above that, their output is only models, not practical products. Julio V. and Jose M.C. [45] present a low-cost platform for a robotics application in education. They use Raspberry Pi 3 as a main board, camera as a main sensor, and Gazebo simulator as a simulation tool. The difficulties of their works are that Raspberry Pi 3 seems to provide very slow computation for image processing while the camera calibration requires expertise. Also, the 3D printable components are not guaranteed to have the correct physical dimension. Pupils lack confidence regarding their ability to assemble the whole system. Last but not least, this project organized for first year or second year students is not applicable in the community.

It is rare to find a free framework for final students to practice the way to be an engineer. It is straightforward to find out an open robotics platform in the market. The famous branch name Lego offers different robotics kits in many versions (Mindstorms RCX, NXT, EV3, or WeDo). References [46,47] represent a platform that is open, low-cost, and easy to assemble for education and experiment. For the initial stage of students, these studies do not contain much technology. References [48,49] are discussed with respect to a robotics platform and teaching course. Some of them are from 3D printed parts, while the others are from an existing hardware system. However, the multi-disciplinary characteristics and collaborative spirit are not mentioned. In the other view, the results of References [50,51] reveal a technique-based platform that is fit for professional researchers. The complex system and fusion data are combined to achieve excellent performance. In this paper, AGV, the symbolizing target of mechatronics, is chosen to be an open platform to develop in multi-disciplinary collaborations for final year students. Each undergraduate student in a specific area contributes his or her works toward the team's success. The collaboration among majors and multidisciplinary applications are validated in experimental discussion.

3. Descriptions of Proposed AGV Model

With the idea of multi-disciplinary applications, the platform should be developed, implemented, and expanded when connected to others. Both hardware and software support learners to enhance their self-development. In order to achieve the mission, an AGV-based structure is investigated with respect to the driving mechanism, appropriate size, load capability, feeding actuators, and line-following sensors.

3.1. Design of an Open Hardware

In Figure 2, the whole vehicle is shown as an overview of the structure. Basically, the robot can be divided into four sections: base layer, middle layer, feeder layer, and lifting layer. The base layer is to bear the body mass and the load. The linear slider leans on the middle layer to orientate the vertical direction. The feeder layer hangs rubber rollers to provide cargo automatically. The top, or lifting, layer is elevated using an electric cylinder and directly contacts the cargo.

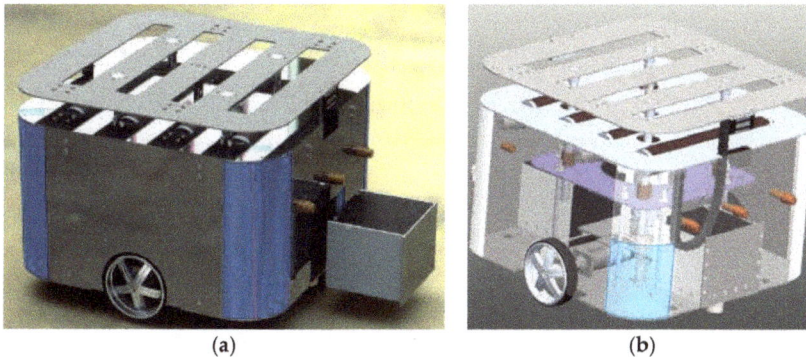

| (a) | (b) |

Figure 2. Overview of proposed autonomous vehicle: (**a**) outside view, and (**b**) inside view.

The driving mechanism consisted of two driving wheels on the side, with castor-type wheels in the front and back. In Figure 3, this structure avoids the slipping phenomenon, guarantees flexible motion, and all wheels are in contact with a certain surface. Two powerful motors were connected with shafts of wheels via gear boxes. The directional movement depended on the difference between the speed of the left wheel and the right wheel. The following path sensors were located at the center of the bottom layer, precisely at the front and the back to provide a bi-directional driving ability. Furthermore, the RFID reader module on this layer helped to navigate via tags, which were placed at each intersection. By reaching a crossroad, the vehicle read the identification numbers on the tags and returned to the host computer. Therefore, the server could track the path followed and generate the proper trajectory.

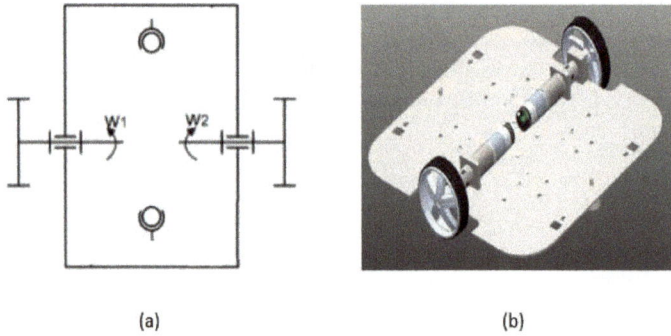

Figure 3. Driving mechanism on bottom layer: (**a**) theoretical design, and (**b**) 3D model.

The internal structure of the bottom of the vehicle is demonstrated in Figure 4. The electric cylinder mainly provided a lifting force to elevate freight, while four slider actuators linearly preserved vertical balance. Under the electric piston, there was a loadcell to evaluate the weight level of goods. As a result, the automated vehicle was able to process proper control signals smoothly. Furthermore, it enhanced the advanced functions via the feeding mechanism. Whenever the mobile vehicle stopped at a destination, without human intervention, it automatically delivered freight to storage. In both the head and tail, the electromagnets were attached to lead the behind carriages. In this case, the overall vehicle-load played the role as a tractor–trailer system.

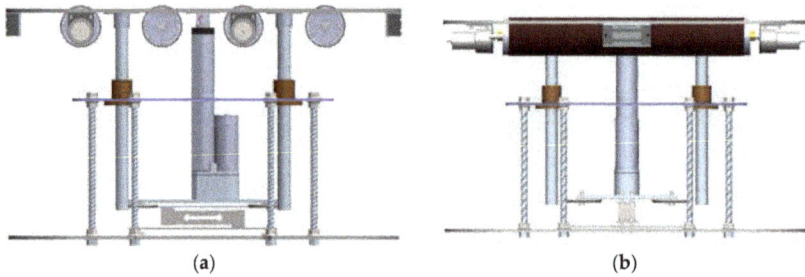

Figure 4. Inside architecture of loading and feeding mechanism: (**a**) side view, and (**b**) front view.

In this system, it was necessary to provide powerful management to process all of peripheral devices. In Figure 5, the connection diagram between the micro-processor and other components is illustrated. The high-performance ARM Cortex M4 32-bit operating at a max frequency of 168 MHz is considered as the core of the CPU module. This embedded processor features a floating-point unit (FPU) single precision that supports all single-precision data processing instructions. Some key specifications, such as 1 Mbyte Flash, 192 Kbytes SRAM, three channels A/D 12-bit, and two channels D/A 12-bit offer a wide range of applications. The two DC servo motors with a power of 60 W and a max speed of 70 rpm were able to drive the whole system. Due to the different speeds among servo motors, the autonomous vehicle can move easily and conveniently. There are several ways to track the reference trajectory. Generally speaking, using a line following sensor or magnetic tracking sensor are two popular methods. To enhance the competition of this system in the marketplace, a line following sensor turns out to be the best choice. The micro-controller could control an electric cylinder to lift cargo up or down with max load of 200 kg. Moreover, this robot was equipped with electric lock which was capable of holding a powerful electromagnetic force to drive followers. Previously, human operators must stand at the start point and target location to pick up freights, which leads to high operating cost. In this design, the mobile robot will be run from the beginning to the end by itself. Thanks to roller actuators, the automated ability of the robot can be improved.

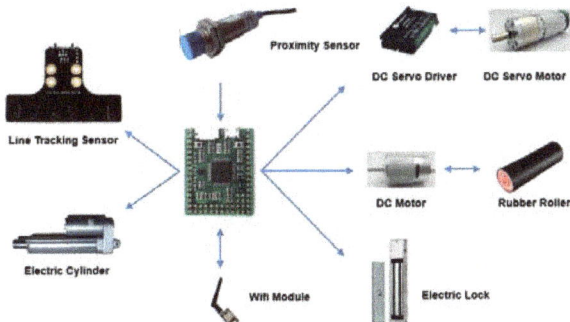

Figure 5. Block diagram of the control system in the proposed system.

The system coefficients and operating information are listed as Table 1. The physical size was large enough to carry various loads while it still maintained the flexible execution. In some cases, the vehicle's suitable height helped it to move underneath shelves. Hence, it could work anywhere in the map regarding dimension. In industry, the working route consists of two kinds: uni-direction and bi-direction. To deal with numerous maps, in this work, the autonomous vehicle could move forward and backward. Furthermore, in some specific situations, such as material transportation in the textile industry, the robot extricates multiple shelves. As a result, the power of the motor and tires of wheels were studied to optimize selection. All electronics components were integrated on the mother control board as in Figure 6.

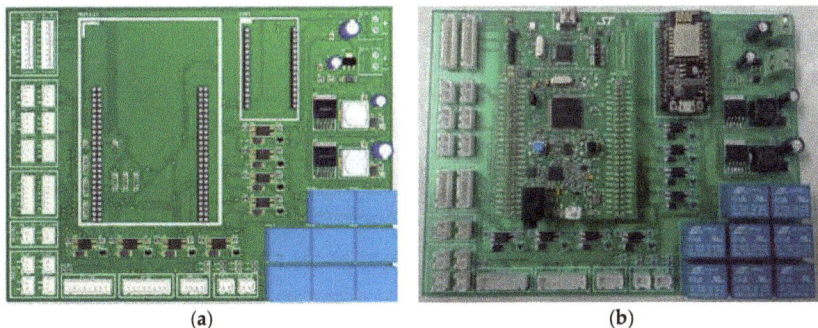

(a)   (b)

Figure 6. Simulation (**a**) and experiment (**b**) of control board in the project.

Table 1. Parameters and specifications of the system.

Item	Description
Physical dimension	650 × 420 × 350 mm
Wheels	4 (2 driving wheels, 2 castor wheels)
Max speed	0.5 m/s
Driving mode	Differential drive
Microcontroller unit	ARM STM32F405
Power	2 battery 12 V DC-48 Ah
Tracking method	Line follower or magnetic sensor
Sensors	Loadcell, proximity, current
Carrying mode	Lift-up, trailer

3.2. Design of an Open Software

The main programming language chosen for this project was C/C++ because most students had studied C/C++ and the software components were written using applications in C/C++ as well.

In Figure 7, the overall structure of the software level is explained in detail. Students could access and control hardware by using the open library. This library consisted of various Application Programming Interface (API) functions categorized into sub-classes, i.e., information (firmware date, type of robot), data (configurations, system parameters), motion (movement, stop), and monitoring (status, sensors). At the initial stage, the graphical user interface was provided to help beginners get acquainted with system. Later, at a higher level, students could customize the graphical user interface depending on their fields. For example, students in mechatronics major could attach more cameras, vibration sensors, or navigation sensors, while those in logistics were able to measure the efficiency of the imports and exports of warehouses. The library sent data to firmware to execute the autonomous robot. The method to control the DC motor is based on Pulse Width Modulation (PWM).

To support the students in programming, a set of API functions is recorded in Table 2. Relying on functional similarity, they were classified into sub-classes. When using this library, the user simply copied *.lib and *.dll into the project's directory. Later, referring to the header file, they just needed to program the syntaxes with the function's name. During the experimental tests, if the devices required calibration or they worked incorrectly, a standard graphical user interface in Figure 8 assisted students to overcome problems.

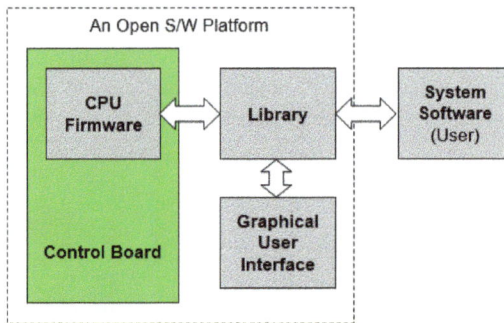

Figure 7. Flowchart of an open software (S/W) platform.

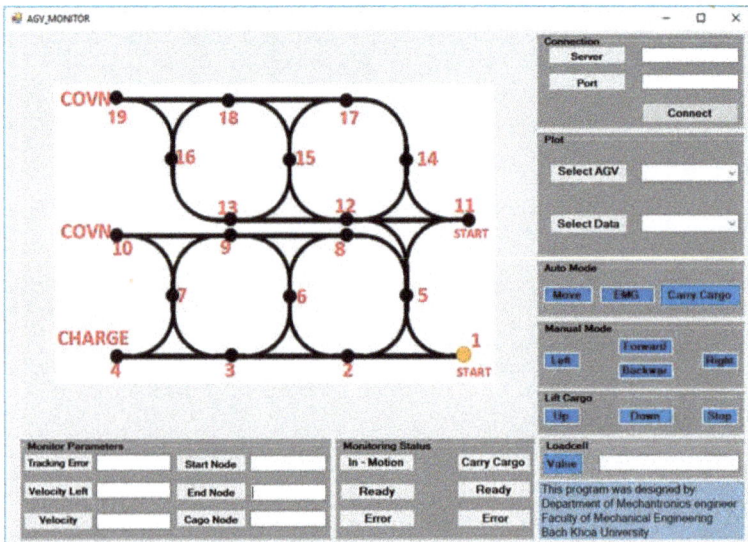

Figure 8. Graphical user interface of an open software platform.

Table 2. List of API function groups.

API Group	Description
Initialization	Initialize and setup connection in middleware
System parameters	Check whether status of connected devices is ready or not
Motion parameters	Establish parameters such as position, velocity, acceleration
Movement	Movement with absolute or relative mode
Status	Exchange data continuously while moving
Stop	Slow-down stop, emergency stop
Energy	Check status of battery

Figure 9 shows the architecture of the firmware including the supervising program and interrupt program. In Figure 9a, the host software in the computer reads the status of the channel and command and communicates with the firmware. Each status of the sensor, real position, and target position read from the variables of interrupt mode are provided to the host software. From the event control stage, the setting values of each item are checked and when the event occurs, the set of actions is controlled correspondingly. In Figure 9b, the interrupt had the cycle of 250 μs out of 4 kHz. In interrupt mode, the value of the encoder was read to assign the real position and each sensor status was also read. In previous sampling times, the calculated output value was generated for the actual movement of the motor. The real position read from the encoder was calculated for the real velocity. In the generator stage, if velocity changes were obtained from the equation of velocity, the amount of changing velocity would be added to the previous position to obtain the current target position. In the current sampling time, the known value of real position, i.e., the calculated value of output from the previous sampling time, was applied to match with the new calculated target position, then the control algorithm was used to compute the new value of the output. The newly calculated value was output on a real amplifier at the next sampling time. Depending on the need, the real position and target position were stored in a buffer. This provided function must be added in the host.

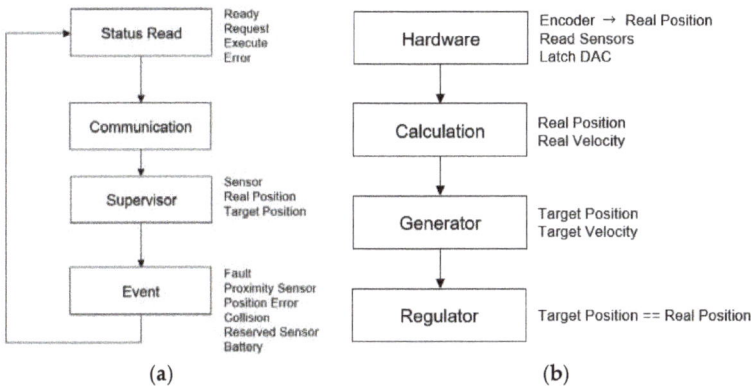

Figure 9. Flowchart in firmware: (**a**) background mode, and (**b**) interrupt mode.

4. Case Studies

To use the platform, it required users to build up additional hardware depending on their applications. In the most essential respect, this system accommodated the movement control by driving the left wheel and right wheel. The autonomous platform changed the directional movement with different velocities between the two wheels. The students stayed at the host computer to monitor the driving trajectory and control it as they desired.

This open platform was used in the EPICS class that was based on PBL and the flipped classroom method for students from different programs working together to design the AGV. Therefore, the organization of classes is also described.

4.1. Implementation of a Vision-Based Application

Nowadays, the image processing approach has been applied in industry widely. The enterprise wants to employ an engineer with vision-based knowledge such that, by using digital cameras, the AGV must capture targets and track to follow human in the shared workspace. As a result, the teaching class should provide an opportunity for students to be familiar with image processing techniques. In this case, students designed the pan-tilt mechanism for two cameras (Figure 10) by themselves. The mechanical structure ensured that two cameras were stable, flat, and operated smoothly. They hung the pan-tilt structure on the autonomous vehicle's front side. Due to the cheap price of the camera, students examined the camera to find intrinsic and extrinsic parameters. To use the image processing techniques, an open source computer vision library called OpenCV (https://opencv.org/) was downloaded and installed. This tool offers functions such as cv2.calibrateCamera(objectPoints, imagePoints, imageSize[, cameraMatrix[, distCoeffs[, rvecs[, tvecs[, flags[, criteria]]]]]]) to calibrate parameters of the camera.

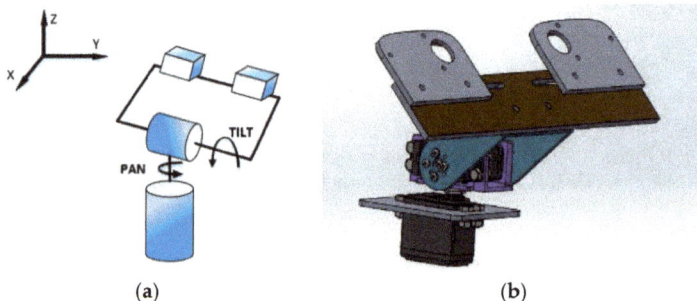

(a)　　　　　　　　　　　(b)

Figure 10. The pan-tilt mechanism for the digital camera: (**a**) theoretical design, and (**b**) 3D design.

Next, student calibrated the camera's system based on an epipolar geometry (Figure 11), a pre-processing stage to estimate the distance from camera to human. The purposes of sub-steps, consisting of stereo calibration and rectification were to discover a rotation and translation matrix between the two cameras and correct the epipolar line of images from cameras located in an axis.

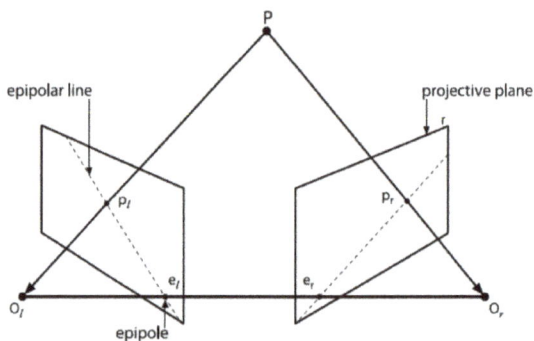

Figure 11. Epipolar plane in a dual-camera system.

Basically, in the flipped class, students were taught to do four stages to gain information in 3D. First, the un-distortion was to reject the radial and tangential distortion due to computation. Then, the rectification was to output an image that was row-aligned (the same direction and location on the y-axis). Later, the finding of features between the left and right image helped to get the disparity of the object. Finally, from the disparity's results, it was easy to calculate the distance to a human, as in Figure 12.

Figure 12. Experiment to estimate the distance statically by using digital cameras.

One of the simplest methods to recognize a target is to use the image threshold of image. Students captured the RGB picture and transferred it to an HSV (Hue, Saturation, Value) format. The dilation and erosion operators assisted in obtaining a smoother image, as in Figure 13. To tie the boundary pixels, edge linking algorithms, for instance, contour tracing, were suggested. During the procedure, the capturing time and processing time from the camera ought to be synchronous. In order to decrease the asynchronization error, the parallel programming, which included multi-thread in Python, was implemented on the computer. The results of the image processing stage were to deliver data for the automated guided vehicle to track a target, as in Figure 14. For convenient control, a group of students designed a graphical interface (Figure 15) by using the API functions in this paper.

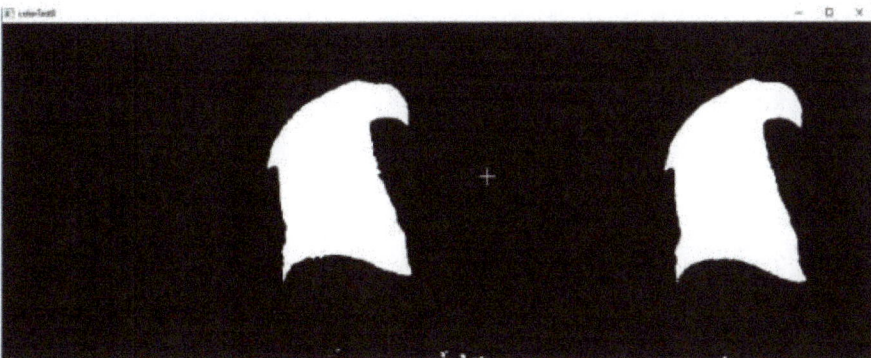

Figure 13. Image after threshold in HSV.

197

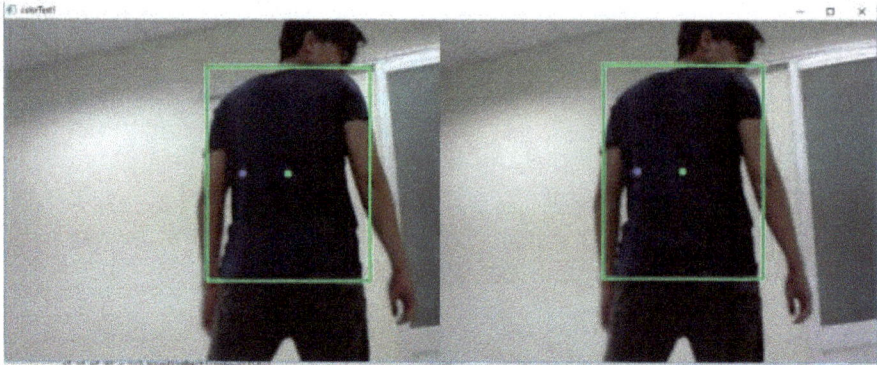

Figure 14. Experimental image that students obtained from digital cameras.

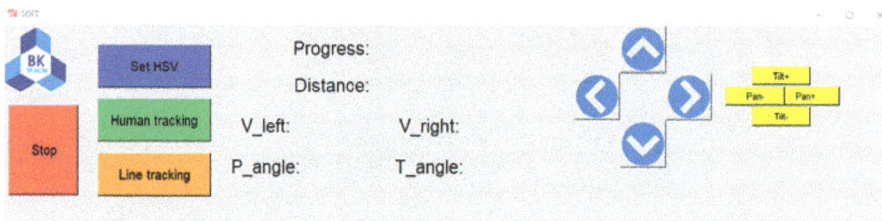

Figure 15. Student-designed interface by using the API functions in the computer.

4.2. Implementation of Control Application

From experiments, undergraduate students acknowledged that the total loads put on the shelf were unknown in advance. On the other hand, in a conventional case of programming, there was no difference among test cases (no load, light load, medium load, full load) in operating mode. Some of them recommended that it was compulsory to implement an intelligent algorithm to distinguish the control signal depending on current status of the load. The loading mechanism for the load cell is demonstrated in Figure 16. Additionally, the control scheme needed to keep the tracking error converge to zero as the autonomous vehicle followed the reference trajectory. The intelligent controller was selected using fuzzy logic under the instructor's guide. In Figure 17, the inputs of fuzzy control were load and tracking error, while the outputs were the control signals to the left and right wheels. This meant that the driving strategy of vehicles relied on the load mode and current following error. In Table 3, the fuzzy rules are established in the sense that, for example, if the load was light and the vehicle deviated on the left side too much, then the fuzzy controller drove max speed on the left wheel and at a much lower speed on the right wheel.

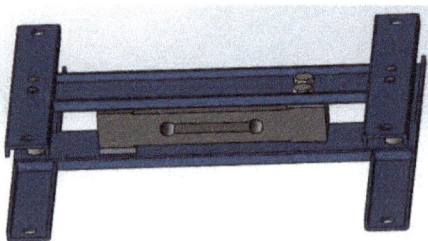
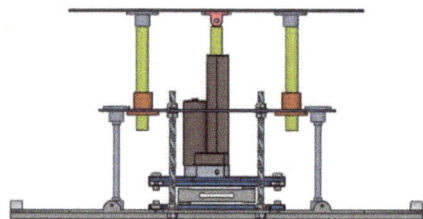

(a)   (b)

Figure 16. Design of attaching mechanism for load cell (**a**) and loading actuator (**b**).

Figure 17. Structure of proposed fuzzy controller.

Table 3. Table of rules in fuzzy sets.

Error Load	VL	L	MID	R	VR
Light	L-MAX R-VL	L-VF R-SL	L-N R-N	L-SL R-VF	L-VL R-MAX
Medium	L-VF R-VL	L-F R-VL	L-SL R-SL	L-VL R-F	L-VL R-VF
Heavy	L-F R-SL	L-N R-VL	L-VL R-VL	L-VL R-N	L-SL R-F

In Figures 18 and 19, most of membership functions are triangular shape and trapezoidal shape because students were familiar with them. In fact, following the instructor's guide, undergraduate learners were able to complete programming in a short amount of time. The sampling time set up in the micro-controller was 500 μs to guarantee the real-time performance. To deal with the short cycle period, the fast computation consisting of singleton outputs was more relevant.

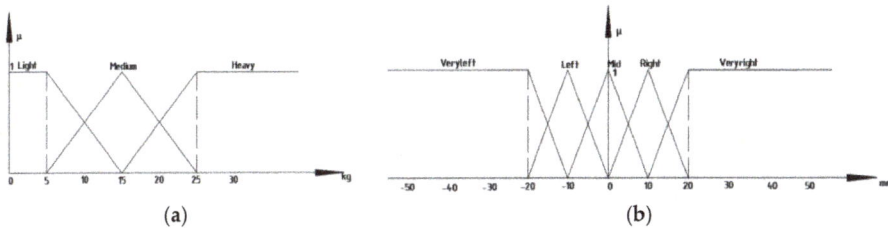

Figure 18. Input membership functions: (a) load, and (b) tracking error.

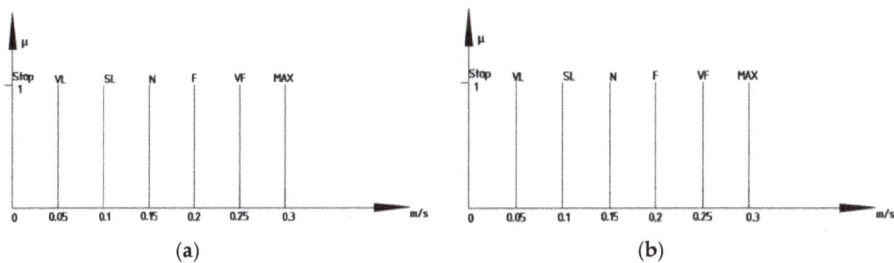

Figure 19. Output membership functions: (a) speed left, and (b) speed right.

The test scenario was planned as given in Figure 20. Initially, the AGV would stay at the start point in a ready status. After the command was released, the host computer scheduled the reference trajectory from the start point to the end point. The cargo or load was located at point C where the AGV travelled. Moving with an unknown load was a good lesson for students in control application. It must keep tracking the following line while keeping the cargo steady. The AGV completed a given task whenever it moved to the end point with the load. Students attained the data in off-line mode and plotted it using Matlab software (R2014a version, MathWorks, Natick, MA, USA), as in Figure 21.

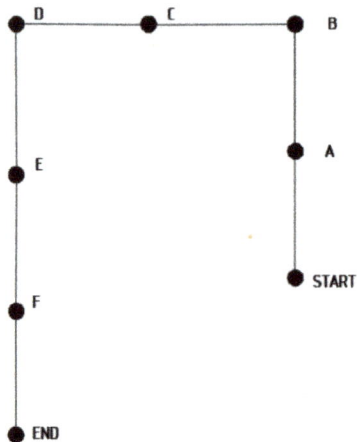

Figure 20. Reference trajectory in the test case.

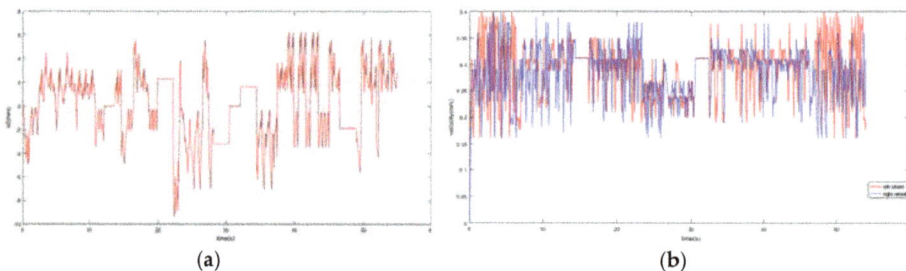

| (a) | (b) |

Figure 21. Experimental results that students obtained: (**a**) tracking error, and (**b**) speeds of wheel.

4.3. Implementation of Logistics Application and EPICS Course

The combination of third and fourth year students of Mechanical Engineering, Mechatronics, Computer Engineering, and Industrial Systems Engineering in the EPICS class began in early 2018 in the BUILD-IT project under the mentoring of lecturers from the University of Arizona, one of the schools that successfully implemented the EPICS class model. This was a course with the assessment based on tangible products and students completed six stages. As a pilot class, there were approximately 20 student groups from 4 universities for the first time and 30 groups from 6 universities for the second time. The lead lecturers of the class were from the Department of Mechatronics and Department of Industrial Systems Engineering, along with support from lecturers in other departments. Lecturers played an instructing role in the process, organizing group activities, guiding assessment, and commenting on every step in class hours. Students needed to work in teams outside the class, learn the relevant knowledge by themselves (from an open platform and flipped classroom documents or videos), and discuss and implement the model design and fabrication steps. After two EPICS classes, students excitedly learned necessary specialized knowledge for their needs. The student groups knew how to divide the work based on the strengths and weaknesses of each member. Students responded well to the learning outcome of the course and the program outcome of the Accreditation Board for Engineering and Technology (ABET) criteria. It was found that the model proposed in this paper was very successful.

The six stages of EPICS included: (1) project identification, (2) specification development, (3) conceptual design, (4) detailed design, (5) delivery, and (6) service maintenance. There were too few class hours to guide the content of course in detail, so the theory lesson for each design stage was uploaded to Google classroom or the LMS (learning management system—Moodle) of our University. For example, Stage 1 included: conduct a needs assessment, identify stakeholders, understand the social context of your community partner, define basic stakeholder requirements, and determine time constraints of the project. The document of these theory lesson included the definition, implementation method, or support tools. Oriented questions and suggestions for implementation processes were also introduced. Specific examples were provided. The checklist was also available for groups to self-assess the process.

In addition to the course's six-step design process, specialized logistic and AGV knowledge on design and fabrication tools were implemented. The logistic system's lesson included two main issues of logistics: warehouse and transport. Transport is used to determine the locations of warehouses or distribution centers and the selection of transport modes, as well as vehicle and resources schedule. The warehouse's lesson provided the theory of warehouse models and decision-making issues in the warehouse from design to operation. The handling equipment was also listed in terms of its characteristics, effects, and deployment conditions. The activities of loading and unloading that related to the AGV was described to help students understand the current situation or its operations in a warehouse. The required operational parameters of AGV that related to goods transportation in a warehouse were also introduced. For each group of students, the warehouse model was selected to propose the requirements of AGV's operational parameters. This affected their AGV's design and fabrication. The supported resource for AGV's design and fabrication were presented in the previous section.

In this course, each student had a unique role to lead the team depending on their background knowledge and the remaining members supported the leader. For example, with the first two phases, it was necessary to identify the needs of the system in general and in the logistics system in particular, where the Industrial Systems Engineering students lead the group. The analysis and conceptual design phase was lead by a Mechatronics student. The detailed design stage and delivery stage was be the primary responsibility of the Mechanical Engineering or/and Computer Engineering students. The test and evaluation step of the system (service maintenance), as well as the writing of the report by English, was done by all of the team members. In each stage, every member should understand basic knowledge, so it was important to have the open source for those who were in charge of each stage to explain to the remaining members. Therefore, students could learn, discuss, survey, measure, design, fabricate, and evaluate the group's products.

5. Discussion

The open platform described in this paper for the multi-disciplinary approach was proposed to surpass the existing limitations of the traditional educational method. From the authors' experiences, a student-centered view is one of the best training methods nowadays. By merging the idea of PBL as a teaching method in a class that was organized by EPICS, the interests of student in learning, professional skills, effective cooperation, and confidence were enhanced significantly as compared to traditional classes. To validate the effectiveness of the proposed approach, we analyzed both direct tools and indirection tools. For direct ones, it was agreed that reports, exercises, oral discussions, and seminars were to be used to carry out rapid reflections from students. We collected data from the beginning lessons to the end of the course. In our research, the final showcase and interview were judged as indirect measuring methods by means of the program manager, stakeholders, students, and board of referees. We believe that our results reflected the effectiveness, feasibility, and reliability of the proposed approach.

5.1. Direct Measurements

Unlike the traditional course where there is only a paper exam at the end of the semester, we used the various tools as Table 4 to assess the student's performance during the class. Several detailed evaluations and activities are provided in Tables A1–A3. From the view of success in the EPICS class, students were required to possess professional skills, for example, engineering skill, soft skill, and management skill, through many kinds of evaluation. Their own a depth knowledge of mechatronics and programming, and the ability to synthesize related information to apply what they learnt was also examined. They also gained experience in timing management, scheduling process, communication, and discussion. Above all, the program brought self-confident and learning interest for students after graduation. The authors believe that in the initiative stage of EPICS in Vietnam, the class was more successful than expected, at least in the contributed awareness of engineers in the community. In addition to conventional lectures in mathematics, physics, mechanical design, or operations research, it was helpful for undergraduate students to prepare their careers in our society when they joined in an EPICS class.

Table 4. Lists of direct assessment tools.

Item	Description	Achieved Skill
Weekly report	Every week, students reported a progress of work to instructors. This kind was similar to a short project report	Planning, time management, design skill, writing skill
Technical report	Whenever students met problems in their project, the reports of understanding techniques, discussion, recommended solutions and saving experiences in future are proposed	Ability to refer engineering tools to solve, teamwork, synthesizing skill, technical thinking
Presentation	A multislide show was organized based on team jobs to share knowledge with others	Self-confident, speaking skill, interpretative skill
Oral survey	Each team met stakeholders, users, and operators in community to enhance the design	Communication skill, social interface, design process
Peer evaluation	Each member in the team should assess their friends	Self-evaluated skill, team management

5.2. Indirect Measurements

To gain the objective feedback, our teams joined in the EPICS showcase, as in Figure 22, after completing the program. In this exhibition, teams were evaluated according to the EPICS final showcase competition rubric, as in Figure 23. The following were worthy of note:

- Overall quality of the proposed model: the products of the project were of a moderate quality in terms of both technique and academic knowledge. Teams verified an understanding of EPICS design process, faced challenges, and solved them. Each member needs to collect data from the logistics warehouse, interview workers, and record their opinions. The results were delivered to stake-holders to apply in their business. Furthermore, teams were willing to co-operate with others to widen their projects' applications.
- Team's demonstration: the explanation in systematic design was confirmed to obey the six stages of the design process of EPICS (project identification, specification development, conceptual design, detailed design, delivery, service maintenance, redesign, and retirement).
- Team's ability: in this case, the team answered questions related to the project directly while they were at the site. Although they lacked the experience of a crowd, it required substantial effort to accomplish.
- Quality of design review: the presentation (involving visual information) focused on the design process was highly evaluated.

Moreover, an evaluation report was developed based on the data collected via the end-of-course and post-event survey tailored to the programming and the curriculum of the EPICS course. The survey was designed to collect lecturers' and students' feedback on their overall experience and the EPICS course contents. The survey was administered using a Qualtrics web link. The language settings of the survey were both English and Vietnamese. The survey questions were designed in English and translated into Vietnamese by the organizers.

Figure 22. Students in EPICS Showcase lead by Dr Thinh and Dr Ha.

Overall quality of the project given its stage in the design process:

- The project design or concept is technically sound and feasible

- Level of detail demonstrates an understanding of the design issues, challenges and constraints

- Safety, maintenance and service issues have been addressed in the design

- Project design was aesthetically pleasing

- Curiosity: the team showed initiative exploring new areas of information

- Connection: Has the team drawn from non-engineering sources to inform their design? (i.e.: customer discovery, researched economic data associated with project, user feedback, etc.)

- Creating value: The team has articulated the value the project can/could deliver to its user and/or community partner

Team's Demonstration of the Design Process:

- User-centered issues are addressed

- Team demonstrates effective communication with community partners and other stakeholders

- Team demonstrates a systematic approach to evaluating design alternatives and making design decisions

- Team demonstrates appropriate flexibility in thinking about the design

Team's Ability to Explain the Project and Answer Questions:

- Team communicated the context and the scope of the project effectively and appropriately

- Team's answers were clear, succinct, and engaging

Quality of the design review presentation:

- Information communicated visually was well organized and relevant

- Team members participated and presented themselves in a professional manner

- Team members were dressed appropriately

- The presentation facilitated constructive feedback as a reviewer

Figure 23. EPICS final showcase competition rubric.

Fifty-three participants of the EPICS final showcase completed the survey, which made up a response rate of 48.6% (53 out of 109). The large majority of those who participated in the survey were students and that made up 81% of the total. Approximately 12% of the respondents were from universities; specifically, 6% were faculty members (lecturers, senior lecturers), 4% were university leaders (Rector/Vice Rector, Dean/Vice Dean, Head/Deputy Head of Department), and 2% were other university personnel (e.g., administrative staff). Meanwhile, those who worked for business/industry and those who worked for government shared the same percentage at 2%. The rest of the respondents (2%) were other non-university personnel. The summary of the survey's results is present in Table 5.

Table 5. Table of survey's results.

	Finding About	Percent of Those Surveyed	
Faculty's overall satisfaction about	course materials and supplies	80%	satisfied and very satisfied
	classroom facilities and equipment	67%	
Faculty's evaluation on contribution of the EPICS course to student development	understanding of design as a start-to-finish process	100%	Somewhat useful and very useful
	an awareness of the customers	100%	
	an ability to collaborate with people from other disciplines and develop an appreciation for cross-disciplinary contributions from individuals	100%	
	an awareness of professional ethics and responsibility	100%	
	an appreciation of the role that their discipline could play in social contexts	100%	
	enhance the ability to identify and acquire new knowledge as a part of problem-solving/design process	67%	
	to communicate effectively, both in oral and written formats	67%	
Students' overall satisfaction	quality of teacher's instruction, course materials, and supplies	88%	satisfied and very satisfied
	course facilities and supplies	86%	
	quality of team activities	80%	
	course duration	73%	
Contribution of the EPICS course to student respondents' professional development	understanding of design as a start-to-finish process	95%	Somewhat useful and very useful
	an awareness of the customers	86%	
	an ability to collaborate with people from other disciplines and develop an appreciation for cross-disciplinary contributions from individuals,	81%	
	an awareness of professional ethics and responsibility	86%	
	an appreciation of the role that their discipline could play in social contexts	95%	
	enhance the ability to identify and acquire new knowledge as a part of problem-solving/design process	92%	
	to communicate effectively both orally and written	81%	
Student respondents' perceptions of preparedness	ranging from collaborating with people from other disciplines and developing an appreciation for cross-disciplinary contributions from individuals	98%	Somewhat prepared and very prepared
	identifying and acquiring new knowledge as part of problem-solving/design process	95%	
	applying professional ethics and responsibility in multidisciplinary projects	95%	
	enhancing oral and written communication skills in multidisciplinary design projects	95%	
	applying disciplinary knowledge and skills to real and ill-defined problems	92%	
	providing services and solutions to address the needs of the community	92%	

While the final showcase competition took place, a discussion with the director program, learners, faculty, and stakeholders was propounded in the interview of VTC10 channel [52]. We now illuminate achievements from EPICS students in correlation with other students. In the director's speech, it was stated that young students wished to join in activities for community. They discovered social problems of neighbor groups and actively understood difficulties in their environment. From knowledge in school, they hoped to suggest practical solutions to overcome these. In other words, undergraduate students improved their social awareness and ethics.

> *"In my perspective, the application of technical knowledge for serving the community is absolutely suited Vietnam, because Vietnam has young and active population, they crave to participate in solving community's issues for the future of Vietnam. More and more the youth join in community serving activities, which is highly positive. What important is to understand the problem and put your heart and soul into it."* Mr Michael Greene, USAID/Vietnam mission director.

In the view of the country director, she supposed that learners were able to create creative and useful ideas to contribute to the community. More and more closed relationships between students and people were consolidated. Likewise, students could be trained with professional preparation, systems thinking and design process skills that are matched with EPICS learning outcomes.

> *"Students don't have to wait for graduation to be able to start their career. Right now, they can create something different and contribute to the society, especially helping their community to solve problems. I think this is a meaningful and interesting program. It's not just about education but also helping them find out solutions to such a problem. They have been working directly from the first stage of problem identification to the final stage of analyzing and solving it."* Ms Phuong, country director, Arizona State university.

To measure what EPICS students learnt, their responses after they completed the course have been recorded. One student mentioned that starting from ideas, she discovered how to bring it to reality. A second student gained direct connection with community via interviews and interaction to update the design. As a result, the gap between theory and practice was decreased. Another student thought that communication, teamwork, discipline knowledge, broader context, and multidisciplinary design techniques are obtained when learners participated during the process of EPICS.

> *"This is a new program. We have learnt a lot from it and we can bring our ideas to life."* —First student

> *"Here we have the opportunity to practice and to be creative, to have direct contacts with the community to interview and interact with people so as to find the right solution, and to develop a project to create something useful for our society."* —Second student

> *"Firstly, the contest is sponsored by Arizona State university. All documents are written in English so I can develop my English skills, one of the most important skills for students. Secondly, I want to develop design skill because it is a must-have skill to serve the community. One more thing is that I want to learn soft skills, teamwork skills in the process of participating in the contest."* —Third student

Finally, the comparison results between EPICS students and other students are illustrated in Table 6. They were appreciated in terms of three rating levels in the aspects of social perception, multidisciplinary design, specialized knowledge, and community-based responsibility. It can be seen that EPICS students were fully aware of practical engineers in the global community and well-prepared for their careers in future.

Table 6. Comparison between EPICS students and other students in three levels (good, fair, medium).

Learning Outcomes	EPICS Students	Other Students
Community awareness	Good	Medium
Multidisciplinary design	Good	Medium
Social entrepreneurship	Fair	Medium
Design process	Good	Fair
Teamwork	Fair	Medium
System thinking	Fair	Fair
Broader context	Good	Medium

6. Conclusions

This research concerned the integration of an open platform in various fields for final year students or pre-engineer learners. It met requirements compelled by Engineering Project in Community Service and demands from industry.

The authors have spent several years joining and teaching in PBL and EPICS classes. We realize that this was a new education approach when combining the EPICS program with practical applications, which involved:

- AGV-based open platform for multidisciplinary majors: The hardware was built up to be similar with an industrial system by manufacturing a tool machine. With an open 3D technical drawing, it was easy to manufacture in workshop anywhere. To simply access the hardware, a free library with numerous API functions was supported to read/write data. Students could develop a user-defined graphical interface in their field. After training sections, students were able to control the whole system.
- Implementation of educational tools: In the market, the existing commercial hardware systems were different and made for specialized industry. Some of them were inappropriate for teaching owing to price, extendibility, and maintenance. By designing additional components, students could extend more in exercises in the proposed platform.
- Collaborative enhancement: One of the primary conditions in this program is an integrated team that involves undergraduate students in mechatronics, computer science, mechanics, and industrial systems. These teams with complementary skills were committed to one shared purpose: to complete goals and try out approaches for which they held themselves mutually accountable. The think-pair-share technique was useful for this kind of team collaboration.

Author Contributions: Conceptualization, H.Q.T.N.; Methodology, M.-H.P. and H.Q.T.N.; Software, H.Q.T.N.; Validation, M.-H.P. and H.Q.T.N.; Formal Analysis, H.Q.T.N.; Investigation, M.-H.P.; Resources, M.-H.P.; Data Curation, M.-H.P.; Writing—Original Draft Preparation, H.Q.T.N.; Writing—Review and Editing, M.-H.P. and H.Q.T.N.; Visualization, H.Q.T.N.; Supervision, M.-H.P.; Project Administration, H.Q.T.N.; Funding Acquisition, M.-H.P.

Funding: This research was partially funded by USAID, BUILD-IT, and Arizona State University through EPICS program in Vietnam.

Acknowledgments: Authors would like to thank USAID, BUILD-IT, and Arizona State University in the EPICS program in Vietnam. Visit our homepage at (https://builditvietnam.org/events/epics2018).

Conflicts of Interest: The authors declare no conflict of interest.

Appendix A

Table A1. Evaluations of Design Process.

	Criteria	Four Driving Wheels	Two Driving Wheels, Two Castor Wheels	One Driving Wheels, Two Castor Wheels	None Driving Wheels, Four Castor Wheels
Electrical subsystem	Schematic	Four DC motor drivers	Two DC motor drivers	One DC motor drivers	No DC motor drivers
	Safety	All connectors tightly and insulated	Most connectors tightly and insulated	Connectors somewhat tight and insulated	No connectors were tight and insulated
	Fashionable	Absolutely reasonable layout	Mostly reasonable layout	Somewhat reasonable layout	Not a reasonable layout
Control program	Program requirements	The control program can: - drive four wheels simultaneously. - turn left or right based on two driving wheels at front side - operate 5-6 tasks of all tasks	The control program can: - drive two wheels simultaneously - turn left or right based on two driving wheels at middle - operate 3-4 tasks of all tasks	The control program can: - only drive one wheel - turn left or right based on one driving wheel at front side and middle - operate 3-4 tasks of all tasks	The control program cannot operate any tasks
	Program structure	Contains sub-functions that fully describe four parts in order: definition, initialization (setup function), main process (loop function), and sub-functions	Contain several sub-functions in order	Contains a few sub-functions in order	Does not contain sub-functions
	Readable	Code must contain: - comments on all initialization, sub-functions, variables at least 30% of comments on process lines of code - variables and functions name are meaningful	Code contains both criteria, comments and variable and function naming but one criterion be wrong	Code contains one of criterion, comments and variable and function naming or code contain both criterion but they all be wrong	Not comment at all variables and functions name are not meaningful
	Algorithm	Algorithm must contain: input, output, definiteness, effectiveness, and termination	Algorithm contains 3-4 criteria in all criterion	Algorithm contains 1-2 criteria in all criterion	Not any criteria
Physical structure	Full load/distance requirement	Full load and distance	Meet <80% of load and distance	Meet <60% of load and distance	Meet <50% of load and distance
	Structure	- Complicated, it is hard to install, adjust, and dismantle - Using large scale of different types materials - Using large scale of unpopular materials	- Fairly easy to install and dismantle - Using many types of different materials - Using many unpopular materials	- Messy structure to install, adjusting and dismantle - Using several types of different materials - Using several unpopular materials	- Easy to manufacture, install, and dismantle - Using not much material types popular materials - Using cheap popular materials
	Designed creation	- Requires self-developed actuators due to limitation of domestic device and finance - Having the ability to upgrade	- The product is based on an existing one, but has some different parts	- It is similar as an existing product - It does not upgrade	- The same as existing product
	Safety operation	- Absolute firmness of mechanism - Operates totally without vibration	- Medium firmness of mechanism - Structure has a little bit of vibration	- Less firm mechanism - Structure has some of vibration	- Unstable structure - Operator requires care to operate

Appendix B

Table A2. Evaluations of Team Presentation.

	4	3	2	1
Oral demonstration	Speak loud enough for everyone to hear	Speak with medium voice and somebody cannot hear	Speak too soft that many in the audience cannot to hear	Mumble their presentation
	Speaking speed reasonable	Speaking speed sometimes too slow or too fast	Speaking speed too slow or too fast	Speaking with a halt
	Combine well with the operation of open platform	Combine pretty well with the operation of open platform	Combine poorly with the operation of open platform	Presentation not combine with the operation of open platform
	Good body language, good eye contact	Good body language, good eye contact but sometime look nowhere	Body language is not good, just focus one side	Shows little care for the audience
Project notebook	Enough information about bill of materials (size, picture, web link, quantity), easy to identify materials	Enough information but some materials that are hard to identify	Lack information about bill of materials and hard to identify materials	Lack information about bill of materials and cannot identify material
	Design drawing similar to the model	Design drawing similar to the model	Design drawing different to the model	Design drawing different to the model
	Assignment task and schedule clearly	Assignment task and schedule is not clearly	Assignment task and schedule is missing some roles	No assignment task and not schedule

Appendix C

Table A3. Evaluations of Team Activities.

	4	3	2	1
Team role/shared leadership	Each role (note taker or time keeper or reporter or facilitator or team member) in team finished their roles excellently and always met the deadline	Each role (note taker or time keeper or reporter or facilitator or team member) in team finished their roles well and meet the deadline most of the time	Each role (note taker or time keeper or reporter or facilitator or team member) in team finished their roles but sometime did not meet the deadline	Each role (note taker or time keeper or reporter or facilitator or team member) in team did not finish their role by the deadline
Participation/collaboration	Member enthusiastically took part and supported others in the project	Member took part and supported others most in the project	Member partly took part and supported others in the project	Member rarely took part and supported others in the project
Work quality/research and information	Member finished their technical tasks excellently (mechanical or electrical or control)	Member finished their technical tasks well (mechanical or electrical or control)	Member partly finished their technical tasks (mechanical or electrical or control)	Member poorly finished their technical tasks (mechanical or electrical or control)

References

1. Oakes, W.; John, S. EPICS: Engineering Projects in Community Service. In Proceedings of the 34th Annual Frontiers in Education, Savannah, GA, USA, 20–23 October 2004.
2. Coyle, E.J.; Leah, H.J.; William, C.O. EPICS: Engineering projects in community service. *Int. J. Eng. Educ.* **2005**, *21*, 139–150.
3. Nasser, H.; Mohammad, R.H. Application of Project-Based Learning (PBL) to the Teaching of Electrical Power Systems Engineering. *IEEE Trans. Educ.* **2012**, *55*, 495–501. [CrossRef]
4. Aaron, M.K.; David, C.J.; Matthew, B.B.; Matthew, E.D. Bioinstrumentation: A Project-based Engineering Course. *IEEE Trans. Educ.* **2016**, *59*, 52–58. [CrossRef]
5. Ho, M.L.; Rad, A.B.; Chan, P.T. Project-based Learning: Design of a prototype semiautonomous vehicle. *IEEE Control Syst. Mag.* **2004**, *24*, 88–91. [CrossRef]
6. Laio, O.S.; Romeu, H.; Eduardo, A.B. Agent-based Simulation of Learning Dissemination in a Project-based Learning Context Considering the Human Aspects. *IEEE Trans. Educ.* **2018**, *61*, 101–108. [CrossRef]
7. Ngo, H.Q.T.; Nguyen, T.P.; Huynh, V.N.S.; Le, T.S.; Nguyen, C.T. Experimental Comparison of Complementary Filter and Kalman Filter Design for Low-cost Sensor in Quadcopter. In Proceedings of the IEEE International Conference on System Science and Engineering, Ho Chi Minh City, Vietnam, 21–23 July 2017; pp. 488–493. [CrossRef]
8. Blumenfeld, P.C.; Elliot, S.; Ronald, W.M.; Joseph, S.K.; Mark, G.; Annemarie, P. Motivating project-based learning: Sustaining the doing, supporting the learning. *Educ. Psychol.* **1991**, *26*, 369–398. [CrossRef]
9. Barron, B.J.S.; Schwartz, D.L.; Vye, N.J.; Moore, A.; Petrosino, A.; Zech, L.; Bransford, J. Doing with understanding: Lessons from research on problem-and project-based learning. *J. Learn. Sci.* **1998**, *7*, 271–311. [CrossRef]
10. Thomas, J.W. A Review of Research on Project-Based Learning. 2000. Available online: https://documents.sd61.bc.ca/ANED/educationalResources/StudentSuccess/A_Review_of_Research_on_Project_Based_Learning.pdf (accessed on 20 December 2018).
11. Stephanie, B. Project-based learning for the 21st century: Skills for the future. *Clear. House* **2010**, *83*, 39–43. [CrossRef]
12. Wiek, A.; Xiong, A.; Brundiers, K.; van der Leeuw, S. Integrating problem-and project-based learning into sustainability programs: A case study on the School of Sustainability at Arizona State University. *Int. J. Sustain. High. Educ.* **2014**, *15*, 431–449. [CrossRef]
13. Kricsfalusy, V.; Colleen, G.; Maureen, G.R. Integrating problem-and project-based learning opportunities: Assessing outcomes of a field course in environment and sustainability. *Environ. Educ. Res.* **2018**, *24*, 593–610. [CrossRef]
14. Wurdinger, S.; Mariam, Q. Enhancing college students' life skills through project-based learning. *Innov. High. Educ.* **2015**, *40*, 279–286. [CrossRef]
15. Baş, G.; Ömer, B. Effects of multiple intelligences supported project-based learning on students' achievement levels and attitudes towards English lesson. *Int. Electron. J. Elem. Educ.* **2017**, *2*, 365–386.
16. Ai, O. Exploring the Potential of CLIL in English Language Teaching in Japan Universities: An Innovation for the Development of Effective Teaching and Global Awareness. *J. Rikkyo Univ. Lang. Center* **2014**, *32*, 39–51.
17. Lastra-Mercado, D. *An Introductory Study of Project-Based Learning (PBL) and Content and Language Integrated Learning (CLIL) in TEFL*; Universidad de Jaén: Jaén, Spain, 2016.
18. Tawfik, A.A.; Lilly, C. Using a flipped classroom approach to support problem-based learning. *Technol. Knowl. Learn.* **2015**, *20*, 299–315. [CrossRef]
19. Rahman, A.A.; Zaid, N.M.; Abdullah, Z.; Mohamed, H.; Aris, B. Emerging Project Based Learning in Flipped Cclassroom: Technology Used to Increase Students' Engagement. In Proceedings of the 3rd International Conference on Information and Communication Technology (ICoICT), Nusa Dua, Indonesia, 27–29 May 2015.
20. Tsai, C.W.; Shen, P.D.; Lu, Y.J. The effects of problem-based learning with flipped classroom on elementary students' computing skills: A case study of the production of ebooks. *Int. J. Inf. Commun. Technol. Educ. (IJICTE)* **2015**, *11*, 32–40. [CrossRef]
21. Shih, W.L.; Tsai, C.Y. Students' perception of a flipped classroom approach to facilitating online project-based learning in marketing research courses. *Aust. J. Educ. Technol.* **2017**, *33*. [CrossRef]

22. Cukurbasi, B.; Kiyici, M. High school students' views on the PBL activities supported via flipped classroom and LEGO practices. *J. Educ. Technol. Soc.* **2018**, *21*, 46–61.

23. Ngo, H.Q.T.; Nguyen, Q.C.; Nguyen, T.P. Design and Implementation of High Performance Motion Controller for 2-D Delta Robot. In In Proceedings of the Seventh International Conference on Information Science and Technology, Da Nang, Vietnam, 16–19 April 2017; pp. 129–134. [CrossRef]

24. Cheryl, Z.Q.; Carla, Z.; William, O. Collaborating Interaction Design into Engineering Projects in Community Service (EPICS). In Proceedings of the 2012 Frontiers in Education Conference Proceedings, Seattle, WA, USA, 3–6 October 2012; pp. 1–6. [CrossRef]

25. Jonathan, P.S.; Panagiotis, K.L. EPICS: A Service Learning Program at Butler University. In Proceedings of the Frontiers in Education 35th Annual Conference, Indianopolis, IN, USA, 19–22 October 2005; pp. 21–25. [CrossRef]

26. Srijoy, D.; Rohan, M. EPICS High: Digital Literacy Project in India. In Proceedings of the 2014 IEEE Integrated STEM Education Conference, Princeton, NJ, USA, 8 March 2014; pp. 1–6. [CrossRef]

27. Srijoy, D.; Rohan, M. EPICS High: STEM's impact on community service. In Proceedings of the 2013 IEEE Integrated STEM Education Conference (ISEC), Princeton, NJ, USA, 9 March 2013; pp. 1–4. [CrossRef]

28. Carla, B.Z.; William, C.O. Learning by Doing: Reflections of the EPICS Program. *Int. J. Serv. Learn. Eng.* **2014**, *9*, 1–32. [CrossRef]

29. Angela, R.B.; Kurtis, G.P.; Christopher, W.S. Measuring the Value Added from Service Learning in Project-based Engineering Education. *Int. J. Eng. Educ.* **2010**, *26*, 535–546.

30. Luis, C.; Marta, R.; Lidon, M.; Teresa, G. University Social Responsibility towards Engineering Undergraduates: The Effect of Methodology on a Service-Learning Experience. *Sustainability* **2018**, *10*, 1823. [CrossRef]

31. Joachim, W.; Nicola, W.S.; Nadia, N.K. Quality in Interpretive Engineering Education Research: Reflections on an Example Study. *J. Eng. Educ.* **2013**, *102*, 626–659. [CrossRef]

32. James, L.H.; Carla, B.Z.; William, C.O. Preparing Engineers for the Workplace through Service Learning: Perceptions of EPICS Alumni. *J. Eng. Educ.* **2016**, *105*, 43–69. [CrossRef]

33. Ghiani, G.; Laporte, G.; Musmanno, R. Introducing logistics system. In *Introduction to Logistics System Planning and Control*; Ross, S., Weber, R., Eds.; John Willey & Son: Chichester, UK, 2004; pp. 1–22, ISBN 0-470-84916-9.

34. Whelan, P.L. Material Handling System and Method for Manufacturing Line. U.S. Patent 4,293,249, 6 October 1981.

35. Kadaba, N. Special Handling Processing in a Package Transportation System. U.S. Patent 6,539,360, 25 March 2003.

36. Thatcher, J.L.; Easterling, A. Automated Order Filling Method and System. U.S. Patent 6,505,093, 7 January 2003.

37. Mountz, M.C. Material Handling System Using Autonomous Mobile Drive Units and Movable Inventory Trays. U.S. Patent 6,895,301, 17 May 2005.

38. Mountz, MC. Material Handling System and Method Using Mobile Autonomous Inventory Trays and Peer-to-Peer Communications. U.S. Patent 6,950,722, 27 September 2005.

39. Ito, T.; Mousavi Jahan Abadi, S.M. Agent-based material handling and inventory planning in warehouse. *J. Intell. Manuf.* **2002**, *13*, 201–210. [CrossRef]

40. Babiceanu, R.F.; Chen, F.F.; Sturges, R.H. Framework for the control of automated material-handling systems using the holonic manufacturing approach. *Int. J. Prod. Res.* **2004**, *42*, 3551–3564. [CrossRef]

41. Chow, H.K.; Choy, K.L.; Lee, W.B.; Lau, K.C. Design of a RFID case-based resource management system for warehouse operations. *Expert Syst. Appl.* **2006**, *30*, 561–576. [CrossRef]

42. Ting, L.; Peijiang, Y.; Tianmiao, W.; Chengkun, W.; Dongdong, C.; Yong, L. An AGV-based Teaching Approach on Experiments of Mechatronics Course. In Proceedings of the 2014 IEEE International Conference on Robotics and Biomimetics (ROBIO 2014), Bali, Indonesia, 5–10 December 2014; pp. 2104–2109. [CrossRef]

43. Hector, P.; Eugenio, V.; Fernando, H. Using JIGSAW-type Collaborative Learning for Integrating Foreign Students in Embedded System Engineering. In Proceedings of the Design of Circuits and Integrated Systems, Madrid, Spain, 26–28 November 2014; pp. 1–6. [CrossRef]

44. Paulik, M.J.; Krishnan, M. An autonomous ground vehicle competition-driven capstone design course. In Proceedings of the 29th ASEE/IEEE Frontiers in Education Conference, San Juan, PR, USA, 10–13 November 1999; Volume 2, pp. 7–12. [CrossRef]

45. Julio, V.; Jose, M.C. PiBot: An Open Low-Cost Robotic Platform with Camera for STEM Education. *Electronics* **2018**, *7*, 430. [CrossRef]

46. Timothy, D.; Nicole, H.; Gautam, B. Design and Development of a Low-Cost Open-Source Robotics Education Platform. In Proceedings of the 50th International Symposium on Robotics, Munich, Germany, 20–21 June 2018; pp. 1–4.

47. Paulo, A.F.R.; Hector, A.; Luiz, C. HeRo: An open platform for robotics research and education. In Proceedings of the Latin American Robotics Symposium and Brazilian Symposium on Robotics, Curitiba, Brazil, 8–11 November 2017; pp. 1–6. [CrossRef]

48. Marsette, V.; Shekar, N.H. Teaching Robotics Software with the Open Hardware Mobile Manipulator. *IEEE Trans. Educ.* **2013**, *56*, 42–47. [CrossRef]

49. Pedro, J.N.; Carlos, F.; Pedro, S. Industrial-Like Vehicle Platforms for Postgraduate Laboratory Courses on Robotics. *IEEE Trans. Educ.* **2013**, *56*, 34–41. [CrossRef]

50. Lorenz, M.; Dominik, H.; Marc, P. PX4: A node-based multithreaded open source robotics framework for deeply embedded platforms. In Proceedings of the 2015 IEEE International Conference on Robotics and Automation (ICRA), Seattle, WA, USA, 26–30 May 2015; pp. 6235–6240. [CrossRef]

51. Marta, M.; Juan, C.G.; Lukasz, K.; Pedro, D.M.; David, P.; Javier, L. Open platform and open software for an intelligent wheelchair with autonomous navigation using sensor fusion. In Proceedings of the 42nd Annual Conference of the IEEE Industrial Electronics Society, Florence, Italy, 23–26 October 2016; pp. 5929–5934. [CrossRef]

52. USAID/Vietnam Director Talks with VTC10 Channel about the Engineering Projects in Community Service Program 2018 in Vietnam. Available online: https://goo.gl/oczFa7 (accessed on 10 July 2018).

electronics

MDPI

Article

PiBot: An Open Low-Cost Robotic Platform with Camera for STEM Education

Julio Vega *,† and José M. Cañas †

Department of Telematic Systems and Computation, Rey Juan Carlos University, Camino del Molino S/N, 28934 Fuenlabrada, Madrid, Spain; jmplaza@gsyc.es
* Correspondence: julio.vega@urjc.es; Tel.: +34-914-888-755
† These authors contributed equally to this work.

Received: 16 October 2018; Accepted: 10 December 2018; Published: 12 December 2018

Abstract: This paper presents a robotic platform, PiBot, which was developed to improve the teaching of robotics with vision to secondary students. Its computational core is the Raspberry Pi 3 controller board, and the greatest novelty of this prototype is the support developed for the powerful camera mounted on board, the PiCamera. An open software infrastructure written in Python language was implemented so that the student may use this camera as the main sensor of the robotic platform. Furthermore, higher-level commands were provided to enhance the learning outcome for beginners. In addition, a PiBot 3D printable model and the counterpart for the Gazebo simulator were also developed and fully supported. They are publicly available so that students and schools without the physical robot or that cannot afford to obtain one, can nevertheless practice, learn and teach Robotics using these open platforms: *DIY-PiBot* and/or *simulated-PiBot*.

Keywords: teaching robotics; science teaching; STEM; robotic tool; Python; Raspberry Pi; PiCamera; vision system

---

1. Introduction

Over the last decade, technology has become increasingly common in the majority of contexts of daily and industrial life. A machine's capacity for taking optimum decisions in real time and simultaneously handling a large quantity of data is undoubtedly far greater than that of a human being. *Industrialization 4.0* [1] involves the integration of complex robotic systems in factories, logistics and what is known as the *Internet of things*, where sophisticated automatons handle an immense quantity of data to take strategic decisions for companies.

In addition to a large computational capacity, these mobile and intelligent robots need a complex sensory system to act intelligently not only in factories but in general robot–human interaction [2]. The fixed automation of well-structured production chains is giving way to an unpredictable world and an unstructured reality, which underlines the need for a wide complementary range of sensors and actuators to attain complete autonomy [3].

Although cameras have not been the most used option in mobile robotics for some years (sonar and/or laser have been more commonly used as sensors), vision is currently the most widely used sensor and will definitely be the most commonly used in the long-term future, because of the possibilities it offers and the processing power of current computers. Cameras are low-cost devices that are potentially a rich source of information.

However, visual capacity in robots, in contrast to that of living beings, is not a simple technique. The main difficulty lies in extracting useful information from the large amount of data that a camera provides. Good algorithms are needed for this task.

Summarizing, the advance of Artificial Intelligence (AI), robotics and automation in society [4], and the future of work and industry, in particular, converge in what is already referred to as the fourth industrial revolution. According to the analysis of the University of Oxford [5] and the professional services of Deloitte [6], almost half of all jobs will be occupied by robots in the next 25 years. Furthermore, as the McKinsey Institute shows in its most recent report on the global economy [7], robots will perform the work of about 800 million employees in 2030.

It is therefore important to incorporate technology, and specifically robotics with vision systems, in the pre-university educational system since, within ten years, todays' youngest students will have to confront a labor market demanding profiles related to automation of systems [8]. From the educational point of view, robotics is a field where many areas converge: electronics, physics (Figure 1 left), mechanics (Figure 1 right), computer sciences, telecommunications, mathematics, etc.

Figure 1. Different robotic prototypes to work in different educational areas.

Hence, robotics is growing in importance in pre-university education, either as a field of knowledge in itself or as a tool to present technology and other subjects [9,10] to young students in an attractive way. Furthermore, it has the power to motivate students, bringing technology closer to young people [11] by using robotics as a tool to present basic concepts of science [12], technology, engineering and mathematics (STEM) [13]. In an almost play-like context, students learn notions, which are difficult and complex to explain or to assimilate in the classic masterclass [14,15].

The implementation of robotics in higher education is already in place. Six states in the U.S. (Iowa, Nevada, Wisconsin, Washington, Idaho and Utah) have announced plans and investments with this aim in mind in the last five months. Likewise, four countries—Canada, Ireland, New Zealand and Romania—have recently announced similar plans, with a total investment of 300 million dollars. Japan, in its *New Robot Strategy Report* [16], makes clear that investing in robotics is fundamental for the growth of the country.

In this educational field, the teaching of robotics itself converges with other disciplines (e.g., programming), where it is used as a teaching tool [17–19].

Another example of the increasing importance of robotics in education are the robotics competitions for teenagers that have appeared in recent years, which encourage interest in this technology. At international level, numerous championships are organized, which bring together students from all over the world to learn, share experiences and enjoy the development of robotic prototypes. It is worth highlighting the RoboCup Junior (http://rcj.robocup.org) [20–22], with tests such as the rescue or the robotic soccer, as well as the First Lego League (FLL) and the VEX Robotics Competitions (https://www.vexrobotics.com/vexedr/competition). In Finland, the SciFest competition (http://www.scifest.fi) attracts students from all over Europe [23] and has agreements with educational institutions in South Africa [24].

In the academic community, several conferences also highlight the role of robotics in education, including the Conference on Robotics in Education (RIE), and the Workshop on Teaching Robotics with ROS (TRROS) within the European Robotics Forum (http://www.eu-robotics.net/robotics_forum). Special editions on education in robotics have also appeared in several scientific journals.

To support this increasing presence of educational robotics, there are many teaching frameworks used to teach robotics to children, ranging from those focused on primary education to more powerful ones designed for secondary education and high school. They are usually composed of a concrete robotic platform, i.e., a robot, which is programmed in a certain language using software tools. Students engage in different exercises, challenges or projects (practice activities). They teach the basic operation of sensors, actuators and the rudiments of programming.

2. Robotic Platforms for STEM Education

Most robots used in commercial educational platforms are proprietary. It is worth mentioning the well-known Lego, which has featured for some years in educational robotics kits, with different versions: Mindstorms RCX, NXT, EV3 and WeDo [15,21].

Arduino boards appeared some years ago, in an effort to work around the closed-platforms limitation, providing cheaper and more adapted robotic platforms. This is a free hardware board that allows a wide variety of low-cost robotic components to be added [21,25–28]. Thus, beginning with a basic and affordable Arduino platform, teachers and students can freely adapt it to their necessities, developing an effective and low-cost robot, as described in [29–32].

Other platforms are Thymio (Figure 2 left) [17,33,34], Meet Edison's or VEX robots (Figure 2 middle and right), and simulated environments such as TRIK-Studio [28,35] or Robot Virtual Worlds (RVW) [36].

Figure 2. Robots Thymio, VEX IQ and VEX CORTEX.

Beyond the robotic hardware platforms, different software frameworks can be found in educational robotics. Lego has its own option, EV3-software. In addition, Lego Mindstorms and WeDo can now be interfaced through Scratch language [32,37], especially version 3.0, which will be officially online from the beginning of next year. Other variants are Blockly [38], Bitbloq or VPL. All of these contain graphic blocks that typically connect in sequence in a graphic editor. Arduino platforms can be programmed with a simple, C++ based, text language using Arduino-IDE. C++ is widely used at university level, but its complexity makes it unsuitable for pre-university students.

Exploring the existing literature, we found many other works that have presented robotic platforms for educational purposes and their underlying philosophy. In [39], the authors focused on a six degrees of freedom (DOF) serial robotic arm as a robotic platform for training purposes. They derived the kinematic and dynamic models of the robot to facilitate controller design. An on-board camera to scan the arm workspace is included.

In [40], Alers and Hu presented the AdMoVeo robotic platform, which was developed for the purpose of teaching the basic skills of programming to industrial design students. This platform lets students explore their creativity through their passion for graphic and behavioral design.

In [26], Jamieson examined whether Arduino was a suitable platform for teaching computer engineers and computer scientists by means of an embedded system course. He described a project-based learning embedded system course that has been taught and identified the topics covered in it compared to the IEEE/ACM recommendations. He finally concluded by saying that students expressed high praise for the Arduino platform and that, compared to previous years, students' final projects were of better quality and more creative.

In [41], the authors presented eBug as a low-cost, open robotics platform designed for undergraduate teaching and academic research in areas such as multimedia smart sensor networks, distributed control, mobile wireless communication algorithms and swarm robotics. This prototype used the Atmel AVR XMEGA 8/16-bit micro-controller.

Miniskybot is presented in [42] as a mobile robot for educational purposes that is 3D-printable on low cost RepRap-like machines, fully open source (including mechanics and electronics), and designed exclusively with open source tools. It is based on an 8-bit pic16f876a micro-controller.

Nevertheless, there is no system, and much less a guided one, which maintains a constant level of motivation and challenge, especially one in which vision plays an important role. In fact, the majority of these kits or robotic platforms on the market focus on doing specific tasks or are designed to arouse the interest of either the youngest students or university students in robotics, but not so that students in pre-university courses acquire correct and complete training in programming, something that is in great demand and is widespread in almost any degree. Although it is true that other kits exist that are more specialized in specific scientific fields [43], our proposed framework goes further and provides all the open tools for both students and teachers [44] required for a complete academic year. Its versatile design puts at their disposal numerous sophisticated algorithms, including vision, with a pleasant and intuitive interface.

In addition, a large gap has been identified between the level of academic training at university level in scientific and technological degrees and the official curriculum implemented at pre-university levels, specifically in science subjects at secondary education level. Thus, the present work proposes to mitigate this gap, developing a complete teaching framework for robotics with vision, which today is non-existent, integrating:

1. A RaspberryPi-based open hardware platform, economically suitable for secondary schools to satisfy the needs of a complete class, but at the same time standardized and powerful, which allows the execution of algorithms of robotics with vision.
2. An open software infrastructure that is simple and intuitive for young students to manage but at the same time is powerful and versatile. It incorporates enough resource libraries, as well as diverse examples, to provide practical exercises in programming robots with vision that are sufficient in both number and complexity to continuously motivate students [45].
3. A wide repertoire of practice activities that can be followed during a complete academic year, including sufficient and properly staggered sessions for students to correctly assimilate the content [46].

3. Design of the PiBot Tool for STEM Education

Following the analysis of the most important available educational robots, this section describes the design of the proposed robot. It leverages some of the new possibilities offered by different technologies and aims to overcome limitations observed in current platforms such as having no cameras or not being usable with programming languages such as Python. Using PiBot with Python is not intended for primary education or first year secondary education, where visual languages such as Scratch are a better starting point [47]. Instead, it is designed for students of secondary education aged over 12 years and even introductory university courses.

Better tools improve learning processes in young learners. The PiBot education tool follows a three-part architecture, as shown in Figure 3: the robot platform, the software drivers and the exercises. The robot and the drivers can be regarded as the infrastructure for the exercises, which can be organized in courses or levels and focus on different aspects of robotics.

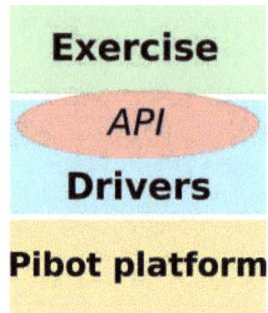

Figure 3. Architecture of the PiBot tool: hardware (platform) and software (drivers and exercise).

The creation of the PiBot tool followed several design principles:

1. Low cost (under 180 euros), to make it affordable for most schools and students.
2. Open: First, the robot hardware should be easy to assemble for students, who can also make the pieces using a 3D printer. Thus, the assembly of a PiBot can be an educational activity and interesting for the makers community. Second, drivers should be open source and publicly available.
3. Compatibility with common sensors and actuators in (Arduino-based) educational robots. In this way, if an Arduino-based robot is already available, the transition to PiBot is affordable; and, in any event, the acquisition of components for PiBot is very simple, given the wide availability of components for Arduino.
4. Including vision in an easy way. Cameras are useful sensors and this platform can give students easy and practical access to vision.
5. Supporting not only the real robot but also a simulated robot. Thus, even without a physical platform, the PiBot tool may be used to teach and learn robotics.
6. Python as a programming language because of its simplicity, expressive power and because it is widely used in higher levels of education and programming.

4. PiBot Robotic Platform

Robots are typically composed of a computer or a microprocessor, several sensors, actuators and some form of connectivity. Sensors provide information about the environment, the computer runs the robot software and actuators allow the robot to do things such as moving or performing actions in the real world.

4.1. Hardware Design

The block diagram of the PiBot hardware is shown in Figure 4. The main computer is a Raspberry Pi 3 controller board (Figure 5 middle). It is more powerful than Arduino processors, maintains low costs, and runs a functional operating system based on Linux; specifically, the Raspbian Stretch distribution. It allows the use of standard development tools on the Linux community and the use of the PiCamera.

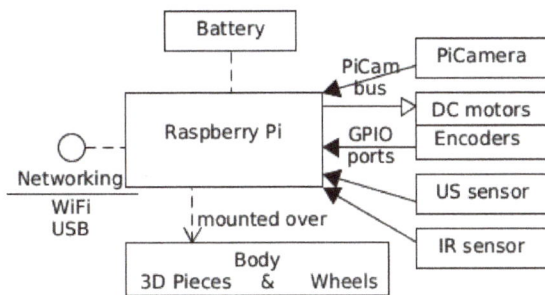

Figure 4. Hardware design of the PiBot robot.

Figure 5. Motors, Raspberry Pi board and PiBot made with 3D printable pieces.

The sensors mounted onboard PiBot are:

- An ultrasound sensor model HC-SR04 (Figure 6 left)
- Infrared sensors
- Motor encoders
- Raspberry PiCamera (Figure 6 right). This is connected to the computer using a dedicated data bus. Its technical details are included in Table 1.

The ultrasonic (US), infrared (IR) and encoder sensors are connected to the Raspberry Pi board through several GPIO ports (General Purpose Input/Output). This protocol allows the connection and control of several devices at the same time and requires a configuration on each port to serve as input and output of data [48].

The actuators mounted onboard PiBot are two DC motors (Parallax Feedback 360° High Speed Servo (Figure 5 left)). They provide movement and differential drive to the PiBot. The motors include encoders and are connected to the main processor through GPIO bus.

Figure 6. Ultrasonic sensor model HC-SR04, IR sensors and Raspberry PiCamera.

Table 1. PiCamera (v2.1 board) technical intrinsic parameters.

PiCamera Params.	Values
Sensor type	Sony CMOS 8-Mpx
Sensor size	3.6 × 2.7 mm (1/4" format)
Pixel count	3280 × 2464 (active px.)
Pixel size	1.12 × 1.12 um
Lens	f = 3.04 mm, f/2.0
Angle of view	62.2 × 48.8 degrees
SLR lens equivalent	29 mm

All these components are assembled into a body made of 3D printable pieces. The 3D printable models of all the chassis pieces are publicly available on the web (https://github.com/JdeRobot/ JdeRobot/tree/master/assets/PiBot). The body also allocates a battery of 10,000 mAh, which provides power to all the electronic onboard devices. An official list of components and some tentative providers are also available at the same webpage so that anyone can buy the components, print the pieces and build a `PiBot`.

4.2. Simulated Robot

The Gazebo simulator (http://gazebosim.org) was selected for simulation of the `PiBot` platform. This is an open source robotic simulator powered by Open Robotics Foundation and the de facto standard in the robotics scientific community. It provides a physical engine so collisions and realistic movements are provided.

The students can program an exercise and run their code seamlessly both on the physical `PiBot` or on the simulated `PiBot` inside Gazebo, at will. The student code lies on top of the `PiBot` API (Application Programming Interface), which is used to obtain sensor readings and to command actuator orders. The API is exactly the same in both cases. In the first one, drivers will be used to connect to the physical devices. In the second one, other drivers will exchange messages with the simulator to implement the same functions.

To support this new robot, a 3D model of the robot was developed (Figure 7). In addition, several plugins were also integrated for the simulation of the onboard camera, the distance sensor (sonar) and IR sensors. IR support was implemented using small cameras. Each IR consists of a 4 × 4 pixel camera and an additional code that computes the virtual IR measurement from the values of these pixels. The movement was also supported with the corresponding Gazebo plugin, which also provides a 2D position sensor (as encoder).

The 3D `PiBot` model and all the developed plugins are publicly available on the web (https://github.com/JdeRobot/JdeRobot/tree/master/assets/gazebo; https://github.com/ JdeRobot/JdeRobot/tree/master/src/drivers/gazebo/plugins/pibot).

Figure 7. PiBot robot simulated in Gazebo.

5. Software Infrastructure

Python was chosen as a programming language to support PiBot because of its simplicity, its expressive power and because it is widely used in higher levels of education and many industries (in conjunction with powerful libraries). It is a text language, interpretative and object oriented. It is easier to learn than other also widely used programming languages, such as C/C++ or Java, and at the same time it is highly powerful. It is a real world language but accessible for pre-university students.

Using the proposed educational tool, the students program their exercises in Python by writing the file *exercise.py*, for example, with a text editor. This program uses the PiBot Application Programming Interface (API) to control the robot, which contains a set of natural methods to read the measurements from the robot sensors (US, IR, and camera) and methods to give commands to the robot actuators (DC motors). The most important API methods are detailed in Table 2.

Two different libraries were developed to support this API. One runs onboard the PiBot Raspberry Pi and a second one communicates with the simulated robot inside Gazebo. As the programming interface is the same in both cases, the student application works interchangeably on the physical platform and the simulated one. The final robot in each case is selected by specifying it in the configuration file.

Using this API, students concentrate on the algorithm they are developing, on the robot's intelligence, avoiding the low level details such as ports, connectivity with the robot, etc., which are stored in the library configuration file.

Table 2. Application Programming Interface (API).

	Actuators	Sensors
Low level methods	RightMotor(V) LeftMotor(V)	readUltrasound readInfrared getImage
High level methods	move(V, W)	getColoredObject(color) getDistancesFromVision getRobotPosition

The API methods can be divided into low and high level methods. The low level methods provide access to a single device, such as readUltrasound, readInfrared or getImage. RightMotor(V) controls the single right motor commanding desired speed, as does LeftMotor(V) for the other motor. The high level methods provide a simpler and more compact way to control the whole robot or two vision functions to get useful information from the image in an easy way. These are described below.

5.1. Drivers for the Real PiBot

To support the real PiBot, two modules were programmed, as shown in Figure 8. One includes the management of the PiCamera and the other deals with GPIO devices (US sensor, IR sensor and motors).

They were programmed in Python using standard libraries available in the Python community. It is publicly available on the web (https://github.com/JdeRobot/JdeRobot/tree/master/src/drivers/PiBot/real). The image processing functionality also relies on OpenCV.

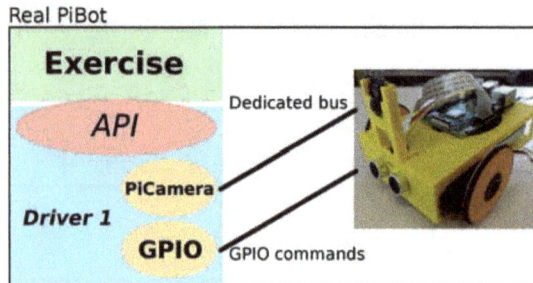

Figure 8. Connection of the library with the real `PiBot`.

5.2. Drivers for the Simulated PiBot

To support the simulated `PiBot` on Gazebo, a specific library was developed, which connects with the previously mentioned plugins (Figure 9), exchanging messages through the ICE communication layer. It achieves sensor readings and camera images through network interfaces built in the JdeRobot project (https://jderobot.org). It is also publicly available on the web (https://github.com/JdeRobot/JdeRobot/tree/master/src/drivers/PiBot/Gazebo).

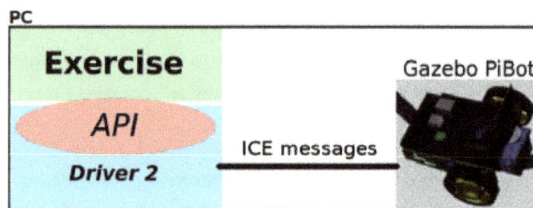

Figure 9. Connection of the library with the simulated `PiBot`.

5.3. Movement Control

Regarding motors, beyond the low level methods RightMotor(V) and LeftMotor(V), a new high level method is provided for simpler control of the robot movements: Move(V,W). As parameters, this method accepts the desired linear speed V and the desired rotation speed W, internally translating them into commands to the left and right motors so that the whole robot moves in accordance with V and W. It takes into account the geometry of the `PiBot` and its wheels.

This function provides general 2D movement control: the `PiBot` may rotate without displacement (setting $V = 0$ and using W) both left or right (depending on the sign of W), may advance in a straight line (setting $W = 0$ and using V) both backwards and forward (depending on the sign of V), and may move in generic arcs advancing and rotating at the same time.

This is a useful speed control when programming reactive behaviors, which is better than position-based control.

5.4. Vision Support

One advantage of the `PiBot` educational tool is its support for the camera. This allows many new exercises with vision and vision-based behaviors. It also introduces the students to computer vision in a simple and natural way. Two functions (getColoredObject(color) and getDistancesFromVision()) have been included so far in the `PiBot` API to easily get useful information from images, because the low level method getImage() and the pixels processing are too complex for high school students. They were implemented and included in a vision library which performs complex image processing, hides all the complexity inside and is simple and intuitive to use. It internally employs OpenCV library, a standard in the Computer Vision community.

First, the high level method getColoredObject(color) accepts the desired *color* as input parameter and filters all the pixels within a range of that color (some of which are already predefined in the library: orange, red or blue) in the current camera image. It delivers as output the position of the colored object inside the image (its mean X and Y value) and its size (the number of detected pixels of that color). It works with single objects, as can be seen in Figure 10.

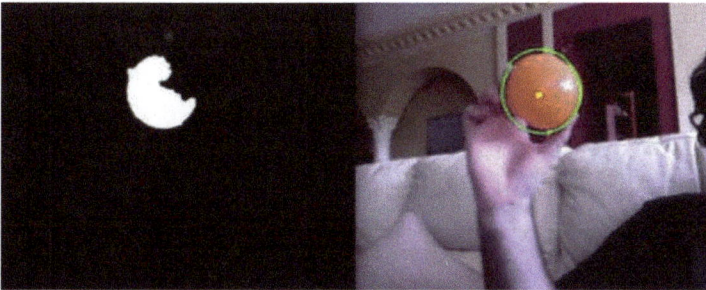

Figure 10. GetColoredObject function for orange color with empirical predefined (H_{min}, H_{max}, S_{min}, S_{max}, V_{min}, and V_{max}) ranges.

It uses HSV color space and OpenCV filtering methods. This function on PiBot API allows for exercises such as Object-Following, which is described in the next section.

Second, the high level method getDistancesFromVision() computes the distance to obstacles in front of the PiBot and provides a depth map from the robot to the surrounding objects. Typically the sonar sensor measures the distances in one direction. Using the camera for the same operation, the angular scope is extended to the camera field of view (around 60 degrees).

The vision library developed contains an abstract model of the camera (pin-hole) and several projective geometry algorithms. The camera parameters are known (K matrix and relative position inside the robot). As the PiBot only has a single camera, no stereo technique can be used for depth estimation. Instead, the implementation of the getDistancesFromVision() method assumes that all objects lie on the floor and the floor surface has a uniform color (ground hypothesis). It sweeps all the columns of the current image from its bottom. When the first edge pixel is found on a column, it is backprojected into 3D space, using ray tracing and the pin-hole camera model. The intersection of this ray with the floor plane is the estimated position of this edge in 3D space, and its distance to the robot is computed. In this way, the 3D point corresponding to each bottom pixel of the obstacle in the image can be obtained (Figure 11).

For instance, the left-hand side of Figure 12 shows the image coming from the camera, with the white floor (the battery was safely ignored as only green pixels were taken into account for explanatory purposes in this test). On the right-hand side, the estimated depths for the green object are displayed as red points and the field of view is also shown as a white trapezoid. The estimated distances are regularly consistent and correct.

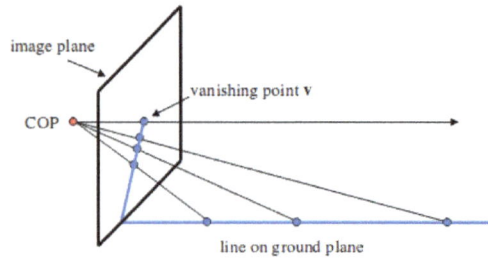

Figure 11. Ground Hypothesis assumes all objects are on the floor.

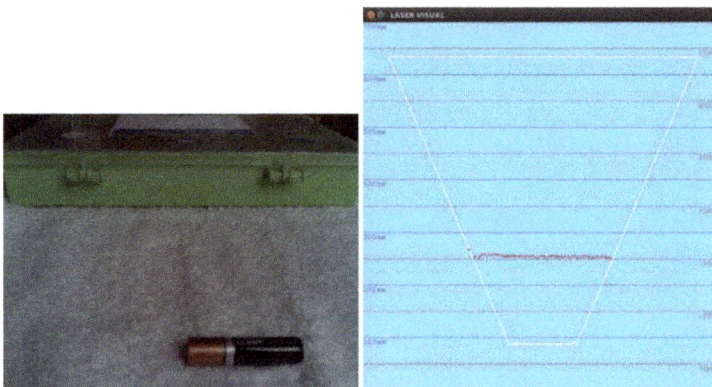

Figure 12. Example of visual sonar reading with 25 cm object shown using 3D scene simulator.

This getDistancesFromVision() function on `PiBot` API allows for exercises such as robot navigation with obstacles. An example is the vision-based obstacle avoidance which is described in the next section.

5.5. Downloading Software for Each Exercise into the Real Robot

Since Raspberry Pi is the brain of the `PiBot`, and unlike current educational robots with an Arduino processor, this board is a fully functional PC (Personal Computer) running a Raspbian Linux OS (Operating System), there are many ways to transfer the Python program to the real robot.

One of the Pi's USB (Universal Serial Bus) ports can be used to insert a memory stick which includes the program to be executed, but the easiest way to transfer it is by connecting the Raspberry Pi to the Internet via either Wi-Fi or Ethernet. When it is connected, its IP (Internet Protocol) address can be found to remotely transfer the program from a PC—which can be running over Windows, Linux or MacOS operating systems—using a graphical SCP client, such as Filezilla, or typing and executing text based commands on a terminal.

Another method to transfer files from the PC to the Raspberry is using Virtual Network Computing (VNC) software. It is only necessary to have a VNC server previously installed on the board and a VNC viewer installed on the PC.

Nevertheless, not only can students remotely transfer their codes but they can also remotely execute them using the SSH protocol both graphically, with tools such as PuTTY, or using commands on a terminal. In addition, because the Raspberry is a full PC, students can even develop the exercises directly on the Raspberry board. It is only necessary to connect it to a screen through the HDMI port, and plug in a keyboard/mouse through the USB ports.

6. Exercises for Students Using PiBot

The program of activities for an introductory course on robotics was prepared. It includes basic exercises with a real robot (learning to deal with its isolated sensors and actuators), basic behaviors (combining both sensors and actuators on a single program) and vision-based behaviors.

It covers the syllabus of the subject of *Programming, Robotics and Technology* in higher education in Madrid (Spain). It also covers the typical topics of any robotics introductory course.

6.1. Basic Exercises

Students can begin assembling different components on the PiBot and review some basic concepts of electronics so they have no problems when connecting the different components, such as infrared or ultrasound sensors (Figure 13).

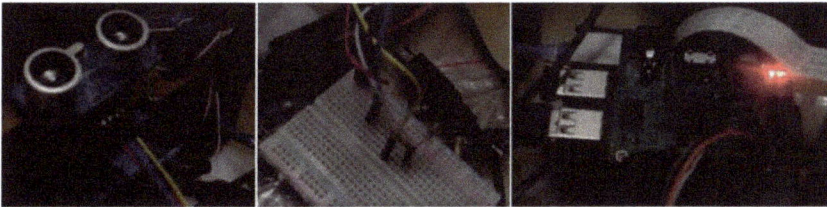

Figure 13. Practice activity with PiBot to handle an ultrasonic sensor.

The exercises here were designed to achieve the objective of *Disassembling objects, analyze their parts and their functions* from the unit *The Technological Process* and all the objectives included in the units of *Electricity*, *Materials for Technical Use* and *Programming*.

6.2. Basic Behaviors

This second step consists of a complete robotics project where students combine everything they have previously learnt. Exploring the existing literature, we found a set of exercises that are frequently used in different teaching frameworks and academic proposals, in an almost cross-curricular manner. One of the classic projects is the follow-line behavior [15,21,28,35], as shown in Figure 14. In this project, the robot has IR sensors pointing to the ground, which is white but has a thick black line. Another is the avoidance of obstacles [21,35], as shown in Figure 15, where the robot has an ultrasound sensor that allows it to detect objects that interfere with its movement. The student's program must then order the motors to stop and turn until they find a free space to go forward again.

These exercises were designed to achieve the objectives included in typical *Hardware and Software* and *Programming* units.

Figure 14. Practice line-tracking task in both real and simulated PiBot platforms using IR sensor.

Figure 15. Navigation practice avoiding obstacles through US in `PiBot`.

6.3. Vision-Based Behaviors

In this group of exercises, the students can solve the previously described exercises but now using vision as the main sensor. Some projects developed are: following a colored object (Figure 16), line tracking or bump-and-go.

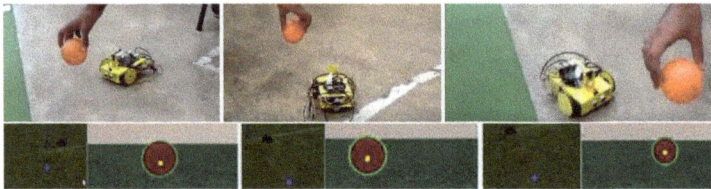

Figure 16. Navigation exercise of following a colored object using vision in both real and simulated `PiBot` platforms.

In July 2018, a Robotics workshop was taught to ten teachers at the Campus of Fuenlabrada of the Rey Juan Carlos University (Madrid) (Figure 17), training them to use the developed framework with `PiBot` as a robotic platform.

Figure 17. Workshop at Rey Juan Carlos University to train teachers for teaching with JdeRobot-Kids framework using `PiBot`.

7. Conclusions

This research focused on incorporating robotics and robots with vision in the classroom to train pre-university students, satisfying the demands imposed by society in the Digital Age and the motivational needs detected in students, who study in a system not currently adapted to the so-called Industrial Revolution 4.0.

Although there are numerous educational robotics kits on the market, most are aimed at very young students. Many robotics kits are based on building robotic platforms with their own programming environments, and do not employ standardized programming languages. A standard language across several robotic platforms is Scratch 3.0, but it is also oriented to elementary level. In addition, typical robotics kits usually have somewhat limited capabilities, which means they tend to generate little mid-term motivation in students (for instance, in students who have already followed an introductory robotics course). Furthermore, given the complexity involved in the processing of images, cameras are not usually included in the educational robotic frameworks despite their great versatility and extensive use in real life applications.

After studying the current market in existing robotics educational kits and conducting an in-depth analysis of what the near to mid-term future holds in terms of the demands of the labor market, the authors (one of whom is an experienced Secondary Education teacher) detected a gap between the level of academic training at university level in scientific and technological degrees and the official curriculum implemented at pre-university levels, specifically in science subjects at secondary education level. Therefore, a complete new educational tool was developed, which includes:

- A robotic platform, the PiBot, based on the free hardware controller board Raspberry Pi 3. This board is mounted over a chassis designed as a 3D printable model, which lets anyone, not only schools, build their own robots at a low cost, following the Do It Yourself (DIY) philosophy. The Raspberry was chosen as the PiBot processor for several reasons: firstly, the inclusion of a camera with its own data bus, the PiCamera, which allows the teaching of Artificial Vision algorithms; secondly, it maintains a low cost robotic platform but with high computational power; thirdly, the inclusion of the GPIO ports on the board, thanks to which various sensors and actuators were connected; and, finally, it is a standardized versatile board.
- A simulated robot for PiBot under Gazebo simulator. Thus, students have the possibility of using both a real and a simulated robot. This provides a valuable learning process through which students can appreciate the differences between the real and the simulated world.
- A software infrastructure developed in Python language, which includes all the appropriate drivers. This facilitated students' programming of the robot, with simple and intuitive functions to handle the different sensors and actuators. At the same time, this infrastructure has great potential due to its handling of a camera as a sensor.
- A wide-ranging set of exercises that serve as a support to students for their progression in learning to program robots with vision. The real or simulated robot is programmed in a simple and powerful language, Python, which is not widely used in educational robotics due to the limitations of the processors managed to date.

Regarding future lines of research, one intended improvement in the short term is to extend the vision support: (a) developing new practical sessions with vision such as the detection and monitoring of people's faces, and materializing a visual memory in the PiBot; (b) seating the camera on a servo so the current vision range can be extended to a wider field of view, thanks to the movement of the camera.

It is also intended to develop the support for the encoders of the PiBot motors, which should allow more position-based sophisticated navigation to be developed.

In addition, in the 2018–2019 academic year, although use of this new tool has already been successful under a pilot plan, with 25 students (15 using the simulated platform from Ceuta, and 10 using the real robot in a workshop), the impact on the educational effectiveness of this new tool will be measured by means of an experiment with actual students and a control group in several schools.

Finally, the authors are also working to support PiBot programming with the popular visual Scratch language, so that younger students can start simple programming of this robot. With the same

`PiBot` platform, they can start learning robotics with Scratch and subsequently move up to Python and engage in more motivating exercises.

Author Contributions: Conceptualization, J.V. and J.C.; methodology, J.V. and J.C.; software, J.V.; validation, J.V. and J.C.; formal analysis, J.C.; investigation, J.V.; resources, J.C.; data curation, J.V.; writing-original draft preparation, J.V. and J.C.; writing-review and editing, J.V. and J.C.; visualization, J.V. and J.C.; supervision, J.C.; project administration, J.C.; funding acquisition, J.C.

Funding: This work was partially funded by the Community of Madrid through the RoboCity2030-III project (S2013/MIT-2748) and by the Spanish Ministry of Economy and Competitiveness through the RETOGAR project (TIN2016-76515-R). The APC was funded by the RoboCity2030-III project (S2013/MIT-2748).

Conflicts of Interest: The authors declare no conflict of interest.

References

1. Schwab, K. *The Fourth Industrial Revolution*; World Economic Forum: Cologny, Switzerland, 2016.
2. Vega, J.; Cañas, J. Sistema de atención visual para la interacción persona-robot. In Proceedigns of the Workshop on Interacción Persona-Robot, Robocity 2030: UNED, Madrid, Spain, 29 September 2009; pp. 91–110, ISBN: 978-84-692-5987-0.
3. Arbel, T.; Ferrie, F. Entropy-based gaze planning. *Image Vis. Comput.* **2001**, *19*, 779–786. [CrossRef]
4. Solove, D. *The Digital Person: Technology and Privacy in the Information Age*; NYU Press: New York, NY, USA, 2004.
5. Frey, C.; Osborne, M. *The Future of Employment: How Susceptible Are Jobs to Computerisation*; University of Oxford: Oxford, UK, 2013.
6. Deloitte. *From Brawn to Brains: The Impact of Technology on Jobs in the UK*; Deloitte LLP: London, UK, 2015.
7. Institute, M. *Jobs Lost, Jobs Gained: Workforce Transitions in a Time of Automation*; McKinsey Global Institute: New York, NY, USA, 2017.
8. UK-RAS White Papers. *Manufacturing Robotics: The Next Robotic Industrial Revolution*; UK-RAS: London, UK, 2016.
9. Rubinacci, F.; Ponticorvo, M.; Passariello, R.; Miglino, O. Robotics for soft skills training. *Res. Educ. Media* **2017**. [CrossRef]
10. Rubinacci, F.; Ponticorvo, M.; Passariello, R.; Miglino, O. Breeding Robots to Learn How to Rule Complex Systems. *Robot. Educ.* **2017**._13. [CrossRef]
11. Rodger, S.H.; Walker, E.L. Activities to attract high school girls to computer science. In Proceedings of the twenty-seventh SIGCSE technical symposium on Computer science education, Philadelphia, PA, USA, 15–17 February 1996.
12. Altin, H.; Pedaste, M. Learning approaches to applying robotics in science education. *J. Balt. Sci. Educ.* **2013**, *12*, 365–377.
13. Mubin, O.; Stevens, C.J.; Shahid, S. A review of the applicability of robots in Education. *Technol. Educ. Learn.* **2013**. [CrossRef]
14. Cerezo, F.; Sastrón, F. Laboratorios Virtuales y Docencia de la Automática en la Formación Tecnológica de Base de Alumnos Preuniversitarios. *Rev. Iberoam. Autom. Inform. Ind. RIAI* **2015**, *12*, 419–431, doi:10.1016/j.riai.2015.04.005. [CrossRef]
15. Jiménez, E.; Bravo, E.; Bacca, E. Tool for experimenting with concepts of mobile robotics as applied to children education. *IEEE Trans. Educ.* **2010**, *53*, 88–95. [CrossRef]
16. Japan-Economic. *New Robot Strategy*; The Headquarters for Japan's Economic Revitalization: Tokyo, Japan, 2015.
17. Magnenat, S.; Shin, J.; Riedo, F.; Siegwart, R.; Ben-Ari, M. Teaching a core CS concept through robotics. In Proceedings of the 2014 Conference on Innovation & Technology in Computer Science Education, Uppsala, Sweden, 23–25 June 2014; pp. 315–320.
18. Merkouris, A.; Chorianopoulos, K.; Kameas, A. Teaching programming in secondary education through embodied computing platforms: Robotics and wearables. *ACM Trans. Comput. Educ.* **2017**, *17*, 9. [CrossRef]
19. Kubilinskiene, S.; Zilinskiene, I.; Dagiene, V.; Sinkevièius, V. Applying Robotics in School Education: A Systematic Review. *Balt. J. Mod. Comput.* **2017**, *5*, 50. [CrossRef]
20. Eguchi, A. RoboCupJunior for promoting STEM education, 21st century skills, and technological advancement through robotics competition. *Robot. Auton. Syst.* **2016**, *75*, 692–699. [CrossRef]

21. Navarrete, P.; Nettle, C.J.; Oliva, C.; Solis, M.A. Fostering Science and Technology Interest in Chilean Children with Educational Robot Kits. In Proceedings of the 2016 XIII Latin American Robotics Symposium and IV Brazilian Robotics Symposium (LARS/SBR), Recife, Brazil, 8–12 October 2016; pp. 121–126.

22. Kandlhofer, M.; Steinbauer, G. Evaluating the impact of robotics in education on pupils' skills and attitudes. In Proceedings of the 4th International Workshop Teaching Robotics, and 5th International Conference Robotics in Education, Padova, Italy, 18 July 2014; pp. 101–109.

23. Jormanainen, I.; Korhonen, P. Science Festivals on Computer Science Recruitment. In Proceedings of the 10th Koli Calling International Conference on Computing Education Research, Koli Calling'10, Koli, Finland, 28–31 October 2010; pp. 72–73.

24. Graven, M.; Stott, D. Exploring online numeracy games for primary learners: Sharing experiences of a Scifest Africa Workshop. *Learn. Teach. Math.* **2011**, *2011*, 10–15.

25. Araujo, A.; Portugal, D.; Couceiro, M.S.; Rocha, R.P. Integrating Arduino-Based Educational Mobile Robots in ROS. *J. Intell. Robot. Syst.* **2015**, *77*, 281–298. [CrossRef]

26. Jamieson, P. *Arduino for Teaching Embedded Systems. Are Computer Scientists and Engineering Educators Missing the Boat*; Miami University: Oxford, OH, USA, 2012.

27. Chaudhary, V.; Agrawal, V.; Sureka, P.; Sureka, A. An experience report on teaching programming and computational thinking to elementary level children using lego robotics education kit. In Proceedings of the 2016 IEEE Eighth International Conference on Technology for Education (T4E), Mumbai, India, 2–4 December 2016; pp. 38–41.

28. Filippov, S.; Ten, N.; Shirokolobov, I.; Fradkov, A. Teaching robotics in secondary school. *IFAC-PapersOnLine* **2017**, *50*, 12155–12160. [CrossRef]

29. Junior, L.A.; Neto, O.T.; Hernandez, M.F.; Martins, P.S.; Roger, L.L.; Guerra, F.A. A low-cost and simple arduino-based educational robotics kit. *Cyber J. Multidiscip. J. Sci. Technol.* **2013**, *3*, 1–7.

30. Plaza, P.; Sancristobal, E.; Fernandez, G.; Castro, M.; Pérez, C. Collaborative robotic educational tool based on programmable logic and Arduino. In Proceedings of the Technologies Applied to Electronics Teaching (TAEE 2016), Seville, Spain, 22–24 June 2016; pp. 1–8.

31. Afari, E.; Khine, M. Robotics as an educational tool: Impact of LEGO mindstorms. *IJIET* **2017**, *7*, 437–442. [CrossRef]

32. Beyers, R.N.; van der Merwe, L. Initiating a pipeline for the computer industry: Using Scratch and LEGO robotics. In Proceedings of the Conference on Information Communication Technology and Society (ICTAS), Umhlanga, South Africa, 8–10 March 2017; pp. 1–7.

33. Mondada, F.; Bonani, M.; Riedo, F.; Briod, M.; Pereyre, L.; Rétornaz, P.; Magnenat, S. Bringing robotics to formal education: The thymio open-source hardware robot. *IEEE Robot. Autom. Mag.* **2017**, *24*, 77–85. [CrossRef]

34. Roy, D.; Gerber, G.; Magnenat, S.; Riedo, F.; Chevalier, M.; Oudeyer, P.Y.; Mondada, F. IniRobot: A pedagogical kit to initiate children to concepts of robotics and computer science. In Proceedings of the RIE 2015, Yverdon-les-Bains, Switzerland, 20–22 May 2015.

35. Stone, A.; Farkhatdinov, I. Robotics Education for Children at Secondary School Level and Above. In Proceedigns of the Conference Towards Autonomous Robotic Systems, Guildford, UK, 19–21 July 2017; pp. 576–585.

36. Witherspoon, E.B.; Higashi, R.M.; Schunn, C.D.; Baehr, E.C.; Shoop, R. Developing computational thinking through a virtual robotics programming curriculum. *ACM Trans. Comput. Educ.* **2017**, *18*, 4. [CrossRef]

37. Plaza, P.; Sancristobal, E.; Carro, G.; Castro, M.; Blázquez, M.; Muñoz, J.; Álvarez, M. Scratch as Educational Tool to Introduce Robotics. In Proceedigns of the International Conference on Interactive Collaborative Learning, Budapest, Hungary, 27–29 September 2017; pp. 3–14.

38. Naya, M.; Varela, G.; Llamas, L.; Bautista, M.; Becerra, J.C.; Bellas, F.; Prieto, A.; Deibe, A.; Duro, R.J. A versatile robotic platform for educational interaction. In Proceedigns of the 2017 9th IEEE International Conference on Intelligent Data Acquisition and Advanced Computing Systems: Technology and Applications (IDAACS), Bucharest, Romania, 21–23 September 2017; Volume 1, pp. 138–144.

39. Manzoor, S.; Islam, R.U.; Khalid, A.; Samad, A.; Iqbal, J. An open-source multi-DOF articulated robotic educational platform for autonomous object manipulation. *Robot. Comput.-Integr. Manuf.* **2014**, *30*, 351–362. [CrossRef]

40. Alers, S.; Hu, J. AdMoVeo: A Robotic Platform for Teaching Creative Programming to Designers. In *Learning by Playing. Game-based Education System Design and Development*; Springer: Berlin/Heidelberg, Germany, 2009; pp. 410–421.

41. Dademo, N.; Lik, W.; Ho, W.; Drummond, T. eBug—An Open Robotics Platform for Teaching and Research. In Proceedings of Australasian Conference on Robotics and Automation, Melbourne, Australia, 7–9 December 2011.

42. Gonzalez, J.; Valero, A.; Prieto, A.; Abderrahim, M. A New Open Source 3D-printable Mobile Robotic Platform for Education. In *Advances in Autonomous Mini Robots*; Springer: Berlin/Heidelberg, Germmany, 2012.

43. Schweikardt, E.; Gross, M.D. roBlocks: A Robotic Construction Kit for Mathematics and Science Education. In Proceedings of the 8th International Conference on Multimodal Interfaces, ICMI '06, Banff, AB, Canada, 2–4 November 2006.

44. Bers, M.U.; Ponte, I.; Juelich, C.; Viera, A.; Schenker, J. Teachers as Designers: Integrating Robotics in Early Childhood Education. *Inf. Technol. Child. Educ. Annu.* **2002**, *2002*, 123–145.

45. Benitti, F. Exploring the educational potential of robotics in schools: A systematic review. *Comput. Educ.* **2012**, *58*, 978–988. [CrossRef]

46. Ainley, J.; Enger, L.; Searle, D. Students in a Digital Age: Implications of ICT for Teaching and Learning. In *International Handbook of Information Technology in Primary and Secondary Education*; Voogt, J., Knezek, G., Eds.; Springer: Boston, MA, USA, 2008; pp. 63–80.

47. Papadakis, S.; Kalogiannakis, M.; Orfanakis, V.; Zaranis, N. Novice Programming Environments. Scratch and App Inventor: A first comparison. In Proceedings of the 2014 Workshop on Interaction Design in Educational Environments, Albacete, Spain, 9 June 2014; pp. 1–7.

48. Balachandran, S. *General Purpose Input Output (GPIO)*; Technical Report ECE 480 Design Team 3; Available on the College of Engineering, Michigan State University Website. Available online: https://www.egr.msu. edu/classes/ece480/capstone/fall09/group03/AN_balachandran.pdf (accessed on 10 August 2018).

electronics

MDPI

Review

Open-Source Electronics Platforms as Enabling Technologies for Smart Cities: Recent Developments and Perspectives

Daniel G. Costa [1,*,†] and Cristian Duran-Faundez [2,†]

1   Department of Technology, State University of Feira de Santana, Feira de Santana 44036900, Brazil
2   Department of Electrical and Electronic Engineering, University of the Bío-Bío, Concepción 4051381, Chile; crduran@ubiobio.cl
*   Correspondence: danielgcosta@uefs.br; Tel.: +55-75-992389000
†   These authors contributed equally to this work.

Received: 18 November 2018; Accepted: 5 December 2018; Published: 7 December 2018

Abstract: With the increasing availability of affordable open-source embedded hardware platforms, the development of low-cost programmable devices for uncountable tasks has accelerated in recent years. In this sense, the large development community that is being created around popular platforms is also contributing to the construction of Internet of Things applications, which can ultimately support the maturation of the smart-cities era. Popular platforms such as Raspberry Pi, BeagleBoard and Arduino come as single-board open-source platforms that have enough computational power for different types of smart-city applications, while keeping affordable prices and encompassing many programming libraries and useful hardware extensions. As a result, smart-city solutions based on such platforms are becoming common and the surveying of recent research in this area can support a better understanding of this scenario, as presented in this article. Moreover, discussions about the continuous developments in these platforms can also indicate promising perspectives when using these boards as key elements to build smart cities.

Keywords: smart cities; Internet of Things; Raspberry Pi; BeagleBoard; Arduino

1. Introduction

The development of different areas of electronics, computing, data acquisition, and communication has created a fertile environment for the rise of the Internet of Things (IoT) [1]. The IoT is expected to deeply change our lives, allowing the creation of ubiquitous, distributed, and reactive systems that can surprisingly alter the way we interact with the world [2,3]. Smart cities are an important application of the IoT landscape, bringing complex challenges in different aspects of city structure and of their inhabitants' behaviors [4,5], but also opening many commercial, industrial, educational, and cultural opportunities [6–9].

The development of efficient communication technologies and cloud data services with decreasing costs, as well as the affordability of powerful open-source hardware platforms, have boosted smart-city initiatives [10,11]. With such platforms, the development time and related costs are considerable reduced, easing the creation of dedicated smart-city applications that can solve punctual problems. On the other hand, in a bigger scope where smart-city systems are designed and deployed by the governments, embedded hardware platforms are also a feasible solution not only during development phases, but also when providing public data: as most platforms are based on open-source technologies, the adoption of public standards is encouraged. Moreover, academic and scientific smart-city applications are inherently propitious to the adoption of open-source electronics platforms to construct IoT devices. Such smart-city scenarios may also rise concerns about interoperability and concurrency [12,13], but the characteristics of open-source electronics boards are favorable when promoting compatibility.

Among the existing open-source electronics platforms, some options are becoming more popular due to many factors. In common, the most popular boards share similar characteristics such as being versatile, powerful (although some models are limited) and cost-effective, being useful to create virtually any kind of device [14,15]. In this context, as most of them are ultimately a computer, respecting the different computational capabilities of the boards, they can be used to create sensors, actuators, and any processing device, presenting themselves as wild-cards for different types of sensor networks, IoT and smart-city applications [16,17].

In recent years, open-source electronics platforms have been used to support IoT applications and thus they can also be employed to support the creation of smart cities. From a different perspective, those platforms are also being taken as pillars for applications that are already being developed around the smart-city principles. Therefore, understanding this scenario can also indicate how smart cities are evolving and what we can expect from the near future.

Figure 1 presents a generic example of a smart city where different systems may be implemented using open-source electronics platforms, solving relevant problems of urban environments.

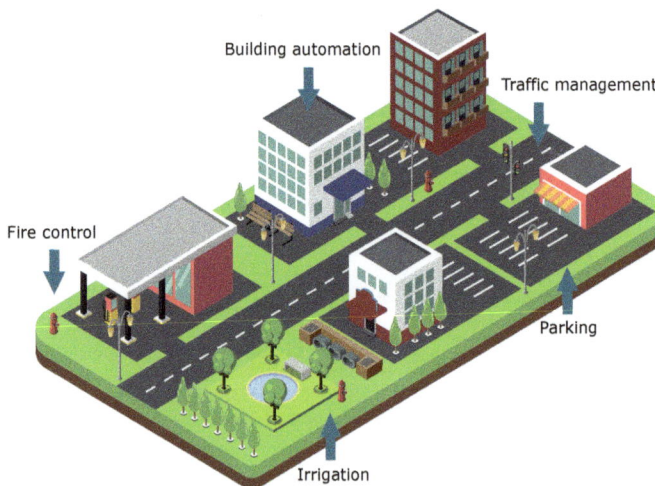

Figure 1. A typical smart city with multiple concurrent systems.

This article surveys recent scientific developments that exploit open-source electronics boards to create IoT and smart-city applications. Moreover, perspectives of this scenario are presented and discussed. As this is a very dynamic area, this article is valuable to support new researches and developments in this field.

The remainder of this paper is organized as follows. Section 2 presents the current availability of embedded open-source platforms that can be used to create IoT devices. Section 3 surveys recent works employing popular open-source boards for IoT and smart-city applications. Perspectives for future developments are presented in Section 4, followed by Conclusions in Section 5 and References.

2. Open-Source Platforms for IoT Development

IoT systems may be designed in different ways, but it is reasonable to expect that sensors, actuators and controller units will be present in most applications. For that, embedded hardware platforms may be used to implement these elements and there are many available options on the market. Among those options, open-source platforms are gaining prominence due to many factors and there are currently some good boards to be chosen. However, the existence of many research and industrial projects that are developed exploiting a few different boards is a good indication of which are the most popular platforms.

There are some characteristics that must be properly considered when choosing the most appropriate open-source platform for a particular IoT application. Those characteristics are described as follows:

- Low cost: To allow massive deployment, especially for physically scattered applications, the chosen platform must be cheap enough to not compromise the design and maintenance of smart-city systems;
- Computational power: Depending on the desired functionality, the board should have enough computational power to execute complex tasks, such as visual data processing and controlling of multiple distributed nodes. Processing and memory resources directly influence the computational power of the board;
- Programming flexibility: The availability of programming languages and libraries is of paramount importance and this characteristic is not necessarily a function of the available hardware capabilities, but also the relevance of the developers community around the platform;
- I/O interfaces. Smart-city applications may use electronic devices to provide relevant data or to take some action, such as cameras, LED, sirens, LCD panels, sensors, among others. The type and number of I/O interfaces in the adopted boards are important parameters when choosing the platform;
- Low energy consumption: As IoT nodes may be powered by batteries, low energy consumption is a relevant characteristic for embedded hardware platforms. Besides the use of batteries, energy efficiency is highly desired in modern cities, encouraging the use of more efficient boards;
- Reduced size: Depending on the expected functions of the board and the deployment place (e.g., lamp posts, traffic lights, transit plates), the size of the adopted board is relevant and should be considered;
- Robustness: Open-source boards should operate for long periods without encountering a hardware failure condition or overheating. If they are deployed at outdoor areas, such remarks are even more important;
- Communication: Different communication technologies may be embedded into the boards, easing the connections of the nodes. Wi-Fi, Bluetooth, Ethernet, LoRa and 4G/5G cellular standards are common communication technologies that may be present;
- Storage: The existence of embedded permanent storage may be helpful in many scenarios. For some boards, permanent storage may only be implemented through portable memory cards;
- Additional resources: LEDs, buttons, RFID (Radio-Frequency IDentification), required power supply (voltage/current), among other characteristics, may influence the choosing of a platform.

There are many open-source electronics platforms but only some of them are being more frequently used to create IoT applications. Among those boards, we selected some of the most popular platforms to be surveyed, considering affordable prices and large use in IoT applications as decision parameters for such selection. The chosen platforms are described in next subsections.

2.1. Raspberry Pi

Raspberry Pi is an affordable tiny and powerful multi-purpose computer that can be used for a variety of applications. When coming to the IoT and the smart-city environment, Raspberry also performs well as can be seen in the increasing number of projects that are exploiting its potentials. Although other hardware platforms have been created for the same purpose, Raspberry Pi is highly used by the industry, technology companies, governments, universities, and hobbyists, supporting the development of different kinds of applications [18,19].

Taking advantage of new electronic technologies and the success of ARM-based processors, the initial Raspberry Pi board was created with more modest purposes. Using cheap components that could be combined in a single board, the initial Raspberry Pi board was originally designed to be an affordable computer to assist students in school. However, it rapidly gained interest from other areas due to its flexibility, computational power, and reduced cost, helping it to become one of the best-selling computers in history.

Raspberry Pi is constructed around an ARM (Advanced RISC Machine) processor, which is the basis for the mobile smartphones industry. Therefore, Raspberry not only takes advantage of the characteristics of the RISC (Reduced Instruction Set Computer) processing platform, but also it was designed when manufacturers of RISC processors were consolidated. Since its initial debut, different models and versions of the original Raspberry board were created, addressing different hardware demands.

The increasing success of Raspberry Pi boards is due to many factors, putting this as the leading open-source hardware development platform. In addition, this hegemony has been reinforced by a series of performance evaluations, which has been conducted in recent years, for different scopes. The work in [20] evaluated a Raspberry Pi 2 B board in terms of graphical processing. The CPU and GPU loading was analyzed according to different visual data processing configurations, and energy consumption was assessed for different testing scenarios. From a different perspective, the work in [21] evaluated a Raspberry board (Raspberry Pi B) as a central element for a networking testbed based on the KODO library [22]. Even using an early model of the Raspberry Pi platform, the results in [21] were encouraging. Energy consumption was also evaluated in [23] but considering particular demands when implementing the RESTful Web Services framework. Similarly, the execution of web services in the Raspberry Pi platform was also addressed in [24], which evaluated load balancing in clusters of Raspberry Pi boards. In fact, all those works evaluated the performance of Raspberry Pi boards in different scenarios, indicating the interest of using this platform for a large set of tasks.

The most popular Raspberry Pi boards are the Raspberry Pi 2 B, the Raspberry Pi 3 B, and the Raspberry Zero W. Raspberry Pi was conceived and it is mostly manufactured in UK.

2.2. BeagleBoard

BeagleBoard is an open-source hardware platform that has the same principles of Raspberry Pi and thus they can often be used interchangeably. However, there are some important characteristics that put BeagleBoard as a viable option with its own particularities [25]. As with the Raspberry Pi family, BeagleBoard is a single-board tiny computer with affordable prices and low energy consumption. The initial use of this board and the related projects put both platforms in an equivalent development field.

When designing IoT applications, BeagleBoard has some particularities that can make it the board to be chosen. In a different way to Raspberry Pi, BeagleBoard board models have digital and analog I/O (similarly to the Arduino family). Other helpful characteristic is the presence of an onboard permanent storage unit in the BeagleBone series, which can be used to store code and sometimes to avoid the use of a microSD card, although there is a slot for those cards.

The work in [25] conducted a deep review of the BeagleBoard platform, discussing and evaluating some performance metrics of different board models.

The most popular BeagleBoard boards are the BeagleBone Black and the BeagleBone Blue. BeagleBoard was conceived and it is mostly manufactured in the USA.

2.3. Arduino

Differently from Raspberry Pi and BeagleBoard, the Arduino platform is an embedded prototyping board that is designed for electronics projects but that does not necessarily operate as a computer. Its use is focused on automation and electronic projects that demand repeated execution of some tasks and thus its software and hardware resources are more limited. However, the simplicity of Arduino finds its place in many automation and control projects and it is common to see Arduino boards composing smart-city systems.

While Raspberry Pi and BeagleBoard are developed around an ARM processor, most Arduino models are developed with a simpler micro-controller manufactured by Atmel, with most of them being of the ATmega family (8-bit, against the 32-bit processors of Raspberry Pi and BeagleBone). In addition, this characteristic has two practical implications: first, an Arduino board is usually cheaper; and second, it has less computational power than a usual Raspberry Pi or a BeagleBoard. Moreover, Arduino operates under a very lower clock speed and it has low RAM memory, but there is a flash memory to store the programs, which are written using special software IDE developed by the Arduino manufacturer. High-level programming languages such as C and Python can be used to program Arduino boards.

Although the reduced computational power may sound badly, the Arduino boards can do many things in terms of electronic control and it is a best-selling platform for its purposes. However, it is often limited for many IoT applications, especially when there is intensive computational processing. Despite that, there are a lot of extension boards (*shields*) to enhance the functions of Arduino, adding flexibility and useful resources. Among those extension boards, the ESP8266 family of microchips is a SoC (System on a Chip) that adds communication capabilities to Arduino and it became common to see many Arduino-based projects exploiting popular chips such as the ESP-01 and ESP-12. For IoT applications, such chips allow Arduino to easily communicate through Wi-Fi.

The Arduino platform was designed and manufactured in Italy and some of its most popular boards are the Arduino Uno, the Arduino Mega and the Arduino Due.

2.4. Open-Source Platforms Comparison

There are different options of open-source hardware platforms, with specifications and constraints that must be properly considered when implementing IoT and smart-city applications. The proper choosing of the most appropriate board is not straightforward, leaving to a heterogeneous scenario with projects and applications developed with different boards or employing combinations of them. However, the proper knowledge of the available off-the-shelf boards can give some light to this matter.

The previously presented open-source electronics platforms have been implemented in different models. Table 1 summarizes some of the most popular models of the presented open-source hardware development platforms (the model revisions and upgrades are not displayed). The "networking" column indicates onboard enabled resources, but additional hardware or dongles can be used to add networking capabilities to the boards.

Table 1. Popular open-source hardware platforms for IoT development.

Platform	Release	CPU	RAM	Networking
Arduino Nano	2008	16 MHz	0.5 KB	-
Arduino Uno	2010	16 MHz	2 KB	-
Arduino Mega ADK	2011	16 MHz	8 KB	-
Raspberry Pi B	2012	700 MHz	256 MB	Ethernet
Arduino Due	2012	84 MHz	96 KB	-
BeagleBone Black	2013	1 GHz	512 MB (4GB storage)	Ethernet
Arduino Yun	2013	16 MHz	2.5 KB	Wi-Fi
Raspberry Pi B+	2014	700 MHz	512 MB	Ethernet
Raspberry Pi 2 B	2015	900 MHz	1 GB	Ethernet
BeagleBone Green	2015	1 GHz	512 MB (4GB storage)	Ethernet
Raspberry Pi Zero	2015	1 GHz	512 MB	-
BeagleBoard-X15	2016	1.5 GHz	2 GB (4GB storage)	Ethernet
Raspberry Pi 3 B	2016	1.2 GHz	1 GB	Wi-Fi, Bluetooth/BLE, Ethernet
BeagleBone Blue	2017	1 GHz	512 MB (4GB storage)	Wi-Fi, Bluetooth/BLE
Raspberry Pi Zero W	2017	1 GHz	512 MB	Wi-Fi, Bluetooth/BLE
Raspberry Pi 3 B+	2018	1.4 GHz	1 GB	Wi-Fi, Bluetooth/BLE, Ethernet
Raspberry Pi 3 A+	2018	1.4 GHz	512 MB	Wi-Fi, Bluetooth/BLE

2.4.1. Software Issues

Most of the available open-source hardware platforms will operate as a computer and thus they require an operating system, excepting the Arduino board (the Arduino Yun board comes with a simplified version of the Linux operating system). In addition, there are many different operating systems for those boards, which are more frequently developed under the open-source premises. This creates a healthy environment for innovative projects, which also takes advantage of the average low costs of the embedded open-source hardware platforms.

Some boards do not have onboard storage, such as Raspberry Pi, while BeagleBone, for example, is shipped with an onboard permanent memory unit. When the board has no permanent storage, it usually employs a microSD card to record the operating system image and all user files. In general, terms, operating systems may be differentiated by performance metrics such as portability, required memory, CPU usage, image size, startup time, integrated tools, among others, and thus it is an important decision when designing smart-city applications. Some manufacturers officially support a particular operating system, such as the Raspbian OS for the Raspberry Pi platform, but even Raspberry can use any of dozens of available operating systems. Following the development trend of the open-source platforms, most operating systems for those boards are developed under open-source licenses.

Besides hardware and operating system characteristics, software and middleware are also relevant parameters that have been addressed in some works. The work in [26] discussed different aspects that are relevant when designing IoT applications, including the specification of the services expected from IoT devices and how they can impact the performance of IoT systems. Common services expected for IoT embedded platforms were also discussed in [27]. In that work, services such as virtualization, storage and use of the Rest API are discussed, which are helpful to add flexibility to IoT devices. In a similar way, the work in [28] also discussed services for IoT devices, but with focus on IoT protocols such as CoAP (Constrained Application Protocol) and MQTT (Message Queue Telemetry Transport). In fact, all those works discussed services that should be provided by IoT devices whatever is the employed hardware. Therefore, the presented considerations are valuable when implementing large-scale IoT and smart-city systems.

When choosing the most appropriate solution to construct IoT systems and ultimately smart-city solutions, all the presented aspects should be properly considered, since the particularities of each project should guide the choosing of the hardware and software technologies. Nevertheless, although there are many software middlewares and hardware platforms to develop IoT systems, and companies and research institutions are playing an important role to create this scenario, the existence of generic open-source multi-purpose development hardware platforms is of paramount importance. Open-source boards are revolutionizing the designing, deployment, operation, and maintenance of IoT and smart-city initiatives and thus they deserve more detailed analysis. In addition, in this scenario, the Raspberry Pi and the BeagleBoard family boards are standing out as promising choices, as can be seen in the increasing number of IoT projects based on them. The work in [15] performed interesting comparisons between the Raspberry Pi and the BeagleBoard platforms relating software and hardware issues.

2.4.2. Hardware Issues

In general, words, IoT development may require high flexibility when attaching new components, since different levels of interactions with the environment may be required. In fact, there are many electronic components that can be attached to the chosen open-source boards to extend their functions. For example, sensing units can be attached to allow monitoring of different information such as temperature, pressure, humidity, and pollution. Cameras and display units can also be easily attached. Raspberry Pi, BeagleBoard and Arduino have different I/O interfaces and there are many companies that develop different types of electronic components to be attached onto those boards. Particularly, the availability of GPIO (General Purpose Input/Output) pins allows the connection to a lot of electronic devices and there are many hardware extensions exploiting them, which significantly facilitates the attachment of new hardware components to the boards. Whatever the case, the number, type, and supported protocols of the I/O pins are important parameters when choosing open-source hardware platforms. In fact, the surveyed boards have different particularities concerning digital and analog I/O pins, which should be properly considered in smart-city projects.

Besides generic I/O, additional interfaces are also relevant, such as the CSI (Camera Serial Interface) standard in the Raspberry Pi platform to allow efficient connection of a camera. Depending on the characteristics of the project, such specialized interfaces may indirectly enhance the performance of the system.

Figure 2 presents three examples of open-source boards. From left to right, the boards are an Arduino Mega ADK, a BeagleBone Black and a Raspberry Pi 3 B.

Figure 2. Some popular open-source electronics boards that can be used in IoT and smart-city applications.

3. Smart Cities and Open-Source Electronics Platforms

The availability of open-source prototyping and development platforms has the potential to deeply change the way IoT applications are created and this trend became even more striking with the release of some popular electronics platforms. Although there is a great appeal around these platforms for DIY (Do-It-Yourself) applications, with a lot of projects in the "Smart Home" domain [19,29–31], different types of IoT applications can be designed. Nowadays, the development of smart-city systems based on such platforms is a reasonable and inexpensive choice for many projects. In addition, the last years have given us some clues of what applications can be created and what can still be envisioned for the near future [32].

Open-source electronics platforms will be typically employed to sense information from the environment or to make specialized processing of such information or other data (e.g., received from the Internet). Concerning sensing functions, we can roughly expect that two different types of information can be gathered: scalar and multimedia. Scalar information such as temperature, pressure, humidity, and luminosity is useful for many applications, and open-source boards can be efficiently used to gather and process such sensed data. However, as such data is usually short, the processing, the transmission and the storage of scalar information are not necessarily a critical issue, opening different opportunities for smart-city applications. On the other hand, multimedia sensing will be performed by cameras and microphones to retrieve large amounts of data, which can also have time restrictions when performing real-time processing. Multimedia-based smart cities are thus more challenging, and they rise additional performance concerns. Nevertheless, it will be natural to see some applications employing scalar and multimedia sensing units at the same board, exploiting continuous developments of more powerful hardware components and more efficient communication standards.

In the context of smart cities, IoT applications are typically designed to solve a particular problem of urban environments. In this sense, smart cities tend to be created by the composition of different concurrent systems [5,33,34]. Considering the state-of-the-art in this area and the use of open-source electronics boards, we selected a group of smart-city common problems to be addressed by such boards, presenting recent works that proposed solutions for them. Although there are many more issues to be addressed by smart-city applications [33], the discussed problems and related approaches are a good indication of the use of open-source boards in this area.

- Parking system: Large cities daily face the prominent problem of parking and the increasing number of cars in urban areas is continuously aggravating this issue. IoT applications may then be used to manage free parking spots, indirectly reducing congestion and car accidents when the time trying to find an available spot is reduced;
- Mobility: Efficient movement of vehicles, accidents prevention, congestion mitigation and (public) vehicles tracking are some important services than can enhance the mobility in urban areas;
- Energy efficiency: Sustainable and efficient consumption of energy in urban areas is one of the challenges of smart-city applications. Most solutions in this area perform dynamical management of the use of energy resources.

The state-of-the-art of IoT and smart-city applications is surveyed in next subsections, but only for works that specify at least one of the considered open-source electronics boards as central elements of the solutions. The selection of the works was made based on this characteristic and the addressing of some of the presented smart-city challenges.

3.1. Parking System

Popular open-source electronics platforms are being used to manage parking systems, following different approaches. The work in [35] employed the Raspberry Pi platform to create an intelligent parking system for smart cities. The authors proposed the use of a Raspberry Pi board to manage the vacant parking spots and to inform drivers about them. Drivers can verify the availability of free parking spots through their smartphones and perform payment when required. A distance (ultrasonic) sensor is used in each parking spot, which is then connected to a NodeMCU/ESP8266 board [36]. A Raspberry Pi 3 B board is used as the central unit of the system, collecting data from all NodeMCU, and delivering information to the users. A similar approach was discussed in [37], which directly connected the ultrasonic sensors (in each parking spot) to a Raspberry Pi board, potentially decreasing the cost of the overall solution (but maybe turning it more complex to handle and more susceptible to failures). Moreover, the work in [37] considered the use of the Raspberry Camera to also detect the presence of cars.

For some solutions, the overall cost is a critical parameter and researchers struggle to develop a cheap system. In this sense and still addressing the problem of smart parking in urban areas, the work in [38] also employed ultrasonic sensors to detect the presence of cars, but instead of monitoring each parking spot individually, the detection of cars is performed only in the entrance and exit gates of each monitored parking area. Doing so, the accounting of the available parking spots is performed considering the total number of spots and the cars flow. In [38], a single Raspberry Pi board was used in each parking area as a central server, computing the available parking spots and delivering such information for the users. In fact, this approach is affordable when there are costs limitations, and it could be easily implemented in many cities.

Parking systems were also developed using the Arduino platform. In [39] a parking system employed Infra-Red (IR) sensors in each parking slot to detect if it is free. Then, an LCD display in the parking area entrance is used to exhibit information about the free spots. Moreover, RFID cards are used to support authorization management, since the parking area may not be public. All sensors and supporting hardware (as a motor to open the gate) are connected to an Arduino board, which makes all processing locally. Similarly, the work in [40] also monitored free parking spots using Arduino boards, which were connected to presence sensors (photosensitive). Additionally, an ESP8266 module was used to allow Wi-Fi communications by Arduino boards, enabling them to send information to an external system to inform users about the current parking system state. Differently, in [41] authors combined Arduino and Raspberry Pi platforms to support efficient parking, leaving the most computation burden to the Raspberry Pi board. As an interesting service, the work in [41] also employed CMOS sensors to detect the cars plates.

Following a different trend of the previously presented works, authors in [42] proposed a generic software solution for easy experimentation of visual sensor networks, which can be used in the context of smart cities. As a generic solution, the proposed approach could be used to perform many different functions. As an example of use, the work in [42] employed the BeagleBoard platform as a use case for parking management. The authors made an experiment using a BeagleBone Black board as a central unit, employing a camera to identify free parking spots. Although the work presented in [42] is generic in the sense that a complete software solution is proposed, the adoption of the BeagleBone Black as the open-source experimentation board also indicates its relevance for IoT applications.

3.2. Mobility

One of the main goals of smart-city systems is to enhance mobility, which can be achieved optimizing the public transportation system, the vehicles flows and the traffic conditions in rush hours. For that, different aspects can be addressed, and open-source electronics boards can be used to support solutions in this area.

A desired service for smart cities is vehicle tracking, since it is important for public transportation enhancement. The work in [43] proposed a bus tracking system, deploying a Raspberry board on each monitored bus. GPS (Global Positioning System) and GSM/GPRS (Global System for Mobile communications/General Packet Radio Service) expansion boards are attached to Raspberry Pi to provide localization information and data communication through the cellular network, respectively. Such services are then used to indicate the online location of the bus. A similar approach was performed in [44], but in that work a pre-determined route is defined for the vehicle trajectories. The idea was to send alert message if the driver picks a different route other than the one previously specified and already stored in the Raspberry Pi storage card. Moreover, auxiliary sensors for gas leakage and temperature were also employed, aiming at the monitoring of the safety conditions of the drivers.

In [45], the Arduino board was used to keep track of moving vehicles using GPS as the localization service. Then, a GSM/GPRS expansion board was employed to transmit positioning information to be displayed on any user smartphone. A similar work was conducted in [46], but now using a XBee expansion board (IEEE 802.15.4) instead of a GSM/GPRS module.

Concerning congestion and traffic jams, a relevant issue in smart cities is traffic-light management. The work in [47] employed a Raspberry Pi board to control the traffic lights to soften traffic jams in harsh hours. As traffic conditions may suddenly change, for example due to a car accident, the Raspberry Pi board could be used to adjust the time programming of traffic lights, but this service was not performed automatically in [47]. In that work, a user accesses the Raspberry board through an Android App and manually controls the traffic lights. From a different perspective, a Raspberry Pi board was used in [48] to measure the traffic flow (counting cars) and to automatically adjust the traffic lights programming, using a camera for that. The traffic light controllers can then transmit information among them using Wi-Fi connections, allowing a more efficient control of a set of traffic lights.

Still following this trend, the work in [49] employed Arduino boards to control traffic lights. The idea is to perform image processing to change the traffic lights programming, changing waiting times according to the detected vehicle flows on the lanes. In the proposed system, processing is performed in a "conventional" computer (webcams are connected to it) and only decisions are transmitted to two Arduino boards, which ultimately control the lights time. The Arduino platform was also used to support the Raspberry-based solution described in [48].

Also addressing the mobility issues in smart cities, but from a different perspective, the work in [50] leveraged RFID cards as a mechanism to manage passengers in a public system, controlled by a BeagleBone Black board. That board also controls advertisement and information displayed on LCD displays, integrating two important services expected from smart-city environments. From the same perspective, data gathered from every monitored bus is displayed on LCD panels in [51], employing a Raspberry Pi board to gather GPS information transmitted by buses (they have a Raspberry Pi board) and to display that information in a bus terminus.

3.3. Energy Management

Open-source boards can also be used to save energy and manage energy consumption in urban areas. The authors in [52] proposed a system to turn on lamp posts only in the presence of cars or pedestrian, instead of keeping the lamps always on during dark times. A Raspberry Pi board was used to control the entire system, being deployed as a single control node. Then, ZigBee communications [53] were performed to control the lamp posts on a defined area. The LEDs in the lamp posts are turned on when it is dark, and cars or pedestrians are detected through presence sensors. Those sensors are not connected to the Raspberry Pi board, but they communicate through Wi-Fi. A similar approach is presented in [54], but it used Arduino boards in all lamp posts, except the coordinator post, which was controlled by a Raspberry Pi board.

Still considering efficient management of lamp posts, the work in [55] employed an Arduino board to control the lights and to monitor the temperature and humidity of a region. As LED lamps may have their light intensity adjusted, authors proposed the use of LDR (Light Dependent Resistors) sensors to reduce light intensity (and energy consumption) when a region is already too bright, potentially achieving efficient lighting on an area. An Arduino board was used in [55] to integrate the sensor units and to perform the adjustments.

An interesting work was presented in [56]. In that work, a neural network was employed to reduce energy consumption in residences, mainly in peak hours, which may affect energy distribution planning in the neighborhoods. As energy consumption in a city is not uniform, and different fares may apply for different periods of the day, the proposed system acts reducing energy consumption when demand is naturally higher. The proposed solution is mainly processed in a Raspberry Pi board.

3.4. Discussions and Comments

There are many works proposing IoT and smart-city applications to solve relevant problems in urban areas. The presented and discussed works are only some examples of feasible projects that have open-source electronics boards as the processing core of the proposed solutions. Table 2 summarizes these smart-city prototypes and projects that are centered around at least one of the presented open-source development platforms.

Table 2. Some smart-city applications based on open-source electronics development platforms.

Work	Year	Model	Sensors	Smart-City Challenge
Vakula and Kolli [35]	2017	Raspberry Pi 3 B	Ultrasonic	Parking system
Ramaswamy [37]	2016	Raspberry Pi B	Ultrasonic, Camera	Parking system
Zadeh and Cruz [38]	2016	Raspberry Pi B+	Ultrasonic	Parking system
Chaudhary et al. [39]	2017	Arduino (model not specified)	IR	Parking system
Huang et al. [40]	2017	Arduino Nano	Presence	Parking system
Kanteti el al. [41]	2017	Raspberry Pi and Arduino (models not specified)	CMOS camera, Ultrasonic	Parking system
bondi et al. [42]	2015	BeagleBone Black	Camera	Parking system
Nalawade and Akshay [43]	2016	Raspberry Pi 3 B	GPS	Mobility
Shinde and Mane [44]	2015	Raspberry Pi B+	GPS, Gas, Temperature	Mobility
Rahman et al. [45]	2016	Arduino Uno	GPS	Mobility
Ibraheem and Hadi [46]	2018	Arduino Uno	GPS	Mobility
hariff et al. [50]	2016	BeagleBone Black	RFID	Mobility
Vakula and Raviteja [51]	2017	Raspberry Pi 3	GPS	Mobility
Misbahuddin et al. [47]	2015	Raspberry Pi B	-	Mobility
Basil and Sawant [48]	2017	Raspberry Pi 3, Arduino Uno	Camera	Mobility
Khushi [49]	2017	Arduino Uno	-	Mobility
Leccese et al. [52]	2014	Raspberry Pi B	-	Energy efficiency
Sunehra and Rajasri [54]	2017	Raspberry Pi 3 B, Arduino Uno	Presence, Ultrasonic	Energy efficiency
Dheena et al. [55]	2017	Arduino Nano	LDR, Temperature, Humidity	Energy efficiency
Mahapatra et al. [56]	2017	Raspberry Pi 3 B	Energy smart meter	Energy efficiency

As can be seen in Table 2, the surveyed works were published in recent years as expected. Although Arduino is a bit older, the Raspberry and the BeagleBoard platforms are still very young, but their potentials as development platforms for smart-city applications are continuously being proved. Their open-source nature, the reduced energy consumption and low acquisition costs are some of the characteristics that are making these boards winning choices.

4. Perspectives and Future Directions

The adoption of open-source electronics development platforms as key elements to construct monitoring and control applications has the potential to accelerate the expected transformations promoted by the IoT era [57], with smart cities, vehicular networks, smart farming, integrated health assistance and other disruptive applications potentially changing the way we perceive and interact with the world. The projects and products being created with such open-source boards are becoming increasingly common and this trend will still be more evident with transformations in the educational system: children are having early contact with these boards, supporting the creation of a maker generation [58,59].

For the particular maturation of the smart-city initiatives, the presented open-source hardware platforms will continue to have an important role. In addition, there are some perspectives in this area that are relevant when envisioning future directions for them. Next subsections discuss some of these perspectives.

4.1. Multimedia Sensing

The development of affordable tiny camera and microphone hardware components to be connected to open-source boards opened new possibilities for monitoring applications. Different manufacturers are continuously releasing new extension boards and specialized electronics components, which has facilitated the designing of smart-city solutions based on multimedia data. The selection of the most appropriate multimedia components is then a relevant project choice.

For Raspberry Pi, the success of the original board encouraged the creation of a specialized camera that could be easily installed, being initially released in 2013. The Raspberry Camera is a low-power high-definition small camera that is shipped with a flexible flat cable to be plugged into the CSI (Camera Serial Interface). For the BeagleBoard and Arduino platforms, some specialized cameras have good performance and affordable prices, such as the CMOS OV7670. Concerning microphones, there are different components and the acquisition cost will be a function of the desired sensibility and audio capturing quality.

Besides choosing the proper multimedia electronic components, the development of smart-city applications based on image, video and audio will also have to be concerned with other issues, since its inherent higher complexity will directly impact storage, processing, and data transmission. Additionally, more complex tasks such as video compressing may also demand high processing power, which may prohibit the use of some resource-constrained boards. One of the trends for this problem is the adoption of development extension boards, as described in Section 4.2. Such extensions may bring additional processing power or even an extra GPU (Graphical Processing Unit), better supporting multimedia-based applications.

Some works have exploited multimedia sensing for IoT applications. The work in [60] investigated the transmission of high-definition video on the Raspberry Pi platform (particularly the Raspberry Pi 2 B), considering time constraints (real-time transmissions). The Raspberry Camera is used to transmit H.264 video streams and authors in [60] concluded that the desired high-definition real-time video transmissions were satisfactorily performed using the Raspberry Pi board and an IoT P2P communications framework. Considering the BeagleBone Black board, the work in [61] proposed a framework to create smart nodes that can handle any kind of data, including audio, image, and video, making easier the development of distributed sensor nodes. For the work in [62], audio sensors were created using a Raspberry Pi 2 B board. In that work, different audio processing tasks were performed in the Raspberry board, aiming at the calculation of different audio noise parameters. Audio processing for noise detection was also performed in [63], which created audio sensing nodes with Raspberry Pi 3 B boards. All these recent works are leveraging the processing capabilities of open-source electronics platforms for multimedia data processing and storage, which may bring significant results for smart-city scenarios.

4.2. Development Extension Boards

One of the perspectives when using open-source electronics platforms is the availability of development extension boards. Such extensions improve the capabilities of the boards, typically adding additional processing power, sensing units, or even facilitating prototyping. Therefore, the choosing of development extension boards may be an important project decision and the constant releasing of new products reinforces the use of open-source boards for IoT and smart-city applications.

Development extension boards may be designed and produced by the same manufacturers of the open-source electronics platforms or even by other companies. Of course, this is possible due to the open-source nature of the platforms. In this way, this "extension boards market" is active and dynamical, constantly bringing novelties that can further support the development of significant applications for smart cities.

The connection of an extension board to an open-source board is performed exploiting the I/O pins. As most of these extensions will attach to the boards in a way that they will be over them, they are referred following this idea: for Raspberry Pi they are HAT (Hardware Attached on Top); for BeagleBoard, they are referred to as *capes*; and for Arduino, they are called *shields*.

Table 3 presents some popular extension boards for the surveyed open-source electronics platforms.

Table 3. Some popular extension boards for open-source electronics boards.

Extension Board	Platform	Resources	Manufacturer
Matrix Creator	Raspberry Pi	FPGA, LED array, Multiple sensors, Accelerometer, 8 microphones, ZigBee	http://www.matrix.one
Matrix Voice	Raspberry Pi	FPGA, LED array, 8 microphones, Wi-Fi/Bluetooth (ESP8266)	http://www.matrix.one
PiJuice HAT	Raspberry Pi	Portable battery	http://www.pijuice.com
Ethernet shield	Arduino	Ethernet communications capability	http://www.arduino.cc
BeagleBone proto cape	BeagleBone family	Prototyping support	http//www.beagleboard.org
GPS Logger Shield	Arduino	GPS	http://www.sparkfun.com
1Sheeld	Arduino	Integration of smartphone's resources with Arduino	http://1sheeld.com
SparkFun Pro RF	Arduino	LoRa communications capability	http://www.sparkfun.com

Among those extension boards, some of them can be very helpful for many smart-city applications. As an example, the Matrix Creator board has several onboard sensor units, which can provide many important sensed information for applications in an integrated way. Other interesting extension board is the 1Sheeld, which integrates smartphones and their sensors to Arduino (also providing software to facilitate this integration). Doing so, for example, the smartphone's camera and microphone can be used by the Arduino board, opening interesting opportunities. New extension boards are still being released and many of them may positively impact the development of IoT and smart-city applications.

Figure 3 presents an example of two development extension boards (HAT) for the Raspberry Pi platform.

Figure 3. Examples of different development HAT for Raspberry Pi.

4.3. New Communication Technologies

The evolving of communications standards has a relevant role in the development of smart cities. Early applications were based on Wi-Fi technologies (more recently supported by ESP modules) but the advent of the IEEE 802.15.4 and ZigBee standard considerably changed this scenario. When extension boards supporting the ZigBee standard (mostly XBee modules) started to be released, a new wave of developments was possible [53,64]. In addition, we can expect such waves when new communication standards are released, once open-source boards are upgraded, and new extensions are released to allow such communications.

Concerning the perspectives in this area, a new communication standard can significantly enrich IoT smart-city applications. In a short term, a new standard, LoRa, has come as an efficient long-range low-power wireless communication standard that fits very well into the IoT demands [65]. Recently, some LoRa expansion boards were released, and new developments in this area are expected.

Other promising technology is 5G cellular networks. With this communication technology, high-bandwidth low-latency communication is possible, with higher communication range than current cellular networks. However, although promising, the introduction of 5G into the IoT world is still slow and other technologies are winning the race. With new releases of extensions boards, LoRa is being more believed to fulfill the IoT dreams and other standards such as Wi-Fi and Bluetooth are continuously evolving to increase bandwidth and reduce latency while keeping energy consumption low. The recent Raspberry Pi 3 B+ and Raspberry Pi 3 A+ models provide onboard support to the IEEE 802.11ac standard, allowing more efficient Wi-Fi communications. Some more years are still required to say what will be the winning technology for smart-city applications.

Therefore, multimedia sensing, real-time decisions and large data processing and storage will be some of the hot-topics to be addressed in smart-city applications and open-source electronics boards will play an important role to support the achievement of these goals. When concerning new communication technologies, the materialization of new standards in extensions boards and revised versions of open-source platforms will be crucial for the evolution of IoT and smart cities.

5. Conclusions

The advent of new hardware and communication technologies in the beginning of this century anticipated a series of innovations in the way information is gathered and processed, and initial developments of wireless sensor networks were first steps in this direction. However, although specialized hardware platforms were created to support sensing networks, such as the MicaZ [66] and the TelosB [67], and they had an important role in the consolidation of sensor networks, the era of open-source electronics platforms is coming to enhance not only wireless sensor networks and their variations, but also to boost the complete IoT landscape. This article is then a valuable resource when trying to understand the transformations we are still living.

This article is an important contribution for new developments in this area, providing information that can guide new researches and projects for IoT applications. The discussions and surveyed works may contribute to create a vibrating atmosphere around open-source electronics platforms, which will help them to become the main development solution for smart cities. In addition, other emerging areas such as smart agriculture, smart retail, smart health, among others, may also benefit from those platforms.

The open-source electronics boards are still evolving, and new products are continuously being released. New platforms such as the NodeMCU [36,68] are still coming to the scene, opening new possibilities for small and big developers. In this context, we can expect a promising future for smart-city initiatives, when manufacturers and governmental agencies are seeing this as a fertile environment for the so eagerly awaited transformations.

Author Contributions: Authors contributed equally to the development of this work. D.G.C. and C.D.-F. contributed with the articles surveys and the discussions about the perspectives of the area. Both authors participated in the writing of the manuscript.

Funding: The APC was funded by the University of the Bío-Bío under grants DIUBB GI 160210/EF and DIUBB 184110 3/R.

Conflicts of Interest: The authors declare no conflict of interest. The founding sponsors had no role in the design of the study; in the collection, analyses, or interpretation of data; in the writing of the manuscript, and in the decision to publish the results.

References

1. Lin, J.; Yu, W.; Zhang, N.; Yang, X.; Zhang, H.; Zhao, W. A Survey on Internet of Things: Architecture, Enabling Technologies, Security and Privacy, and Applications. *IEEE Internet Things J.* **2017**, *4*, 1125–1142. [CrossRef]
2. Talari, S.; Shafie-khah, M.; Siano, P.; Loia, V.; Tommasetti, A.; Catalão, J.P.S. A Review of Smart Cities Based on the Internet of Things Concept. *Energies* **2017**, *10*, 421. [CrossRef]
3. Peixoto, J.P.J.; Costa, D.G. Wireless visual sensor networks for smart city applications: A relevance-based approach for multiple sinks mobility. *Future Gener. Comput. Syst.* **2017**, *76*, 51–62. [CrossRef]
4. Zanella, A.; Bui, N.; Castellani, A.; Vangelista, L.; Zorzi, M. Internet of Things for Smart Cities. *IEEE Internet Things J.* **2014**, *1*, 22–32. [CrossRef]
5. Costa, D.G.; Collotta, M.; Pau, G.; Duran-Faundez, C. A Fuzzy-Based Approach for Sensing, Coding and Transmission Configuration of Visual Sensors in Smart City Applications. *Sensors* **2017**, *17*, 93. [CrossRef] [PubMed]
6. Glasmeier, A.K.; Nebiolo, M. Thinking about Smart Cities: The Travels of a Policy Idea that Promises a Great Deal, but So Far Has Delivered Modest Results. *Sustainability* **2016**, *8*, 1122. [CrossRef]
7. Allam, Z.; Newman, P. Redefining the Smart City: Culture, Metabolism and Governance. *Smart Cities* **2018**, *1*, 4–25. [CrossRef]
8. Molina, B.; Palau, C.E.; Fortino, G.; Guerrieri, A.; Savaglio, C. Empowering smart cities through interoperable Sensor Network Enablers. In Proceedings of the 2014 IEEE International Conference on Systems, Man, and Cybernetics (SMC), San Diego, CA, USA, 5–8 October 2014; pp. 7–12.
9. Fortino, G.; Savaglio, C.; Zhou, M. Toward opportunistic services for the industrial Internet of Things. In Proceedings of the 2017 13th IEEE Conference on Automation Science and Engineering (CASE), Xi'an, China, 20–23 August 2017; pp. 825–830.
10. Costa, D.G.; Duran-Faundez, C.; Andrade, D.C.; Rocha-Junior, J.B.; Just Peixoto, J.P. TwitterSensing: An Event-Based Approach for Wireless Sensor Networks Optimization Exploiting Social Media in Smart City Applications. *Sensors* **2018**, *18*, 1080. [CrossRef]

11. Santana, E.F.Z.; Chaves, A.P.; Gerosa, M.A.; Kon, F.; Milojicic, D.S. Software Platforms for Smart Cities: Concepts, Requirements, Challenges, and a Unified Reference Architecture. *ACM Comput. Surv.* **2017**, *50*, 1–37. [CrossRef]

12. Fortino, G.; Savaglio, C.; Palau, C.E.; de Puga, J.S.; Ganzha, M.; Paprzycki, M.; Montesinos, M.; Liotta, A.; Llop, M., Towards Multi-layer Interoperability of Heterogeneous IoT Platforms: The INTER-IoT Approach. In *Integration, Interconnection, and Interoperability of IoT Systems*; Gravina, R., Palau, C.E., Manso, M., Liotta, A., Fortino, G., Eds.; Springer International Publishing: Cham, Switzerland, 2018; pp. 199–232.

13. Costa, D.G.; Guedes, L.A. Exploiting the sensing relevancies of source nodes for optimizations in visual sensor networks. *Multimedia Tools Appl.* **2013**, *64*, 549–579. [CrossRef]

14. Tavade, T.; Nasikkar, P. Raspberry Pi: Data logging IOT device. In Proceedings of the 2017 International Conference on Power and Embedded Drive Control (ICPEDC), Chennai, India, 16–18 March 2017; pp. 275–279.

15. Kruger, C.P.; Hancke, G.P. Benchmarking Internet of things devices. In Proceedings of the 2014 12th IEEE International Conference on Industrial Informatics (INDIN), Porto Alegre, Brazil, 27–30 July 2014; pp. 611–616.

16. Nikhade, S.G. Wireless sensor network system using Raspberry Pi and zigbee for environmental monitoring applications. In Proceedings of the Smart Technologies and Management for Computing, Communication, Controls, Energy and Materials (ICSTM), Chennai, India, 6–8 May 2015; pp. 376–381.

17. Patil, N.; Ambatkar, S.; Kakde, S. IoT based smart surveillance security system using raspberry Pi. In Proceedings of the Communication and Signal Processing (ICCSP), Chennai, India, 6–8 April 2017; pp. 0344–0348.

18. Costa, D.G. On the Development of Visual Sensors with Raspberry Pi. In Proceedings of the 24th Brazilian Symposium on Multimedia and the Web, WebMedia '18, Salvador, Brazil, 16–19 October 2018; ACM: New York, NY, USA, 2018; pp. 19–22.

19. Patchava, V.; Kandala, H.B.; Babu, P.R. A Smart Home Automation technique with Raspberry Pi using IoT. In Proceedings of the Smart Sensors and Systems (IC-SSS), Bangalore, India, 21–23 December 2015; pp. 1–4.

20. He, Q.; Segee, B.; Weaver, V. Raspberry Pi 2 B+ GPU Power, Performance, and Energy Implications. In Proceedings of the 2016 International Conference on Computational Science and Computational Intelligence (CSCI), Las Vegas, NV, USA, 15–17 December 2016; pp. 163–167.

21. Paramanathan, A.; Pahlevani, P.; Thorsteinsson, S.; Hundeboll, M.; Lucani, D.E.; Fitzek, F.H.P. Sharing the Pi: Testbed Description and Performance Evaluation of Network Coding on the Raspberry Pi. In Proceedings of the 2014 IEEE 79th Vehicular Technology Conference (VTC Spring), Seoul, Korea, 18–21 May 2014; pp. 1–5.

22. Pedersen, M.V.; Heide, J.; Fitzek, F.H.P. Kodo: An Open and Research Oriented Network Coding Library. In *NETWORKING 2011 Workshops*; Casares-Giner, V., Manzoni, P., Pont, A., Eds.; Springer: Berlin/Heidelberg, Germany, 2011; pp. 145–152.

23. Nunes, L.H.; Nakamura, L.H.V.; de F. Vieira, H.; de O. Libardi, R.M.; de Oliveira, E.M.; Estrella, J.C.; Reiff-Marganiec, S. Performance and energy evaluation of RESTful web services in Raspberry Pi. In Proceedings of the 2014 IEEE 33rd International Performance Computing and Communications Conference (IPCCC), Austin, TX, USA, 5–7 December 2014; pp. 1–9.

24. Maduranga, M.W.P.; Ragel, R.G. Comparison of load balancing methods for Raspberry-Pi Clustered Embedded Web Servers. In Proceedings of the 2016 International Computer Science and Engineering Conference (ICSEC), Chiang Mai, Thailand, 14–17 December 2016; pp. 1–4.

25. Nayyar, A.; Puri, V. A Review of Beaglebone Smart Board's-A Linux/Android Powered Low Cost Development Platform Based on ARM Technology. In Proceedings of the 2015 9th International Conference on Future Generation Communication and Networking (FGCN), Jeju, Korea, 25–28 November 2015; pp. 55–63.

26. Samie, F.; Bauer, L.; Henkel, J. IoT technologies for embedded computing: A survey. In Proceedings of the 2016 International Conference on Hardware/Software Codesign and System Synthesis (CODES+ISSS), Pittsburgh, PA, USA, 2–7 October 2016; pp. 1–10.

27. Heo, Y.J.; Oh, S.M.; Chin, W.S.; Jang, J.W. A Lightweight Platform Implementation for Internet of Things. In Proceedings of the 2015 3rd International Conference on Future Internet of Things and Cloud, Rome, Italy, 24–26 August 2015; pp. 526–531.

28. Hejazi, H.; Rajab, H.; Cinkler, T.; Lengyel, L. Survey of platforms for massive IoT. In Proceedings of the 2018 IEEE International Conference on Future IoT Technologies (Future IoT), Eger, Hungary, 18–19 January 2018; pp. 1–8.

29. Sicari, S.; Rizzardi, A.; Miorandi, D.; Coen-Porisini, A. Securing the smart home: A real case study. *Internet Technol. Lett.* **2018**, *1*, e22. [CrossRef]

30. Costantino, D.; Malagnini, G.; Carrera, F.; Rizzardi, A.; Boccadoro, P.; Sicari, S.; Grieco, L.A. Solving Interoperability within the Smart Building: A Real Test-Bed. In Proceedings of the 2018 IEEE International Conference on Communications Workshops (ICC Workshops), Kansas City, MO, USA, 20–24 May 2018; pp. 1–6.

31. Gunputh, S.; Murdan, A.P.; Oree, V. Design and implementation of a low-cost Arduino-based smart home system. In Proceedings of the 2017 IEEE 9th International Conference on Communication Software and Networks (ICCSN), Guangzhou, China, 6–8 May 2017; pp. 1491–1495.

32. Pradhan, M.A.; Patankar, S.; Shinde, A.; Shivarkar, V.; Phadatare, P. IoT for smart city: Improvising smart environment. In Proceedings of the 2017 International Conference on Energy, Communication, Data Analytics and Soft Computing (ICECDS), Chennai, India, 1–2 August 2017; pp. 2003–2006.

33. Vidiasova, L.; Kachurina, P.; Cronemberger, F. Smart Cities Prospects from the Results of the World Practice Expert Benchmarking. *Procedia Comput. Sci.* **2017**, *119*, 269–277. [CrossRef]

34. Al Nuaimi, E.; Al Neyadi, H.; Mohamed, N.; Al-Jaroodi, J. Applications of big data to smart cities. *J. Internet Serv. Appl.* **2015**, *6*, 25. [CrossRef]

35. Vakula, D.; Kolli, Y.K. Low cost smart parking system for smart cities. In Proceedings of the 2017 International Conference on Intelligent Sustainable Systems (ICISS), Palladam, India, 7–8 Deember 2017; pp. 280–284.

36. Barai, S.; Biswas, D.; Sau, B. Estimate distance measurement using NodeMCU ESP8266 based on RSSI technique. In Proceedings of the 2017 IEEE Conference on Antenna Measurements Applications (CAMA), Tsukuba, Japan, 4–6 December 2017; pp. 170–173.

37. Ramaswamy, P. IoT smart parking system for reducing green house gas emission. In Proceedings of the 2016 International Conference on Recent Trends in Information Technology (ICRTIT), Chennai, India, 8–9 April 2016; pp. 1–6.

38. Zadeh, N.R.N.; Cruz, J.C.D. Smart urban parking detection system. In Proceedings of the 2016 6th IEEE International Conference on Control System, Computing and Engineering (ICCSCE), Batu Ferringhi, Malaysia, 25–27 November 2016; pp. 370–373.

39. Chaudhary, H.; Bansal, P.; Valarmathi, B. Advanced CAR parking system using Arduino. In Proceedings of the 2017 4th International Conference on Advanced Computing and Communication Systems (ICACCS), Coimbatore, India, 6–7 January 2017; pp. 1–5.

40. Huang, K.Y.; Chang, S.B.; Tsai, P.R. The advantage of the arduino sensing system on parking guidance information systems. In Proceedings of the 2017 IEEE International Conference on Industrial Engineering and Engineering Management (IEEM), Singapore, 10–13 December 2017; pp. 2078–2082.

41. Kanteti, D.; Srikar, D.V.S.; Ramesh, T.K. Intelligent smart parking algorithm. In Proceedings of the 2017 International Conference On Smart Technologies For Smart Nation (SmartTechCon), Bangalore, India, 17–19 August 2017; pp. 1018–1022.

42. Bondi, L.; Baroffio, L.; Cesana, M.; Redondi, A.; Tagliasacchi, M. EZ-VSN: An Open-Source and Flexible Framework for Visual Sensor Networks. *IEEE Internet Things J.* **2016**, *3*, 767–778. [CrossRef]

43. Nalawade, S.R.; Akshay, S.D. Bus tracking by computing cell tower information on Raspberry Pi. In Proceedings of the 2016 International Conference on Global Trends in Signal Processing, Information Computing and Communication (ICGTSPICC), Jalgaon, India, 22–24 December 2016; pp. 87–90.

44. Shinde, P.A.; Mane, Y.B. Advanced vehicle monitoring and tracking system based on Raspberry Pi. In Proceedings of the 2015 IEEE 9th International Conference on Intelligent Systems and Control (ISCO), Coimbatore, India, 9–10 January 2015; pp. 1–6.

45. Rahman, M.M.; Mou, J.R.; Tara, K.; Sarkar, M.I. Real time Google map and Arduino based vehicle tracking system. In Proceedings of the 2016 2nd International Conference on Electrical, Computer Telecommunication Engineering (ICECTE), Rajshahi, Bangladesh, 8–10 December 2016; pp. 1–4.

46. Ibraheem, I.K.; Hadi, S.W. Design and Implementation of a Low-Cost Secure Vehicle Tracking System. In Proceedings of the 2018 International Conference on Engineering Technology and their Applications (IICETA), Al-Najaf, Iraq, 8–9 May 2018; pp. 146–150.

47. Misbahuddin, S.; Zubairi, J.A.; Saggaf, A.; Basuni, J.; A-Wadany, S.; Al-Sofi, A. IoT based dynamic road traffic management for smart cities. In Proceedings of the 2015 12th International Conference on High-capacity Optical Networks and Enabling/Emerging Technologies (HONET), Islamabad, Pakistan, 21–23 December 2015; pp. 1–5.

48. Basil, E.; Sawant, S.D. IoT based traffic light control system using Raspberry Pi. In Proceedings of the 2017 International Conference on Energy, Communication, Data Analytics and Soft Computing (ICECDS), Chennai, India, 1–2 August 2017; pp. 1078–1081.

49. Khushi. Smart Control of Traffic Light System using Image Processing. In Proceedings of the 2017 International Conference on Current Trends in Computer, Electrical, Electronics and Communication (CTCEEC), Mysore, India, 8–9 September 2017; pp. 99–103.

50. Shariff, S.U.; Swamy, J.C.N.; Seshachalam, D. Beaglebone black based e-system and advertisement revenue hike scheme for Bangalore city public transportation system. In Proceedings of the 2016 2nd International Conference on Applied and Theoretical Computing and Communication Technology (iCATccT), Bangalore, India, 21–23 July 2016; pp. 781–786.

51. Vakula, D.; Raviteja, B. Smart public transport for smart cities. In Proceedings of the 2017 International Conference on Intelligent Sustainable Systems (ICISS), Palladam, India, 7–8 December 2017; pp. 805–810.

52. Leccese, F.; Cagnetti, M.; Trinca, D. A Smart City Application: A Fully Controlled Street Lighting Isle Based on Raspberry-Pi Card, a ZigBee Sensor Network and WiMAX. *Sensors* **2014**, *14*, 24408–24424. [CrossRef]

53. Wang, H.; Dong, L.; Wei, W.; Zhao, W.; Xu, K.; Wang, G. The WSN Monitoring System for Large Outdoor Advertising Boards Based on ZigBee and MEMS Sensor. *IEEE Sensors J.* **2018**, *18*, 1314–1323. [CrossRef]

54. Sunehra, D.; Rajasri, S. Automatic street light control system using wireless sensor networks. In Proceedings of the 2017 IEEE International Conference on Power, Control, Signals and Instrumentation Engineering (ICPCSI), Chennai, India, 21–22 September 2017; pp. 2915–2919.

55. Dheena, P.P.F.; Raj, G.S.; Dutt, G.; Jinny, S.V. IOT based smart street light management system. In Proceedings of the 2017 IEEE International Conference on Circuits and Systems (ICCS), Thiruvananthapuram, Indian, 20–21 December 2017; pp. 368–371.

56. Mahapatra, C.; Moharana, A.K.; Leung, V.C.M. Energy Management in Smart Cities Based on Internet of Things: Peak Demand Reduction and Energy Savings. *Sensors* **2017**, *17*, 2812. [CrossRef]

57. Kurkovsky, S.; Williams, C. Raspberry Pi as a Platform for the Internet of Things Projects: Experiences and Lessons. In Proceedings of the 2017 ACM Conference on Innovation and Technology in Computer Science Education, ITiCSE '17, Bologna, Italy, 3–5 July 2017; pp. 64–69.

58. Adams, J.C.; Brown, R.A.; Kawash, J.; Matthews, S.J.; Shoop, E. Leveraging the Raspberry Pi for CS Education. In Proceedings of the 49th ACM Technical Symposium on Computer Science Education, SIGCSE '18, Baltimore, MD, USA, 21–24 February 2018; pp. 814–815.

59. Pattichis, M.S.; Celedon-Pattichis, S.; LopezLeiva, C. Teaching image and video processing using middle-school mathematics and the Raspberry Pi. In Proceedings of the 2017 IEEE International Conference on Acoustics, Speech and Signal Processing (ICASSP), New Orleans, LA, USA, 5–9 March 2017; pp. 6349–6353.

60. Jennehag, U.; Forsstrom, S.; Fiordigigli, F.V. Low Delay Video Streaming on the Internet of Things Using Raspberry Pi. *Electronics* **2016**, *5*, 60. [CrossRef]

61. Chianese, A.; Piccialli, F.; Riccio, G. Designing a Smart Multisensor Framework Based on Beaglebone Black Board. In *Computer Science and Its Applications*; Park, J.J.J.H., Stojmenovic, I., Jeong, H.Y., Yi, G., Eds.; Springer: Berlin/Heidelberg, Germany, 2015; pp. 391–397.

62. Noriega-Linares, J.E.; Navarro Ruiz, J.M. On the Application of the Raspberry Pi as an Advanced Acoustic Sensor Network for Noise Monitoring. *Electronics* **2016**, *5*, 74. [CrossRef]

63. Peckens, C.; Porter, C.; Rink, T. Wireless Sensor Networks for Long-Term Monitoring of Urban Noise. *Sensors* **2018**, *18*, 3161. [CrossRef]

64. De Oliveira, K.V.; Castelli, H.M.E.; Montebeller, S.J.; Avancini, T.G.P. Wireless Sensor Network for Smart Agriculture using ZigBee Protocol. In Proceedings of the 2017 IEEE First Summer School on Smart Cities (S3C), Natal, Brazil, 6–11 August 2017; pp. 61–66.

65. Pasolini, G.; Buratti, C.; Feltrin, L.; Zabini, F.; De Castro, C.; Verdone, R.; Andrisano, O. Smart City Pilot Projects Using LoRa and IEEE802.15.4 Technologies. *Sensors* **2018**, *18*, 1118. [CrossRef]

66. Ali, N.A.; Drieberg, M.; Sebastian, P. Deployment of MICAz mote for Wireless Sensor Network applications. In Proceedings of the 2011 IEEE International Conference on Computer Applications and Industrial Electronics (ICCAIE), Penang, Malaysia, 4–7 December 2011; pp. 303–308.

67. Guevara, J.; Vargas, E.; Brunetti, F.; Barrero, F. Open architecture for WSN based on runtime reconfigurable systems and the IEEE 1451. In Proceedings of the 2013 IEEE SENSORS, Baltimore, MD, USA, 3–6 November 2013; pp. 1–4.

68. Araújo, A.; Kalebe, R.; Girão, G.; Filho, I.; Gonçalves, K.; Melo, A.; Neto, B. IoT-Based Smart Parking for Smart Cities. In Proceedings of the 2017 IEEE First Summer School on Smart Cities (S3C), Natal, Brazil, 6–11 August 2017; pp. 31–36.

MDPI

St. Alban-Anlage 66

4052 Basel

Switzerland

Tel. +41 61 683 77 34

Fax +41 61 302 89 18

www.mdpi.com

Electronics Editorial Office

E-mail: electronics@mdpi.com

www.mdpi.com/journal/electronics

www.ingramcontent.com/pod-product-compliance
Lightning Source LLC
Chambersburg PA
CBHW051725210326
41597CB00032B/5609